24th International Conference on Vacuum Technique and Technology (VTT 2017)

Journal of Physics: Conference Series Volume 872

St. Petersburg, Russia
6 – 8 June 2017

ISBN: 978-1-5108-4424-7
ISSN: 1742-6588

Printed from e-media with permission by:

Curran Associates, Inc.
57 Morehouse Lane
Red Hook, NY 12571

Some format issues inherent in the e-media version may also appear in this print version.

Printed by Curran Associates, Inc. (2017)

For permission requests, please contact the Institute of Physics
at the address below.

Institute of Physics
Dirac House, Temple Back
Bristol BS1 6BE UK

Phone: 44 1 17 929 7481
Fax: 44 1 17 920 0979

techtracking@iop.org

Additional copies of this publication are available from:

Curran Associates, Inc.
57 Morehouse Lane
Red Hook, NY 12571 USA
Phone: 845-758-0400
Fax: 845-758-2633
Email: curran@proceedings.com
Web: www.proceedings.com

Table of contents

Volume 872

24th International Conference on Vacuum Technique and Technology
6–8 June 2017, St. Petersburg, Russian Federation

Accepted papers received: 30 June 2017
Published online: 10 July 2017

Preface

Papers

IOP Conf. Series: Journal of Physics: Conf. Series **872** (2017) 011001 doi:10.1088/1742-6596/872/1/011001

Main Results of the 24th International Conference on Vacuum Technique and Technology (VTT2017)

A A Lisenkov[1] **and D K Kostrin**[2]

[1] Institute of Problems of Mechanical Engineering, Russian Academy of Sciences, 199178, Saint Petersburg, Russia
[2] Saint Petersburg Electrotechnical University “LETI”, 197376, Saint Petersburg, Russia

E-mail: lisran@yandex.ru, dkkostrin@mail.ru

Abstract. In this preface the main results of the 24th International Conference on Vacuum Technique and Technology (VTT2017) that was held on 6–8 June 2017 in Saint Petersburg, Russia are discussed.

In the period from 6 to 8 June 2017 at the Saint-Petersburg Electrotechnical University “LETI” and All-Russian Research Institute of Metrology took place the 24th International Conference on Vacuum Technique and Technology (figure 1). The Organizers of the Conference were: Russian Academy of Sciences, Institute of Problems of Mechanical Engineering RAS, Saint Petersburg Electrotechnical University “LETI”, Saint Petersburg Polytechnic University, All-Russian Research Institute of Metrology and Russian Scientific and Technical Vacuum Society.

Figure 1. Emblems of the Saint-Petersburg Electrotechnical University “LETI” (left) and All-Russian Research Institute of Metrology (right).

The aim of the Conference was to study the results of the research in physics of vacuum, vacuum measuring, mass spectrometry and leak detection. Were discussed the issues of producing vacuum, vacuum equipment and development of the new technological processes. Special attention was paid to the use of vacuum technology in the processes of films and coatings formation using plasma and related methods, study of the properties of coatings and methods of their research, new composite coatings, including nanomaterials, as well as new areas of their use, and the development of modern equipment and processes. Were analyzed the problems of industry cooperation and collaboration in the framework of the Federal Target Programs and the formation of the General Strategy of the development of vacuum equipment and technologies as an independent high-tech industries.

In this regard, the main sections of the Conference were:

1. Vacuum technique. Vacuum measurement. Mass spectrometry.
2. Vacuum ion-plasma methods of surface treatment.
3. Nanomaterials and nanotechnologies.
4. Physical processes of coating formation, properties and methods of their evaluation.

In the course of work of the Conference were presented 87 oral reports, including 13 plenary, and 21 posters. Among the speakers were professors, PhD students, students and representatives of industry. The Conference was attended by representatives of different regions of Russia and Republic of Belarus, Kazakhstan, Ukraine and Vietnam (figure 2). The total number of participants was 115.

Figure 2. Some photos from the Conference.

All of the presented scientific results and developments correspond to the level of World analogues, and in some cases surpass them in the performance and technological capabilities. Most of the developments are protected by patents for inventions; some of them were also presented at the XII International Exhibition of Vacuum Equipment "VacuumTechExpo 2017" (Moscow, 11–13 April 2017).

All messages of the Conference participants draw the scientific and practical interest. In general discussions questions, comments and opinions were presented. It should be noted the high professionalism of the speakers and the high ethics of communication parties during the exchange of the opinions and answers to the questions.

During the Conference were held "round tables" devoted to the analysis of the reports and the work of the Conference, also they touched upon the issues of training of the students and postgraduates. It was noted that currently young people pay more attention to the technical education and after graduation find jobs at the enterprises of electronic industry.

In order to summarize the Conference results participants noted that the diversity and number of the submitted reports are the signs of the constant technical development of this field of science and technology. However, against this background, participants, technical specialists with deep theoretical knowledge and extensive practical experience, expressed concern about the absence of a strategic plan for the development of vacuum technology in the Country in the near future.

The Conference participants decided to:

1. Consider the work of the Conference as successful.

2. State that the Conference is an effective link between the industrial enterprises, educational and academic institutions located in different regions of Russia and abroad.

3. Keep the format of this Conference and held the next one in June 2018.

4. Draw the attention of the Russian Scientific and Technical Vacuum Society and its President on the need for consolidation of specialists of this direction, their organizations and, subject to the comments made, further promoting the achievements and relevance of tasks associated with the development of vacuum technology.

IOP Conf. Series: Journal of Physics: Conf. Series **872** (2017) 011002 doi:10.1088/1742-6596/872/1/011002

Peer review statement

All papers published in this volume of *Journal of Physics: Conference Series* have been peer reviewed through processes administered by the proceedings Editors. Reviews were conducted by expert referees to the professional and scientific standards expected of a proceedings journal published by IOP Publishing.

IOP Conf. Series: Journal of Physics: Conf. Series **872** (2017) 012001 doi:10.1088/1742-6596/872/1/012001

Some problems of the vacuum technique development

L N Rozanov[1] **and A N Sokolov**[2]

[1] Saint Petersburg Polytechnic University, 195251, Saint Petersburg, Polytechnicheskaya st. 29, Russia
[2] JSC Corporation "Comet", 194021, Saint Petersburg, Shatelena st. 7, Russia

E-mail: lnrozanov@mail.ru

Abstract. In this work the problems of obtaining ultra-high vacuum without warming of the vacuum chamber are discussed. Numerical calculations have shown the possibility of obtaining within one hour the degree of coverage 10^{-2} in conventional vacuum chambers and $2\cdot10^{-3}$ in the chambers of special shape.

1. Introduction

The heyday of vacuum technique in the second half of the last century is over, but the possibilities of evolutionary development still exist. They are associated with the increasing use of vacuum technique in industry and scientific research. It is possible to identify ways of its further development on the basis of currently existing problems. The list of these problems in a large part is subjective and depends on the area of interest of the user. Let us consider some of them.

Vacuum physics offers a lot of methods of vacuum systems calculation, but it cannot provide the necessary for practice accuracy. This leads during the design of vacuum systems to do the calculations with a large margin, leading to a significant rise in the cost of the projects.

In the technique of producing vacuum pressure limit of scroll pumps is significantly worse than that of rotary vane oil pumps. Turbomolecular pumps typically can't have a direct exhaust into the atmosphere and have reserves to improve their pressure limit.

Combined vacuum converters have poor metrological characteristics and lower limit of measurement. Metrological basis for calibration of vacuum gauges and leak detectors needs significant modernization, and centralized verification and calibration of analyzers of partial pressure is absent.

Leak detectors have become so expensive devices that only large firms can afford them. The mass-spectrometer leak detectors have problems with the measurement of flows of more than 10^{-4} $m^3\cdot Pa/s$. Use of partial pressures analyzers can be in addition addressed not only to measuring partial pressures, but also for the problem of leakage detection. Moreover, the cost of partial pressure analyzers is several times less than the cost of leak detectors. However, during the direct use of partial pressure analyzers as leak detectors, there are certain difficulties with the pumping of helium.

2. Obtaining ultra-high vacuum without warming of the vacuum chamber

Let us consider the problem of obtaining ultra-high vacuum without warming of the vacuum chamber. Requirement for the ultra-high vacuum is obtaining of clean surfaces free from adsorbed gases. In the middle of the last century it was believed that to obtain ultra-high vacuum the chamber should be warmed up to 450 °C. This value was associated with the softening temperature of the glass, which was necessarily present in the design of vacuum systems. Then, this number declined to 200 °C, providing desorption of the water vapor. Let us analyze is it possible at present to further reduce this temperature to room temperature, i.e. to exclude vacuum installation heating.

Let us define what minimum degree of coverage can be currently obtained without heating of the vacuum chamber for one hour of pumping. We choose water vapor as the main polluter of the surface. Let us conduct theoretical analysis, considering that pumping is carried out using an ideal pump, the pressure limit of which is equal to zero.

The amount of gas adsorbed on the internal surfaces of vacuum systems is determined by the isotherms of adsorption [1–3]. Frendlich equation [2] for isotherms of water vapor adsorption at room temperature on the surface of electropolished stainless steel can be written in the form:

$$\theta = dp^m / a_m, \tag{1}$$

where θ – degree of coverage; $a_m = 0.116$ m^3·Pa/m^2 – specific amount of adsorbed gas in a monolayer; $d = 0.0176$, $m = 0.25$ – constant coefficients [1]; p – pressure, Pa.

The degree of coverage strongly depends on temperature (figure 1), which makes heating of a vacuum unit an effective means of its reducing.

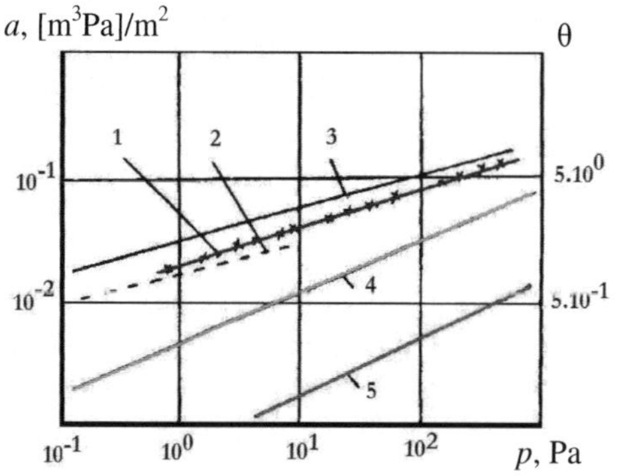

Figure 1. Dependence of the amount of the absorbed gas (a) and the degree of coverage (θ) of water vapors on stainless steel from pressure at different temperatures: 1 – 298 K [2]; 2 – 298 K [1]; 3 – 273 K; 4 – 350 K; 5 – 450 K.

Let us use a mathematical model of pumping of vacuum chambers with absorbent walls [4], which implies the equation for calculating the time of pumping of the water vapors:

$$t = 2.3v\lg\frac{p_i}{p_f} + \frac{dmf}{1-m}\left(p_f^{m-1} - p_i^{m-1}\right), \tag{2}$$

where t – pumping time, s; $v = V_c/S$, s; $f = F_c/S$, s/m; $S = S_m K_u(1 - p_m/p_f)$, m^3/s; V_c – volume of the vacuum chamber, m^3; F_c – area of the inner surface of the chamber, m^2; S – effective pumping speed, m^3/s; S_m – maximum pumping speed, m^3/s; K_u – coefficient of the pump usage; p_m – pump pressure limit, Pa; p_i – initial water vapors pressure, Pa; p_f – final pressure of the water vapors, Pa: d, m – coefficients from the Frendlich equation for adsorption isotherms. Pumping curves calculated using equations (1) and (2) for the pressure and the degree of coverage are shown in figure 2.

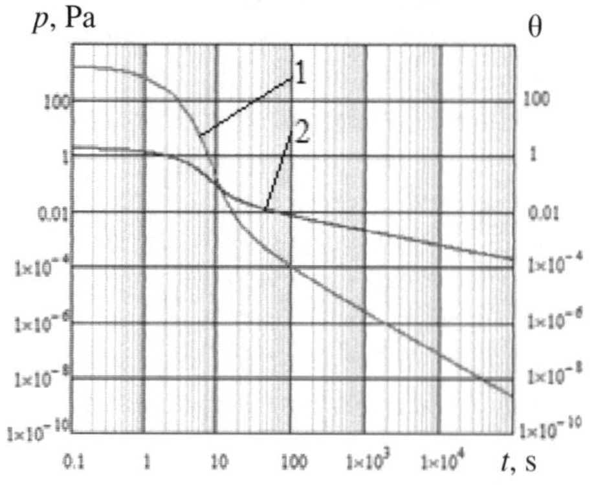

Figure 2. Curves of water pumping at 298 K for $v = 1$ s, $f = 20$ s/m: 1 – pressure; 2 – amount of absorbed gas.

IOP Conf. Series: Journal of Physics: Conf. Series **872** (2017) 012001 doi:10.1088/1742-6596/872/1/012001

Amount of absorbed gas during pumping decreases more slowly than pressure. To reduce the degree of coverage by 10 times it is necessary to reduce the pressure by 10^3 times (figure 2). The first term in the formula (2) takes into account the time of pumping from the volume of the chamber, and the second – time of pumping of the gases adsorbed on the surface. Effect of the chamber volume on the pumping curve is shown in figure 3.

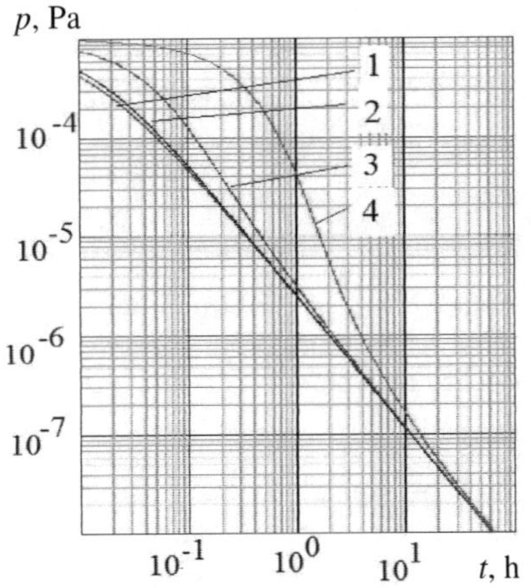

Figure 3. Curves of pumping of the vacuum chamber, made from electropolished stainless steel (d = 0.176; m = 0.25; f = 37.5 s/m; $p_i = 10^{-3}$ Pa), for water vapors at different values of the ratio of chamber volume to effective pumping speed of the pump $v = V/S$: 1 – 1 s, 2 – 10 s, 3 – 100 s, 4 – 1000 s.

If the following condition is met:

$$2.3 v \lg \frac{p_i}{p_f} << \frac{dmf}{1-m}\left(p_f^{m-1} - p_i^{m-1}\right), \qquad (3)$$

pumping curve does not depend on the volume of the chamber. In a typical vacuum system, the influence of the volume of the working chamber for pumping curve becomes insignificant after one hour after the start of the pumping process. Influence of the initial pressure on the pumping curve is shown in figure 4.

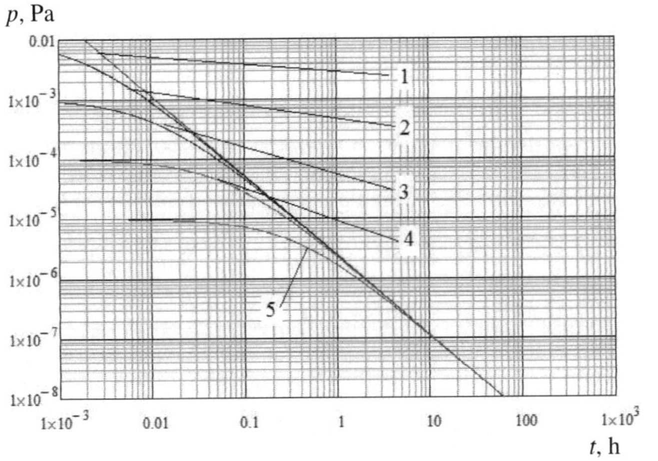

Figure 4. Curves of pumping of the chamber from electropolished stainless steel (d = 0.176; m = 0.25; f = 37.5 s/m; v = 1.25 s) for water vapors at different values of the initial pressure p_i: 1 – 10 Pa; 2 – 10^{-2} Pa; 3 – 10^{-3} Pa; 4 – 10^{-4} Pa; 5 – 10^{-5} Pa.

Provided that

$$p_i / p_f > 10^2, \qquad (4)$$

pumping curve becomes independent of the initial pressure of pumping. Usually the condition (3) is fulfilled within one hour after the start of the pumping process. Thus, we can assume that after the time of pumping more than one hour occurs a regular mode, in which the pumping curve in a logarithmic scale asymptotically approaches a straight line, which is described by the following equation:

$$p(t)=\left[\frac{dmF}{3600S\left(1-m\right)t}\right]^{\frac{1}{1-m}}=A_1t^{-B},\qquad(5)$$

where t – pumping time, h; $A_1=\left[\frac{fdm}{3600\left(1-m\right)}\right]^{\frac{1}{1-m}}$, $B=\frac{1}{1-m}$.

Using the experimental coefficients A_1 and B (5), which are determined from the regular mode, the coefficients of the Frendlich adsorption isotherms (1) could be found:

$$d=3600\frac{(1-m)A_1^{1-m}}{mf},\quad m=\frac{B-1}{B}.\qquad(6)$$

Degree of coverage from (1) and (5) can be calculated by the following equation:

$$\theta=\frac{d\left[dm\frac{f}{3600(1-m)}\right]^{\frac{m}{1-m}}}{a_m}.\qquad(7)$$

Management of the degree of coverage, using the parameters a_m, m and d is difficult, because they are the properties of the material. The f parameter is a characteristic of the vacuum system and can be selected more freely. The dependence of the degree of coverage from the various parameters of equation (7) shows (figure 5) the possibility of obtaining degree of coverage in the range $10^{-2}\ldots10^{-3}$.

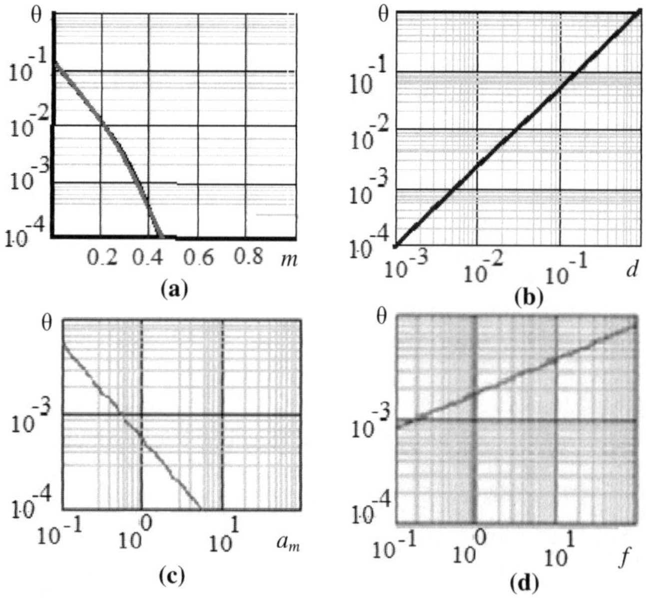

Figure 5. Dependences of the degree of coverage from: **(a)** – m at d = 0.0176, a_m = = 0.116 $m^3\cdot Pa/m^2$, f = 20 s/m; **(b)** – d at m = 0.25, a_m = = 0.116 $m^3\cdot Pa/m^2$, f = 20 s/m; **(c)** – a_m at m = 0.25, d = = 0.0176, f = 20 s/m; **(d)** – f at m = 0.25, a_m = 0.116 $m^3\cdot Pa/m^2$, d = 0.0176.

Value of $\theta = 10^{-3}$ ($2\cdot10^{-9}$ Pa), if f = 0.2 s/m. For the vacuum chamber in a form of a cube (figure 6(a)), through one of the sides of which within one hour pumping by an ideal pump is conducted value f = 0.5, that from figure 2 corresponds to $\theta = 2\cdot10^{-3}$ (10^{-7} Pa). For the pump, closed with a plug (figure 6(b)) value f = 0.1, this corresponds to $\theta = 9\cdot10^{-4}$ ($5\cdot10^{-8}$ Pa).

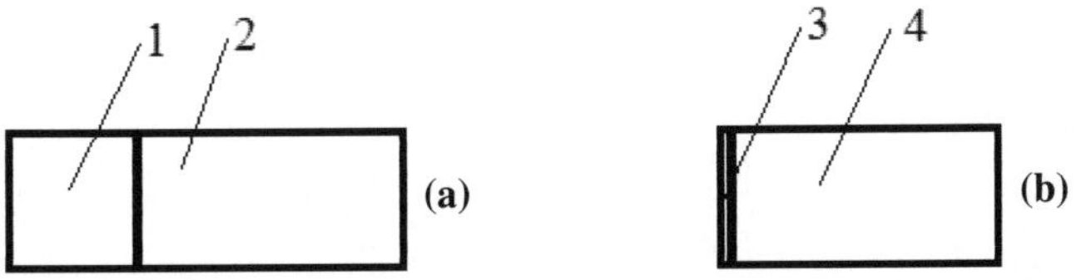

Figure 6. Vacuum chambers: 1 – cubic cell; 2 – pump; 3 – plug; 4 – pump.

Theoretical results (5) and (6) can be checked on the experimental setup (figure 7), including a vacuum chamber made of stainless steel 12KH18N10T with electropolished inner surface of volume $V = 50$ l and internal surface area $F = 1.875$ m^2; spiral Edwards XDS35i pump with a pumping speed of 10 l/s; turbomolecular pump Edwards nEXT 300 with pumping speed 300 l/s; quadrupole mass spectrometer EXtorr XT300M; vacuum valve Du-100 with metal seal.

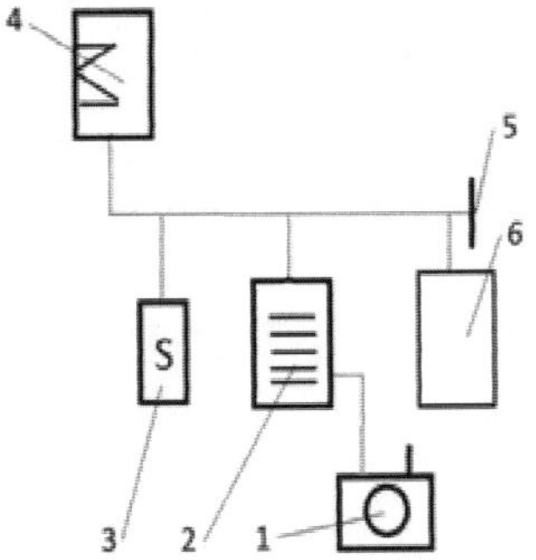

Figure 7. Experimental setup: 1 – spiral pump; 2 – turbomolecular pump; 3 – mass-spectrometer; 4 – cryosorbtion pump; 5 – entrance; 6 – inlet system.

The vacuum chamber was sealed using metal seals. Effective pumping speed was 1 l/s for spiral pump and 50 l/s for turbomolecular pump. Thus, for this installation $v = 1$ s and $f = 37.5$ s/m.

Mass spectrum of residual gases (figure 8) for the experimental setup (figure 7) consists of hydrogen, methane, water, nitrogen and carbon monoxide, oxygen, argon and carbon dioxide, but the main component, as expected for non-heated systems, is the water vapors, pressure of which at the end of pumping was $1.5 \cdot 10^{-6}$ Pa.

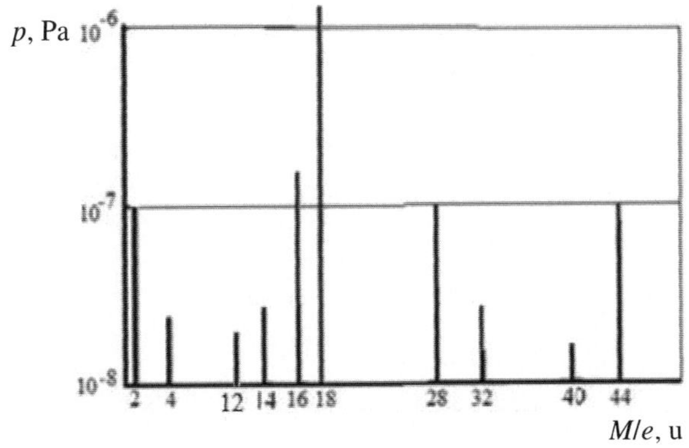

Figure 8. Mass spectrum of residual gases at the end of pumping process.

The results of calculations and experimental data (figure 9) correspond to each other, which confirm the possibility of using equations (5) and (7) to determine the extent of coverage of the internal surfaces of the vacuum chamber by water molecules.

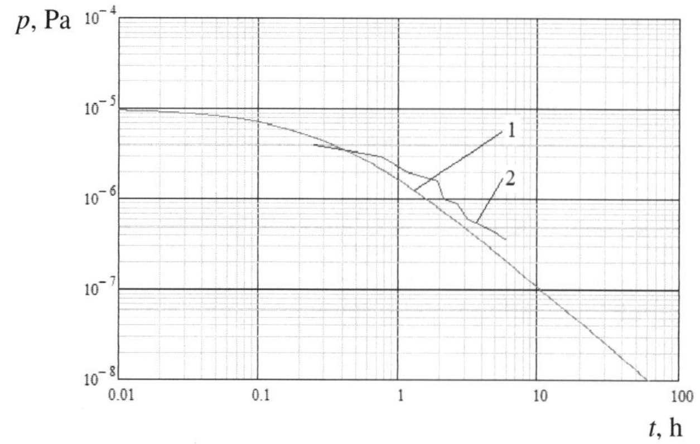

Figure 9. Pumping curve for the experimental setup at $v = 1$ s, $f = 37.5$ s/m: 1 – according to equation (5); 2 – experimental results.

After a brief purge of the atmosphere into the chamber and pumping by a spiral pump to the limit pressure for 850 s, the water vapor pressure was 10^{-5} Pa. After one hour of pumping with a turbo molecular pump was achieved a water pressure of $2 \cdot 10^{-6}$ Pa, which corresponds to the degree of coverage $\theta = 5.7 \cdot 10^{-3}$.

3. Conclusion

Obtaining of the ultra-high vacuum ($\theta = 10^{-2}$) without warming of the vacuum chamber from stainless steel and in the absence of the technological gas release is possible by means of modern vacuum pumps for a one hour high vacuum pumping. A further reduction of the degree of coverage ($\theta = 10^{-3}$) can be achieved by the reduction of *f* factor or increasing the duration of pumping.

References

[1] Jilnin V S, Jilnina L N and Kuzmin A A 1974 *Electronic technique. Electrovacuum and gas-discharge devices* **4** 74–80
[2] Rozanov L N and Akimov Yu D 1985 *Journal of applied chemistry* **6** 1381–4
[3] Rozanov L N 2007 *Vacuum technique* (Moscow: Higher school)
[4] Rozanov L N 2016 *Journal of Physics: Conference Series* **729** 012001

IOP Conf. Series: Journal of Physics: Conf. Series **872** (2017) 012002 doi:10.1088/1742-6596/872/1/012002

Microfocus X-ray tubes with increased power

A A Podimsky[1], N N Potrakhov[2], N Y Staroverov[2], V V Klonov[2] and I A Larionov[2]

[1] JSC "Svetlana-X-Ray", 198095, Saint Petersburg, Russia
[2] Department of electronic instruments and devices, Saint Petersburg Electrotechnical University "LETI", 197376, Saint Petersburg, Russia

E-mail: kzhamova@gmail.com

Abstract. The designs and fundamentals of the three types of microfocus X-ray tubes for increased power and voltage are described, including those with a rotating anode, a stationary internal anode, and an anterior-impaired anode. Areas of application of these tubes are indicated.

Improvement of the electronic component base in terms of increasing the degree of integration and reducing the size of microcircuits, as well as passive *RLC* components led to the creation of new technologies for their mounting on printed circuit boards [1–3].

To ensure quality control of printed circuit boards mounting directly in production conditions, in the early 2000s, specialized X-ray units were developed [4, 5].

At present, such an installation generally includes: a radiation source based on a collapsible X-ray tube with a constant pumping out with a microscopic focal spot and an image receiver on the basis of a flat panel X-ray detector located in a special chamber for radiographic work. When controlling printed circuit boards, two methods of obtaining an X-ray image are commonly used: contact and projection [6].

The leading manufacturers of X-ray inspection systems currently design and manufacture plants only based on microfocus X-ray tubes with constant pumping. The advantage of this design is the possibility of repeated replacement of the cathode and anode target. This makes it possible to operate the tube at the maximum loads – emissive to the cathode and thermal to the target, but it causes a number of disadvantages in comparison with the tubes sealed off from the vacuum (pumping) system:

- large dimensions, weight and complexity of the construction, connected, for example, with the use of a metal-ceramic insulator unit designed for the total working voltage of the tube, as well as vacuum-tight mechanical collapsible connections of a metal tube of a tube with replaceable units;
- presence in the tube of an integral constructive element – a specialized pumping system;
- need for high-voltage tube training after the replacement of individual units, which imposes additional requirements on the generator device of the power source of the tube, and the subsequent alignment of its electron-optical system;
- high cost.

However, three-electrode metal-glass tubes with a long-term anode tube from a vacuum cylinder of a thin-film target of the BS series, which were developed and mass produced in the OKB RP of the LOEP "Svetlana" RP, were made simpler in design and convenient in operation and also cheaper, starting from the mid- X of the last century [7].

This significantly limits the scope of microfocus radiography, implemented on sealed-off tubes. As is known intensity I, generated by the X-ray tube, connected with the parameters of X-ray tube by the following expression:

$$I = kizU^2, \tag{1}$$

where I – intensity of radiation, generated by the X-ray tube; k – coefficient of proportionality; z – the atomic number of the material of the anode target; i – current of the tube; U – voltage on the tube.

In accordance with expression (1), the radiation intensity can be increased and by increasing a tube current, and the voltage on it. Therefore, at creation of new generation of microfocus X-ray tubes have been used both directions.

As is known, on the tubes with a rotating anode current may be several orders of magnitude higher than in tubes with a fixed (stationary) anode. Therefore, during the development of "high current" microfocus tube as base chosen tube 15-40BD46-150, which is a kind of standard tubes in a class with a rotating anode. Modernization undergone two nodes of basic design – cathode and anode. As a result, in the cathode assembly has only one cathode, and to enable adjustment of the focal spot size the both outputs of cathode was electrically insulated from the focusing electrode. The composition of the modernized assembly includes: the cathode holder 1, directly heated cathode and the focusing electrode. The cathode 3 is made as a cylindrical spiral, two outputs of which are mounted in the holder, and both of outputs has ceramic insulators. Focusing electrode fixed on the cathode holder 1 (figure 1).

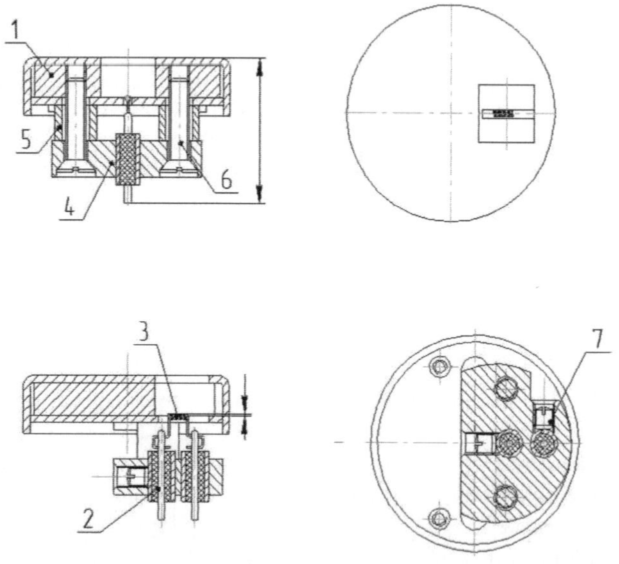

Figure 1. Cathode head of X-ray tube 0.75BD63-150.

To adjust the focal spot size the control voltage U_{adj} (negative polarity with respect to the cathode) is supplied to the focusing electrode. Depending on the magnitude of the voltage, the effective focal spot size will vary from a maximum 0.6×0.6 mm with a control voltage equal to zero to the minimum 0.1×0.1 mm with a control voltage to 100 V.

When upgrading anode node were tightened requirements to the value of bounce at node in the longitudinal and lateral (radial bounce) directions, and also to the magnitude of vibration and noise. Studies have shown that one of the causes of these effects – the lack of rigidity of the bearing housing and of the axis of the assembly rotation in the specified directions. In this basic design of the tube the radial clearance is comparable with the desired size of the focal spot. As a result of analysis and numerical estimates of allowable rotation speed of the anode assembly was chosen unseparatored ball-bearing assembly with a temperature-compensating gap type ZYS99.

The construction of the first domestic X-ray tube with a rotating anode with smooth adjustment of effective focal spot size 0.7BD64-150 shown in figure 2.

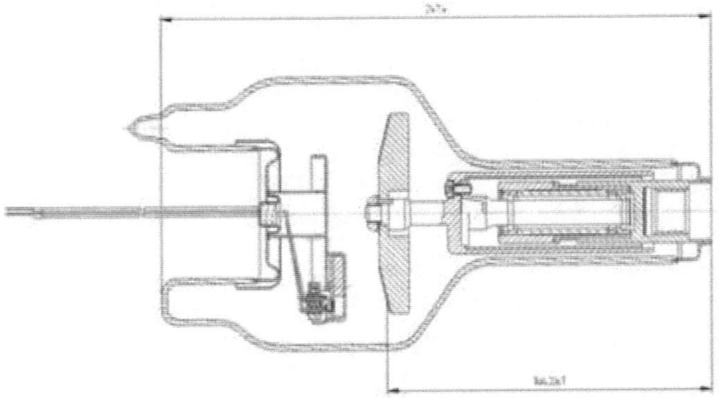

Figure 2. Microfocus X-ray tube with a rotating anode 0.7BD63-150: max. voltage 150 kV; max. current 12 mA; size of the focal spot at a bias voltage 0 V – 0.6×0.6 mm; size of the focal spot at a bias voltage 55V – 0.1×0.1 mm; max. rotation rate of the anode 9000 rot/s.

In the development of microfocus X-ray tube at a higher voltage with a massive target located inside the vacuum container, was chosen tube design 1.2BPK21-200. Modernization has undergone the cathode assembly of basic design. Firstly, in order to provide the micron size of focal spot, instead of the cathode filament in the form a cylindrical spiral was used filament in the form of V-shaped loop. Secondly, in order to enhance the electron beam focusing capability by independent adjustment of the filament current and voltage on the focusing electrode, a second output of the cathode is also insulate from the focusing electrode. For this, both the cathode outputs is mount in the holder by screws, as in the tube with a rotating anode on ceramic insulators.

The construction of the first domestic microfocus X-ray tube with a massive internal anode voltage of 200 kV 0.2BPM64-200 is shown in figure 3.

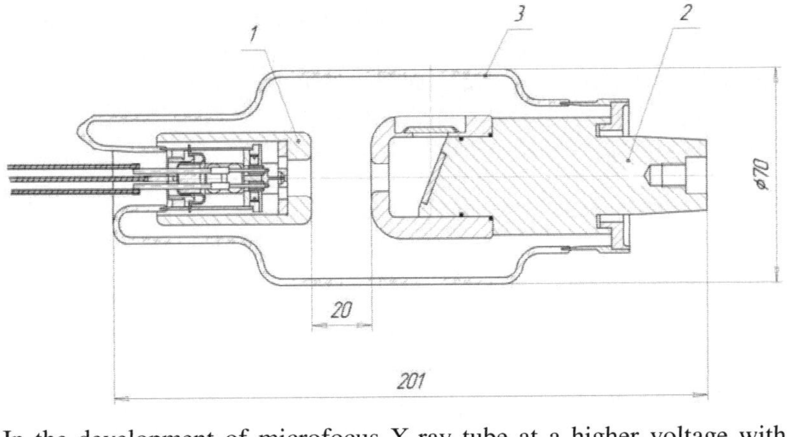

Figure 3. Microfocus X-ray tube with a massive internal anode 0.2BPM64-200: max. voltage 200 kV; max. current 1 mA; min. size of the effective focal spot 0.1 mm; diameter 70 mm; length 210 mm.

In the development of microfocus X-ray tube at a higher voltage with the external target from the vacuum cylinder as a base was choosen tube BS16(I) – maximum operating voltage of 135 kV. During modernization of the base construction in order to increase the operating voltage to 150 kV by minimizing the electric field intensity in the high voltage gap of the tube were refined dimensions of the outside diameter of the focusing electrode of the cathode node and inner diameter of the anode screen. With the same purpose in the assembly of the tube was introduced the operation autofrettaging abrasive medium.

Furthermore, in order to increase heat removal from external target and increase its loading capacity the anode material of the tube instead of stainless steel is replaced by copper. The design of microfocus X-ray tube with the external target from the vacuum cylinder for voltage 150 kV BS16(III) – shown in figure 4.

All three tubes of the new generation are designed for operation as part of a specially developed X-ray sources and devices: 0.7BD63-150 in the source IRD46 intended for medical diagnosis, 0.2BPM64-200 in a monoblock unit RAP-200M, BS16(III) in a monoblock unit RAP-150M.1 intended, in particular, for microfocus X-ray CT scanners family MRKT-01.

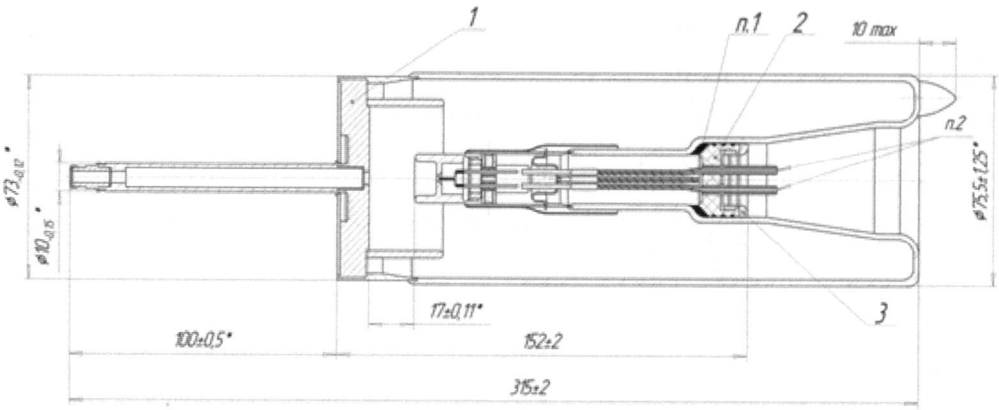

Figure 4. Microfocus X-ray tube with the external target BS16(III): max. voltage 150 kV; max. current 150 μA; min. size of the real focal spot 20 μm; dimensions of the balloon – length 252 mm, diameter 75 mm; dimensions of the external anode – length 100 mm, diameter 12 mm.

According to the preliminary assessment the implementation of these technical means will allow essentially extend the scope of microfocus X-ray radiography in socially important fields of medical diagnostics, in the monitoring of the products and electronic components of the micro- and nanotech in all sorts of research.

Acknowledgment
The work was supported by the Russian Science Foundation in the framework of the project "Creating portable systems for microfocus radiography with the purpose of operational control of the microstructure, physicochemical properties and determining the residual life of aircraft parts and assemblies made of composite materials". Number of project 15-19-00259.

References
[1] Pirogova Y V 2005 *Proyektirovaniye i tekhnologiya pechatnykh plat* (Moscow: Forum, Inra-M)
[2] Bakanov G F, Sokolov S S and Sukhodol'skiy V Yu 2007 *Osnovy konstruirovaniya tekhnologii radioelektronnykh sredstv* (Moscow: Akademia)
[3] Koledov L A 2008 Tekhnologiya i konstruktsii mikroskhem, mikroprotsessorov i mikrosborok (Saint Petersburg: Lan')
[4] Shmakov M 2006 *Tekhnologii v elektronnoy promyshlennosti* **4**
[5] Garanin A 2012 *Elektronika. Nauka. Tekhnologiya. Biznes* **6** 170–5
[6] Artem'yev B V and Bukley A A 2013 *Radiatsionnyy kontrol'* (Moscow: Spektr)
[7] Ivanov S A , Ioffe Y K, Kiriyenko S V et al. 1987 *Obzory po elektronnoy tekhnike. Ser. 4* **4**

Characteristics of radial ion-plasma accelerators

S M Bondarenko, C V Korolev, Yu B Movsesyants and P M Tyuryukanov

All-Russian Electrotechnical Institute, Krasnokazarmennaya str. 12,
Moscow, 111250 Russia

E-mail: tyuryukanov@vei.ru, svulm@list.ru

Abstract. The characteristics of two-stage ion-plasma accelerators are presented. These accelerators are based on a discharge in a transverse highly-inhomogeneous magnetic field and can form radially converging and divergent flows. It is shown that for a radially convergent flow the width of the ion acceleration zone is limited by the condition of a transition through the ion sound point in the vicinity of the plasma ion emitting boundary.

1. Introduction

The use of accelerators that form a radially convergent (IRD) and radially divergent (IRR) quasineutral ion fluxes, is promising in vacuum ion-plasma deposition technologies for the inner and outer surfaces of extended cylindrical parts. Two-stage ion-plasma accelerators (IU) [1] were developed in the All-Russian Electrotechnical Institute on the basis of high-current and high-voltage discharges in a transverse strongly inhomogeneous magnetic field.

Their characteristics have been thoroughly studied in [2]. The parameters of the IU can be regulated over a wide range (ion energy 0.1 to 40 keV and current density up to 100 mA/cm^2). This allows to carry out various technological operations: cleaning, etching, coating and implantation in a continuous mode for a single injection of the part into a vacuum chamber on the same accelerator. These IEDs are used to create tribological pairs with a low coefficient of friction and increased wear resistance. The inner surface of the cylinders and the outer edge of the piston rings were irradiated with molybdenum and argon ions at an energy of 30 keV and a current density of 1 mA/cm^2. High-strength hydrophobic and dusty DLC (diamond like carbon) films are obtained on the surface of high-voltage insulators made of ceramics and silicone rubber.

The properties of the deposited hydrocarbon coatings on the substrates of various materials (Mo, Cu, Al, Si, SiO_2 and silicone rubber) have been determined by X-ray photoelectron spectroscopy (XPS), Reflected Electron Emission (EIS) spectroscopy, Reflected Electron Emission (ESR) spectroscopy and spectroscopy of characteristic Energy losses (SCPE) [3]. Investigations of the elemental composition of the samples were carried out at REC Nanotechnologies of the Moscow Power Engineering Institute. The XPS method made it possible to diagnose a small admixture of oxygen in the surface, in addition to carbon, which is more than 97 %. The exception is rubber, where the oxygen content reaches 29 %. Studies of a series of samples by the SPCE method were carried out at an energy of 3 keV. Spectrometric analysis showed the presence of a characteristic peak.

The loss of energy (4.9 eV) clearly indicates the presence of hydrogen on the surface of the samples under study. This peak is present in the spectra for all 6 samples. This allows us to consider the data of the coating hydrocarbon. The thickness of the DLC film is 5–15 nm.

2. Experimental results and their discussion

The device and principle of operation of radial control devices are described in [1]. Ions are drawn by a radial electric field from the active zone of a high-current discharge in a transverse magnetic field through an annular gap in the cathode electrode. The magnetic field created by permanent ring magnets reaches a maximum in the gap region and amounts to 0.07 T for the IGRD and 0.2 T for the IRIS. The gap has a width of 2 mm, the radius of the cathode for the IRS is 20 mm and for the IRIS 110 mm. These dimensions are set by the possible dimensions of the machined parts. The gas dynamic conditions and the ion current to the collector in these accelerators will be quite different. Plasma-forming gas is fed through a damping gap in the annular anode to the zone of high-current discharge. It maintains a constant pressure, for example for argon 0.3 Pa. Then, through the annular gap in the cathode, the gas enters the accelerating gap, where the pressure is 0.03 Pa and further into the vacuum chamber, which is pumped out by a turbomolecular pump with a capacity of 1000 l/s.

Figures 1 and 2 show the characteristic dependencies of the current on the collector I_c on the magnitude of the accelerating voltage U_a for different currents of the high-current discharge of the first stage of the accelerator for the IRIS and the IGDR, respectively. For each of the accelerators for small values of the ion current, there is a region of a floating potential which, as the accelerating voltage increases, becomes a mode of forming a space charge layer that limits the current. On the current-voltage characteristics, the inflection point is observed. Above it, the rate of current increase with increasing voltage decreases. Figure 3 shows a photograph of the plasma stream that flows from the first stage of the accelerator to the chamber under the potential of the cathode. It is seen that the flux is limited in radius. With increasing current, the minimum radius that a plasma stream can reach, decreases, remaining finite.

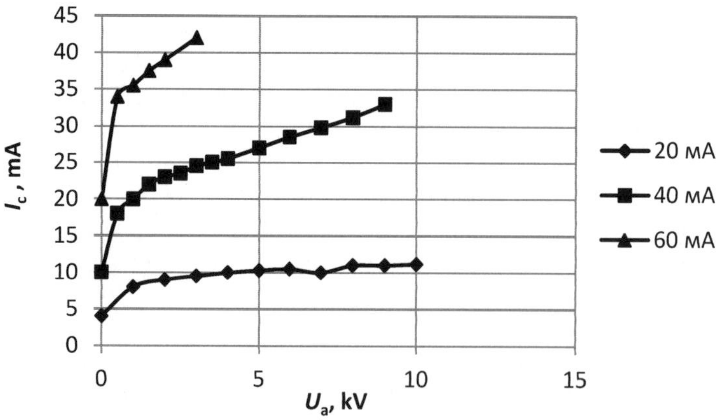

Figure 1. Volt-ampere characteristics of the IRIS at different currents of high-current discharge (0.1 Pa, Ar).

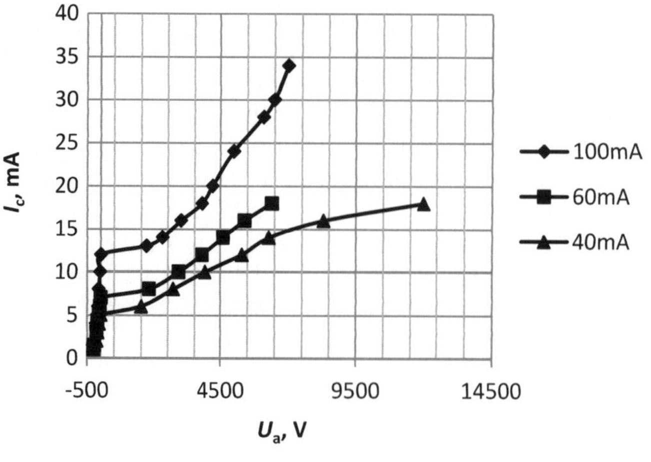

Figure 2. Volt-ampere characteristics of the ICR for different currents of high-current discharge (0.03 Pa, Ar).

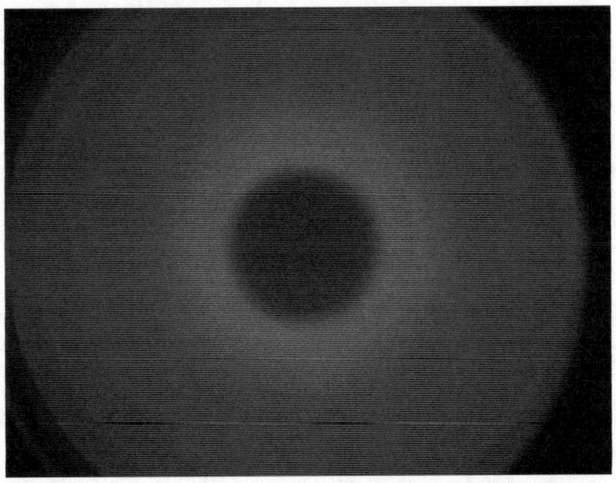

Figure 3. The exposure of the plasma flows to the IRIS (0.1 Pa, Ar).

The equilibrium states of the azimuthally symmetric collisionless ion flux compensated by Boltzmann electrons are described by self-consistent equations of hydrodynamics and Poisson, which can be represented as:

$$\left(\frac{r_D}{r_1}\right)^2 \frac{d}{d\rho}\left(\rho\frac{dy}{d\rho}\right) = \frac{1}{\sqrt{1+2\left(\frac{c_s}{V_0}\right)^2 y}} - \rho e^{-y};\quad V^2 = V_0^2\left(1+2\frac{c_s^2}{V_0^2}y\right);\quad j = \frac{1}{\rho}j_0;\quad j_0 = en_0V_0,$$

boundary conditions: $y(1) = 0$, $y(\rho_1) = \varphi_0$, $\rho_1 = r_2/r_1$,
where $\rho = r/r_1$, $y(\rho) = e(\varphi_0 - \varphi)/T_e$, $\varphi = \varphi(\rho)$ is the potential distribution in the accelerating gap, $\varphi_0 = \varphi(1)$, r_1 is the potential and radius of the flame emitter of ions, R_2 is the radius of the collector, $r_D = (\varepsilon_0 T_e)/(e^2 n_0)$ is the Debye radius, n_0 and T_e are the plasma concentration and electron temperature at the emitting boundary, $c_s^2 = T_e/M$ is the ion sound velocity, e and M are the charge and mass of the ion, E_0 is the dielectric constant of the vacuum. The condition $\rho_1 \geq 1$ corresponds to the ERI, and $\rho_1 \leq 1$ is the IRIS. The ion current density depends on ρ and for IRIS increases as approaching the axis $r = 0$. From the quasineutrality condition of the plasma near the emitter $r \approx r_1$ we obtain

$$\rho = \frac{e^y}{\sqrt{1+2\left(\frac{c_s^2}{V_0^2}\right)y}}.$$

It follows that ρ has a minimum $\rho_m = \exp(y_0)/(c_s/V_0)$ for $y_0 = (c_s^2 - V_0^2)/2c_s^2$ and therefore, if quasi-neutrality is observed, a radially divergent ion flux can be formed in the supersonic flow $V_0 \geq c_s$, and radially convergent under the subsonic condition $V_0 < c_s$. In this case, the quasineutral flow zone is bounded along the radius $\rho_1 \geq \rho \geq \rho_0$. Numerical calculations of potential distributions and volt-ampere characteristics for two types of accelerators are in satisfactory agreement with experiment, but require further development, taking into account the non-one-dimensionality of the flow and the presence of a magnetic field in the drift space.

3. Conclusion

It is established that on the basis of a discharge in a highly inhomogeneous magnetic field and a two-step acceleration scheme, it is possible to form both radially divergent and radially converging ion-plasma flows. The parameters of the ion beam can be controlled within a wide range. This opens the possibility to perform several technological operations on one accelerator and to modify the surface of extended cylindrical parts. The theoretical model above is consistent with the experimental results for both the IRIS and the IGRD.

References

[1] Movsesyants Yu and Tyuryukanov P 2011 *Engineering physics* **3** 21–5

[2] Tyuryukanov P, Nicholas A, Fetisov I and Tolstoy I 1982 *Plasma Physics* **6** 1220–7
[3] Afanasiev P, Ermilov A, Ivanov D et al. 2014 *Surface. X-ray, synchrotron and neutron techniques* **3** 74–8

IOP Conf. Series: Journal of Physics: Conf. Series **872** (2017) 012004 doi:10.1088/1742-6596/872/1/012004

Features of technique for leak test of gas-filled dischargers

S A Bushin

Federal State Unitary Enterprise "Dukhov Research Institute of Automatics (VNIIA)", Sushchevskaya st. 22, Moscow, Russia

E-mail: vniia4@vniia.ru

Abstract. The paper gives an overview of the issues related to the technique for leak test of miniature gas-filled dischargers developed and manufactured by VNIIA.

Creating and maintaining tightness in vacuum and gas-filled devices is one of the main conditions for maintaining operational parameters that determine their operability and guaranteed shelf life. The high requirements for the standard of leak tightness ($1.4\cdot10^{-13}$ Pa·m^3/s) of one of the lines of gas-filled devices produced by VNIIA led to the need to develop a special measuring procedure and to build an automated prototype of a vacuum machine for the final leak test (MFLT) to implement this procedure [1]. The technique for estimating the degree of leakage of dischargers is based on the mass spectrometric method, which is used as the basis for applying the accumulation in the development.

The partial pressure of the working gas on the MFLT is measured with QMG 220 PrismaPlus M2 quadrupole mass spectrometer. Under low pressure, its ion source can operate for a relatively long period of time without the need for any intervention in terms of routine maintenance. Nevertheless, after four years of operation of the mass spectrometer (~10 000 h), signs began to show indicating that the failure of the filament was a matter of just a few months.

Negative trends in the work of the mass spectrometer were not immediately apparent, but appeared gradually; typically after the discharge (transfer) of the accumulated portion, the dynamics of the change in the controlled response of the intensity at a given mass number deviated from the typical relation. This subsequently manifested in the form of significant distortions of the "zero" line for practically the entire group of recorded mass numbers from 1 to 45 (see figure 1). Previously almost absent, i.e. not typical, changes of pressure over time after the transfer of a gas portion in the form of a so-called "potential well" with a level of the trough extending to values below the background ones, began to be observed from one discharge to another, regardless of the accumulation time. As a result of indirect measurements, a minus sign appeared in the value of the measured leakage flux Q_i.

So that the received mass-spectrometric information, which has undergone non-typical changes, was not lost and was not useless given no opportunity to operatively replace the cathode, it became necessary to try to solve this problem in the future, since the probability of its repetition could not potentially be excluded during operation in future. The appearance of the above-mentioned problem required additional studies, mathematical modeling, and debugging certain points of the measurement technique. A mathematical model was developed in the form of equation that approximately describes the process of the gas portion transfer and pressure change during its expansion in the volume of the measuring system (in current terms), using elements of regression analysis.

IOP Conf. Series: Journal of Physics: Conf. Series **872** (2017) 012004 doi:10.1088/1742-6596/872/1/012004

The function of the output current signal of the mass analyzer I = f (t) can be represented as the following equation:

$$I = \Delta I_0 \left[1 - \exp\left(-\frac{S_{ef} t_{[i]}}{V_\Sigma + V_{ch}} \right) \right] \exp\left(-\frac{S_{ef} t_{[i]}}{V_\Sigma + V_{ch}} \right) + \frac{Q S t_{[i]}}{V_\Sigma + V_{ch}}, \quad (1)$$

where ΔI_0 – the increment of ion current of 20 M/e, [A]; S_{ef} – "pseudo pumping" rate of 20 M/e (S_{ef} = Varia), [m^3/s]; Q – flow of leakage from device [Pa·m^3/s]; S – coefficient of absolute sensitivity of mass spectrometer, [A/Pa]; V_Σ – total volume of measuring system, [m^3]; V_{ch} – volume of accumulation chamber, [m^3]; $t_{[j]}$ – interval (1st) of time (1.7...2 s), determines pressure equalization time in two volumes (V_Σ and V_{ch}) after gas portion is transferred, [s]; $t_{[i]}$ – current time from transfer to end of recording during measurements (before opening of the turbomolecular pump valve), [s].

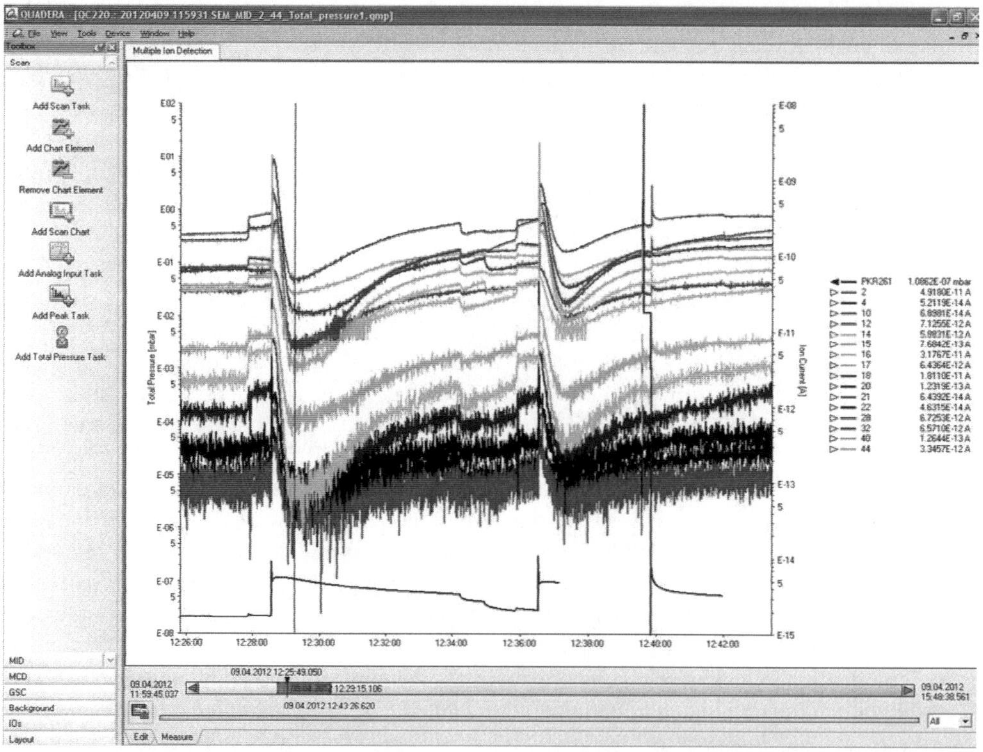

Figure 1. Dynamics of changes in mass numbers during leak test (two gas portions were consistently "dumped") with characteristic distortions – "potential wells".

It should be considered that during the first seconds of the discharge of gas portion, the surge in pressure is dominant due to the lack of complete isotropy (uniformity of distribution) of pressure when it is redistributed in the volume and due to the tribological factor (friction of the valve-saddle surfaces). The process of "pseudo-pumping", described by the expression

$$\Delta I_0 \left[1 - \exp\left(-\frac{S_{ef} t_{[i]}}{V_\Sigma + V_{ch}} \right) \right] \exp\left(-\frac{S_{ef} t_{[i]}}{V_\Sigma + V_{ch}} \right),$$

and the linear function (gas flow from the leak), described by the expression, also take place (due to the continuity of the flow), but their influence during the first few seconds is relatively small. It should be specially noted that the last expression most accurately describes the process of increasing the partial pressure of the working gas in the system due to the flow through the leak, and the desired value of the estimate of the leakage flow, Q, is the more accurate the closer to the "tail" part the sample is taken from the points used to determine it.

The principal nature of the studied function is known in advance based on the construction of the physical model, therefore, it is required to obtain only some numerical parameters that are part of the expression of the function. These include the flow, Q, and S_{ef} and ΔI_0 parameters, which are auxiliary and do not require high accuracy: the acceptable format of the result is "X.X".

The problem of determining the required parameters (Q, S_{ef}, ΔI_0) can be tried to resolve in different ways, for example, using the nonlinear least-squares method through construction of a system of normal equations. However, in order to avoid the "inconvenience" in solving the smoothing problem using (1), in which some of the parameters are nonlinear, the problem can be solved by setting a number of initial values of Q, S_{ef}, ΔI_0 parameters [2]. For each of them, the sum of the squared deviations of Y_k function from the initial measured values I_k with the search for the minimal discrepancy $|Y_k - I_k|^2$ is found.

Since for estimating the flow Q based on (1), the value of the background Q_b (determined from the change in the current signal of the analyzer during the time interval tt_1, tt_2, see figure 2) is not included explicitly in the variables involved in the calculation of the leakage flux, the value of the background flux is necessary for the final computations of the estimate of the degree of leakage Q', determined from the expression:

$$Q' = (Q - Q_b), \text{ [Pa·m}^3\text{/s]}. \quad (2)$$

The value of Q_b is calculated from the known equation of the following form:

$$Q_b = \frac{\Delta I_b V_\Sigma}{S\tau}, \text{ [Pa·m}^3\text{/s]}, \quad (3)$$

where (with other equal parameters, previously given), ΔI_b – difference between signals of mass spectrometer I_{tt2} and I_{tt1} at points of time tt_1 and tt_2, [A]; τ – accumulation time for background (tt_2 – tt_1), [s]. The results obtained in this case can have a "negative" value, but not by current increment, but already by the estimation of leakage flow Q'.

Because of the peculiarities in the operation of the ion source of the mass spectrometer, as well as the "contamination" with water vapor, as previously noted, the pumping process is delayed (3 to 8 h). In view of this, it was necessary to provide another way of obtaining useful information, similar to the case when the background characteristic Q_b exceeds the permissible values (Q').

As practice has shown, the most effective way to "not lose" the information received was the process of combining the operations of accumulation and measurement, i.e. the supply of test gas from the monitored device is not limited to the volume of the accumulation chamber. The gas, flowing "directly" through the leak with the open valves of the accumulation chambers ensuring the integration of the test object and the measuring volume, accumulates in the volume of the measuring system from the moment of shutting down the turbomolecular pump valve (tt_1) for a predetermined time interval. At the same time, the nature of the current function does not change, and there are no manifestations of "distortion" that preceded the transfer of the accumulated portion to the volume of the measuring system. In this way, the operation for accumulating and transferring a portion of gas from the storage chamber is replaced by accumulation with registration in the combined volume, including the vacuum cavity of the measuring system and the storage cavity.

The leakage flow of the monitored device(s) (n = 1...7) is determined by equation (4)

$$Q = \frac{\Delta I (V_\Sigma + n V_{ch})}{S\tau} - Q_b, \quad (4)$$

where [the terms of the equation are the same as the parameters in (3)] ΔI – difference of I_{tt4} and I_{tt3} signals at time tt_4 and tt_3 (more precisely, similar to tt_4 and tt_3, since for in this case, there will be only two time counts), [A]; τ – accumulation time for background (tt_4 – tt_3), [s].

Thus, in addition to the implemented software algorithm of data processing ("positive", "zero" current increment), two more are added for the case with a "negative" increment:

1. Using (1) according to the above algorithm (without repeating the operation associated with the transfer of the accumulated portion, i.e. only resorting to analytical calculations);

2. Using (3) with additional procedures for separate registration of the background of the measuring system and the total leakage flow from the device along with the background of the system through accumulation in the combined volume.

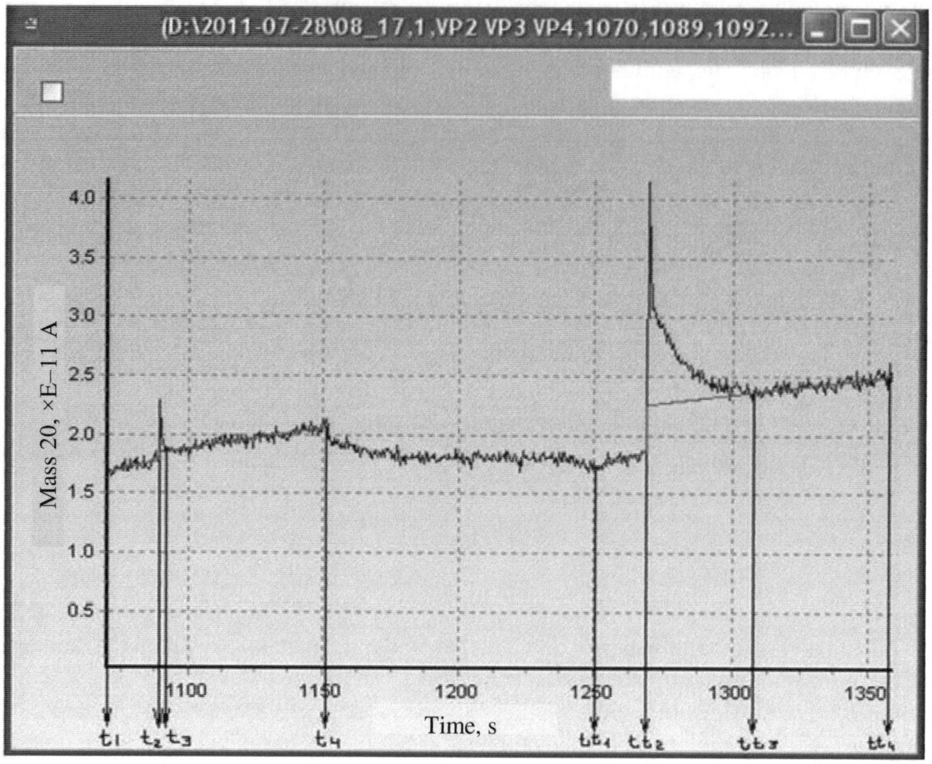

Figure 2. Graph-scan characterizing measuring process during implementation of "Express Analysis" project.

In figure 2: t_1 – start time of sample background record; t_2 – end time of sample background record; t_3 – start time of sample measurement; t_4 – end time of sample measurement; tt_1 – start time of background record of group of valves of accumulation chambers (or one); tt_2 – end time of background record of group of valves of accumulation chambers (or one); tt_3 – start time of measurement of group of valves in accumulation chambers (or one); tt_4 – end time of measurement of group of valves of accumulation chambers (or one).

However, here it should be noted that the option using additional mathematical processing at the final stage of development of special software was still rejected because of its comparative complexity. Therefore, the method used and implemented by the software algorithm was the combination of volumes.

Speaking about the methodological improvements and their usefulness, it is obvious, however, that the further steps taken by the tester due to the emerging uncertainty regarding the time of further operation of the cathode should be aimed at replacing the cathode (cathode node) of the mass spectrometer in order not to aggravate the situation with measurements, since the accuracy of the received data will be low in future.

Thus, it is necessary to take into account the operational characteristics of the mass spectrometer during operation and it is a forced procedure for the operator controlling the leak testing process. Undoubtedly, the mentioned difficulties require certain skills of the operators when operating this analytical equipment, despite the availability of automation. Nevertheless, in case of cathode-related events close to failure, the technique for final leak test of the dischargers, as a separate option, there is

IOP Conf. Series: Journal of Physics: Conf. Series **872** (2017) 012004 doi:10.1088/1742-6596/872/1/012004

a possibility implemented by software to avoid "loss" of data obtained with hardware "distortions" by using an additional method of obtaining useful information, despite the above-mentioned problems.

References

[1] Bushin S A 2012 *Vacuum Technique and Technology* **2** 98–9
[2] Wentzel E S 1964 *Theory of Probability* (Moscow: Science)

A new type of vacuum sensors with sensitive elements based on multicomponent oxide nanomaterials with fractal structures

A A Karmanov[1], I A Averin[1], I A Pronin[1], N D Yakushova[1], S E Igoshina[1], V A Moshnikov[2] and G V Vishnevskaya[1]

[1] Department of Nano- and Microelectronics, Penza State University, 440026, Penza, Russia
[2] Department of Micro- and Nanoelectronics, Saint Petersburg Electrotechnical University "LETI", 197376, St. Petersburg, Russia

E-mail: nano-micro@mail.ru

Abstract. A new type of vacuum sensors with sensitive elements based on two- and three-component oxide nanomaterials with fractal structures is proposed. It is shown that a sensory response for two-component oxide systems based on SiO_2-SnO_2 is determined by their quantitative composition – mass fraction of tin dioxide. An increase in sensitivity to changes in ambient pressure for three-component oxide nanomaterials has been demonstrated.

At present, to measure low pressures, a variety of devices different inoperational and dimensional characteristics are used, e.g., thermocouple, capacitance, ionization, and other types of vacuum gauges [1]. However, they have a number of drawbacks that limit the scope of their practical applications (e.g., large weight and size, energy consumption, narrow operating pressure range, etc.). Vacuum sensors based on nanomaterials are one of the developing branches of nanoengineering. Today, this subject is represented by few publications in the world highly-rated literature. Most of research in this area is based on the idea of replacing the traditional sensitive element of a gauge with a nanostructured one with dimensional effects. In particular, it is proposed to use the effect of field emission of carbon nanotubes [2], or nanomaterials based on zinc oxide [3], whose sensitivity mechanism to reduce the pressure below the atmospheric one is not fully understood at present. The nanotechnological direction associated with the creation of vacuum sensors based on nano- and microelectromechanical systems is actively developing [4].

Nanomaterials based on wide band gap semiconductor oxides with *n*- and *p*-type electrical conductivity (e.g. SnO_2, ZnO, TiO_2, In_2O_3, etc.), as well as multicomponent oxide systems (e.g. SiO_2-SnO_2, SiO_2-SnO_2-ZnO) are classical for the purposes of gas sensing. To synthesize such nanomaterials, various physical and chemical methods are used, which along with the technological modes of a wide range production, depend on their structure, qualitative and quantitative composition, and a number of other parameters. A promising method for obtaining nanomaterial data is a sol-gel technology that allows the synthesis of hierarchical structures of a fractal type spatial organization [5]. Within the framework of this technology, gas-sensitive nanomaterials based on a two-component silicon dioxide-tin dioxide system with a high sensory response to reducing gases, in particular, ethanol vapors have been developed. Structures with a percolation cluster appearing and disappearing upon the emergence of the analyzed gas, and as a consequence, having ultrahigh sensitivity [6], are also proposed. It should also be noted that multicomponent oxide nanomaterials with fractal structures have an undisclosed potential in the measurement of low pressures. In [7] it was demonstrated that for two-component oxide systems based on tin dioxide-silicon dioxide, a high sensory response to a decrease in pressure below the atmospheric one is characteristic, depending on the structure type of the nanomaterial to be synthesized (quasi-spherical aggregates, labyrinthine, percolation net).

The aim of this work was to study the sensory response of sensitive elements of vacuum sensors based on two- and three-component oxide nanomaterials with fractal structure, depending on their qualitative and quantitative composition. The synthesis was carried out within the framework of the sol-gel technology, using filmforming sols based on the following components: tetraethoxysilane ($Si(OC_2H_5)_4$, TEOS), ethyl alcohol, distilled water, hydrochloric acid as a catalyst, tin (II) chloride dehydrate ($SnCl_2 \cdot 2H_2O$) and zinc chloride ($ZnCl_2$) as alloying components. The mass fraction of dopants was selected on the basis of the following conditions: 1) the formation of an ultrathin silica mesh (matrix) with embedded modifiers (tin and zinc heterototoms) of the "guest-host" type; 2) the formation of a conducting cluster that permeates the entire volume of nanostructures.The sol was coated on the substrates of monocrystalline oxidized silicon of 5×5 mm^2 by centrifugation at the table rotation speed of 4000 rev/min. The annealing was carried out at a temperature of 600 °C for 30 min in the air. Planar silver contact pads were formed by thermal evaporation in the vacuum. The main feature of the synthesized two- and three-component metal oxide nanomaterials is a fractal type of structural organization (figure 1).

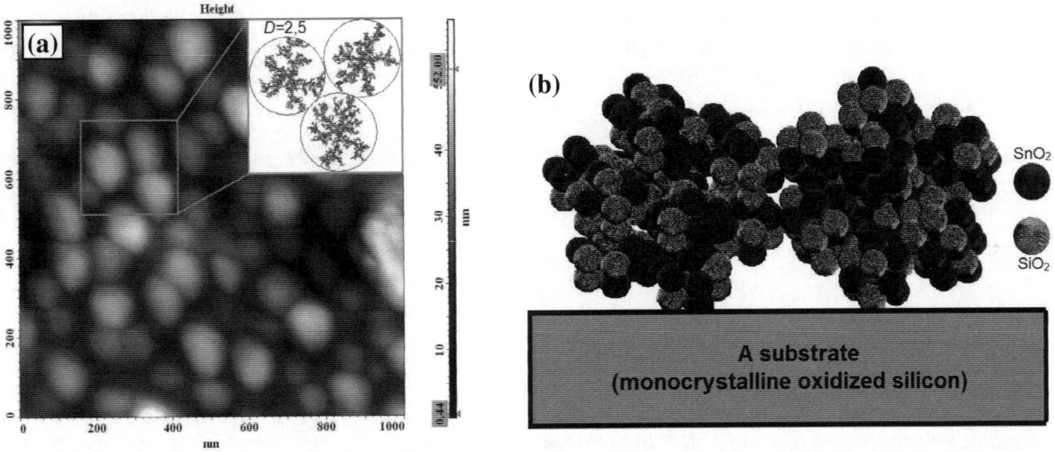

Figure 1. An AFM image of two-component metal oxide nanomaterials of SiO_2-SnO_2 composition with fractal structures **(a)** and the model of sensitive elements of vacuum sensors on their base **(b)**.

Figure 1 shows that the synthesized nanomaterials have a developed surface, formed by clusters of a quasi-spherical form, to describe which it is appropriate to apply either the model of Witten–Sander fractal (box in figure 1(a)), or a three-dimensional Julien fractal (figure 1(b)). However, regardless of the used model, it is well established and confirmed by the experimental data that the multicomponent metal oxide nanomaterials with fractal structures have a high porosity due to a high concentration of macro, micro and mesopores [8]. In turn, the high porosity leads to active gases' interaction with the surface and the volume of the investigated nanomaterials, and their desorption at pressure decrease, lower than the atmospheric one, forms the basis of the offered vacuum sensors operation.The greatest contribution to the sensory response of such sensitive elements at temperatures close to room temperature is made by the adsorption/desorption of oxygen in the form of O_2^-.

Figure 2 shows the relative change in resistance (R/R_0) of sensitive elements of vacuum sensors based on the two-component oxide nanomaterials SiO_2-SnO_2 as a function of pressure.

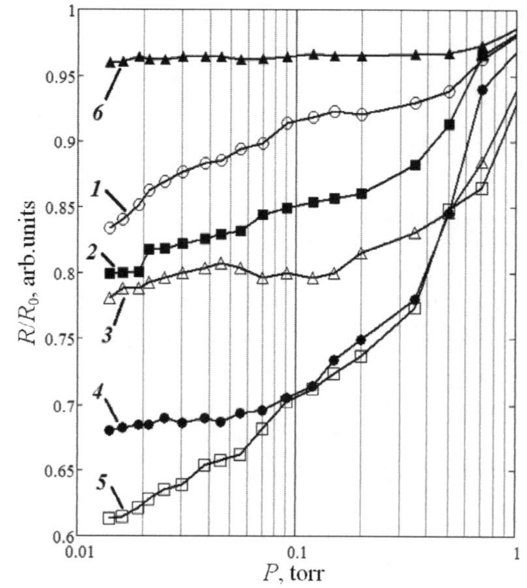

Figure 2. Relative change in resistance of sensitive elements of vacuum sensors based on the two-component oxide nanomaterials SiO_2-SnO_2 at different mass fraction of tin dioxide: *1* – 50 wt. %; *2* – 60 wt. %; *3* – 70 wt. %; *4* – 80 wt. %; *5* – 85 wt. %; *6* – 90 wt. %.

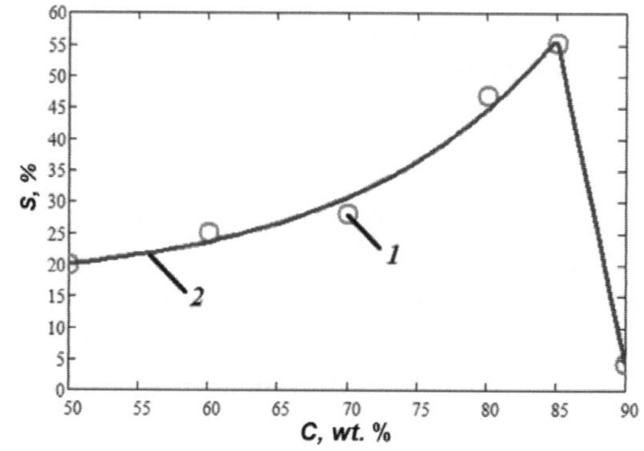

Figure 3. Dependence of sensory response of sensitive elements of vacuum sensors based on the two-component oxide nanomaterials SiO_2-SnO_2 on the mass fraction of tin dioxide: *1* – experimental data; *2* – approximation.

The analysis of experimental data shows that the resistance of the sensitive elements of the developed vacuum sensors decreases monotonically with the pressure decrease. The mass fraction of tin dioxide does not affect shown dependencies (curves 1–6 in figure 2), but it determines the value of the sensor response, which can be calculated using the following formula

$$S = \left| \frac{R_0 - R}{R} \right| \cdot 100\ \%,$$

where R_0 is the initial resistance at the pressure chosen for the reference point; R is the resistance of the sample at a given value of the measured pressure.

Figure 3 shows the sensor response dependence of the sensor vacuum sensing elementson the basis of two-component oxide nanomaterials on the mass fraction of tin dioxide (the measured pressure is taken equal to 0.014 mm Hg). An increase of the alloying component content in the nanomaterial leads to an increase of sensitivity until it reaches its maximum of 55.3 % at 85 wt. % of SnO_2. A further increase of the tin dioxide mass fraction has the opposite effect, which, apparently, is due to the inability of formation of ultrathin silica matrix at low silica dioxide content. It has been proved that when the mass fraction of tin dioxide increases, an increase of the total porosity occurs, since the presence of a modifying inorganic additive in the ash promotesthe pore formation and loosening of the inorganic polymer structure [9].

The use of three-component oxide systems based on SiO_2-SnO_2-Me_xO_y (where Me is the metal *d*-atoms, e.g. Zn) with fractal structures opens great possibilities for controlling the vacuum gauges' characteristics. Figure 4 shows the relative change in resistance of sensitive elements of vacuum sensorsbased on SiO_2-SnO_2-ZnO, and based on SiO_2-SnO_2, for comparison.

It can be seen from figure 4(a) that the resistance of the sensing elements of the proposed vacuum sensors decreases monotonically with the pressure decrease, and the qualitative composition of the multicomponent oxide nanomaterials has practically no effect on the form of the presented dependencies. However, the sensory response for the three-component oxide nanomaterials (SnO_2 and

ZnO mass fractions are 50 and 10 wt. %, respectively) is significantly higher than for the two-component oxide system (tin dioxide mass fraction is 50 wt. %).

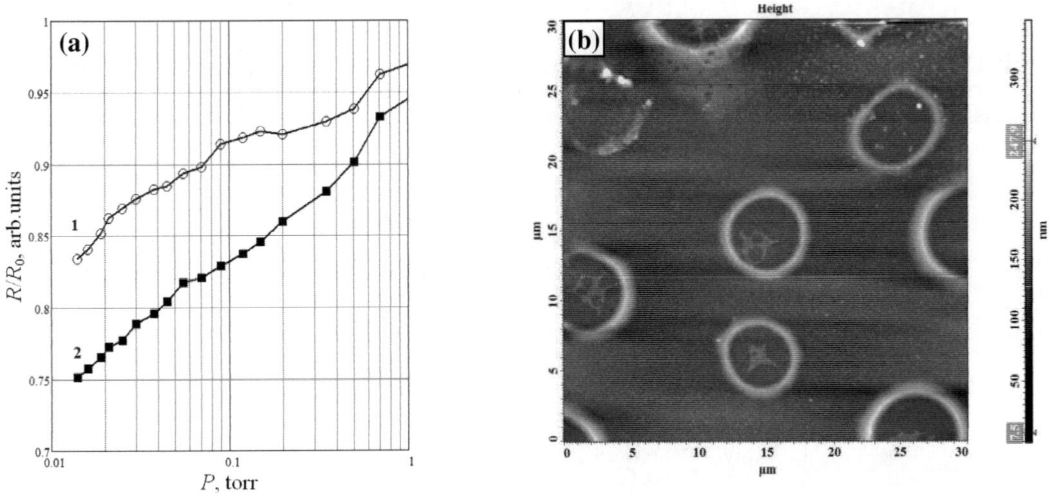

Figure 4. Relative change in resistance of sensitive elements of vacuum sensorsbased on the multicomponent oxide nanomaterials with various fractal structures: *1* – SiO_2-SnO_2; *2* – SiO_2-SnO_2-ZnO **(a)** and an AFM image of the surface morphology of nanomaterials based on SiO_2-SnO_2-ZnO **(b)**.

The analysis of the AFM images presented in figure 4(b) shows that a three-dimensional porous structure with a high concentration of macro and micropores is characteristic for nanomaterials based on SiO_2-SnO_2-ZnO. Apparently, zinc oxide acts as a loosening structure of an inorganic polymer. An increase in the sensory response of the three-component oxide systems can be explained by their high porosity, which determines the concentration of the adsorption/desorption centers, and as a consequence, the sensitivity.

Acknowledgments
The work is financially supported by the Ministry of Education and Science of the Russian Federation within the framework of the project part of the state assignment for Penza State University No. 16.897.2017/4.6, and the Foundation for Assistance to Small Innovative Enterprises in Science and Technology ("UMNIK" program, contract No. 6700 GU2015).

References
[1] Deulin E A and Gatsenko A A 2012 *Vacuum Technique and Technology* **22(1)** 3–12
[2] Kim S J 2005 *Technical Physics Letters* **31(7)** 597–9
[3] Zheng X J, Cao X C, Sun J, Yuan B, Li Q H, Zhu Z and Zhang Y 2011 *Nanotechnology* **22(43)** 435501
[4] Randjelovic D V, Frantlovic M P, Miljkovic B L, Popovic B M and Jaksis Z S 2014 *Vacuum* **101** 118–24
[5] Pronin I A and Goryacheva M V 2013 *Surface and Coatings Technology* **235** 835–40
[6] Karpova S S, Moshnikov V A, Maksimov A I, Mjakin S V and Kazantseva N E 2013 *Semiconductors* **47(8)** 1026–30
[7] Averin I A, Igoshina S E, Moshnikov V A, Karmanov A A, Pronin I A and Terukov E I 2015 *Technical Physics. The Russian Journal of Applied Physics* **60(6)** 928–32
[8] Moshnikov V A, Gracheva I E, Kuznezov V V, Maximov A I, Karpova S S and Ponomareva A A 2010 *Journal of Non-Crystalline Solids* **356** 2020–5
[9] Kanunnikova O M, Mikhailova S S and Muravyev A E 2006 *Glass Physics and Chemistry* **32(2)** 228–33

Calculation of a vacuum system of the installation for cleaning the surface of metal rolling by a cathode spot of a vacuum arc

V G Kuznetsov[1], **T A Kurbanov**[2] **and D K Kostrin**[3]

[1] Institute of Problems of Mechanical Engineering, Russian Academy of Sciences, 199178, Saint Petersburg, Russia
[2] LLC “Magistral of technologies”, 198320, Saint Petersburg, Russia
[3] Department of electronic instruments and devices, Saint Petersburg Electrotechnical University “LETI”, 197376, Saint Petersburg, Russia

E-mail: kvg-ipme@yandex.ru

Abstract. In this work are presented the installations for cleaning the surface of rolled products (wire and ribbon) from scale and technological lubricant with gateway systems of open type. The calculation of gateway devices and the optimal selection of pumping systems are shown.

In recent years in Russia is actively developing a scientific and applied direction of research associated with the modification of the material surfaces using the cathode spots of a vacuum arc discharge. In many works was shown the ability to carry out hardening of the surface layers of materials [1], to restore the shape of plastically deformed surfaces [2], to form on the surface of a regular microrelief [3], to carry out the surface passivation [4], to prepare the surface for flaw detection [5], etc. However, the most significant scientific and applied results were obtained using the cathode spots of a vacuum arc discharge for the tasks of surface cleaning [6] from such contaminants as mill scale, rust, oil film, etc. Figure 1(a) shows the installation for cleaning the surface of a rod with a diameter of 6.5 mm from the scale and figure 1(b) shows the installation for cleaning the surface of a metal strip with width of 12 cm for the thresholds of cars from a technological lubricant.

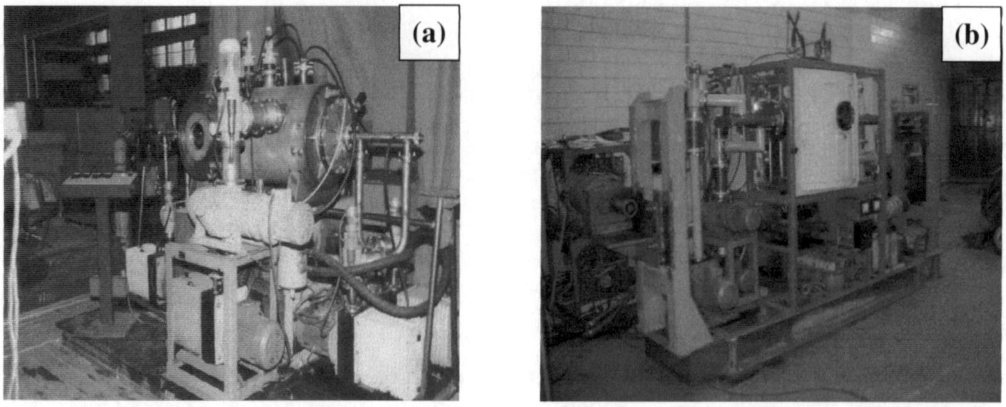

Figure 1. Vacuum arc installations for cleaning rods from scale (**a**) and metal strips from technological lubricant (**b**).

The main directions of improving the reliability, performance and efficiency of the technology and equipment for the vacuum arc cleaning are the improvement of management systems of the cathode spots, optimization of the design and geometry of the cathode-anode nodes and the gateway systems, optimal choice of the vacuum pumps, solution of the problems associated with the disposal of cleaning products from the vacuum system, optimization of the thermal processes on the electrodes, etc.

Figure 2 shows a simplified schematic diagram of the vacuum arc installation for the continuous surface cleaning of a strip metal.

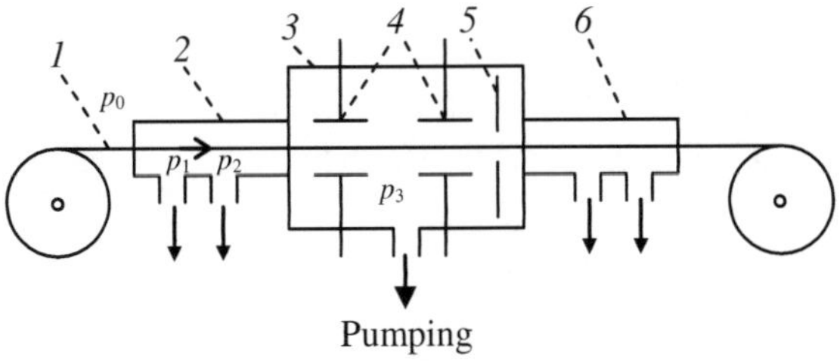

Figure 2. Schematic diagram of the vacuum arc installation for the continuous surface cleaning of metal rolling.

Product (tape) *1* from the unwinding drum, situated at the atmospheric air, through the entrance gateway *2* is continuously supplied into the vacuum chamber *3*. In a vacuum chamber between the product, which acts as the cathode, and the anodes *4* using the ignition electrode (not shown in the diagram) is ignited a vacuum arc discharge. Using a system of electrodes *5* discharge on the cathode surface opposite each of the anodes is localized in a zone of a certain area. The cleaned product from the vacuum chamber through a gateway is pulled to the atmosphere and is wound on a drum. For maintaining in a vacuum chamber of a required pressure pumping using vacuum pumps is conducted not only from the chamber but also from the gateways. In figure 2 is conventionally depicted a three-stage gateway system.

For an optimal design of a gateway system is usually taken the criterion of a minimum total pumping speed of all the vacuum pumps used in the system. This criterion is proportional in the first approximation to the cost of manufacturing a gateway system.

For any of the intermediate chamber of the *i*-th gateway of the installation, it can be written the equation of material balance

$$Q_i = Q_{ip} + Q_{i+1}, \tag{1}$$

where Q_i – total flow of leakage and gas release in the *i*-th chamber; Q_{i+1} –flow of gas leaving into the next (*i*+1) chamber; Q_{ip} – gas flow pumped by the pump from the *i*-th chamber.

This equation can be rewritten in a different form

$$Q_i = S_i p_i + U_{i+1}(p_i - p_{i+1}), \tag{2}$$

where S_i – effective pumping speed of the pump in the chamber; p_i – pressure in the *i*-th chamber; p_{i+1} – pressure in the (*i*+1)-th chamber; U_{i+1} – conductivity of the pipe connecting *i*-th and (*i*+1)-th chambers.

Written equations (1), (2) are true for all the gateway chambers, except for the middle working chamber. For this chamber the equation (2) takes the following form

$$Q_c = 2Q_c + Q_t = S_c p_c, \tag{3}$$

where Q_c – leakage from the gateways; Q_t – technological gas release; S_c – effective pumping speed of the pump in the working chamber.

Equation (2) for *k* gateways can be represented in the form

$$S_c p_c = U_k(p_{k-1} - p_k), \tag{4}$$

where S_c – pumping speed of the pump in the chamber without regard to the own gas release and the availability of the output gateway. For simplicity let us assume that

$$p_i >> p_{i+1};\ U_i >> U_{i+1}, \tag{5}$$

i.e. the gateway system is efficient enough, then

$$S_i = \frac{p_{i-1}}{p_i} U_i,\ i = 1, 2, \ldots, k. \tag{6}$$

The sum of all the pumping speeds of the vacuum pumps in the gateway system, as follows from (1)

$$C = \sum_{i=1}^{k} S_i = \sum_{i=1}^{k} \frac{p_{i-1}}{p_i} U_i. \tag{7}$$

Differentiating (7) by p_i, we can get the minimizing function $C = f(p_i)$

$$\frac{\partial C}{\partial p_i} = \frac{U_{i+1}}{p_{i+1}} - \frac{p_{i-1} U_i}{p_i^2} = 0. \tag{8}$$

Hence the system of equations for selecting an optimum pressure in the gateway chambers is

$$p_i = \sqrt{\frac{U_i}{U_{i+1}} p_{i-1} p_{i+1}}. \tag{9}$$

Defining a constant length of a gateway system

$$L = \sum_{i=1}^{k} l_i$$

in addition to equation (7), differentiating it under these conditions by l_i, can be found the minimum conditions of the function $C = f(l)$ depending on the relative length of the gateway channels. In [7] are recommended the optimum pressures (figure 2) and the relative length of the gateway channels, for example, for three levels of a gateway system

$$p_1 = 0.5\sqrt[7]{p_0^6 p_3};\ p_2 = 0.4\sqrt[7]{p_0^4 p_3^3};\ l_1 = 4/7;\ l_2 = 2/7L;\ l_3 = 1/7L, \tag{10}$$

where p_0 – inlet pressure in a gateway system (atmospheric); p_3 – output pressure from the gateway system (in the chamber); p_1 – pressure in the first stage of the gateway; p_2 – pressure in the second stage of the gateway (figure 2).

The number of the levels of a gateway system should be chosen depending on the pressure in the working chamber. If the product for cleaning is a tape, for calculating the conductivity of the gateway channels we need to use the formula for the conductivity of two parallel narrow rectangular slots in the viscous flow regime of gas flow

$$U_i = \frac{610 p_{av}}{l_i}(a_1^3 b_1 + a_2^3 b_2), \tag{11}$$

where a, b – dimensions of the short and long sides of the rectangle; p_{av} – average pressure in the gateway channel; l_i – length of the i-th channel of the gateway. For example, the average pressure in the first gateway channel $p_{av1} = (p_0 + p_1)/2$.

The calculation of conductivity in a molecular-viscous mode can be approximately carried out, summing up the conductivity in the viscous and molecular flow regimes. As the calculation formula for the conductivity of a narrow rectangular slit in the molecular regime can serve the following expression

$$U_i = 27 \lg\left(\frac{l}{a}\right)\frac{a^2 b}{l}. \tag{12}$$

The flow of gas passing through a gateway channel is

$$Q_i = U_i (p_{i-1} - p_i). \tag{13}$$

The effective pumping speed in the gateway channel is determined by formula

$$S_i = Q_i / p_i, \tag{14}$$

the calculation using which allows to select the pump needed to pump a given gateway channel.

Let us consider an example of the design calculations for a gateway system presented above, taking into account the given recommendations. Working pressure in the chamber will be taken equal to $p_3 = 0.1$ Torr. At first approximation let us consider that the gas flow in all the gateways is viscous. The total length of the gateway chambers will be taken equal to $L = 21$ cm. Then, in accordance with the relations (10): $L_1 = 12$ cm, $L_2 = 6$ cm, $L_3 = 3$ cm, $p_1 = 195$ Torr, $p_2 = 7$ Torr.

For cleaning of the tape cross section of the gateway channel is a rectangle. The cross section of the tape itself is also a rectangle with sides of 0.7 mm and 120 mm. Combination of the fields of tolerances of mating parts leads to the formation of two rectangular slits with dimensions $a = 0.2$ mm and $b = 120$ mm (considering the gap between the tape and the gateway on each side equal to 0.2 mm).

To calculate the conductivity of the first gateway channel let us use the formula for the conductivity of a narrow rectangular slit in the viscous flow regime of a gas flow. The average pressure in the first gateway:

$$p_{av1} = \frac{p_1 + p_2}{2} = 478 \text{ Torr.}$$

The conductivity of two parallel identical slits is equal to

$$U_1 = 2\frac{610 p_{av}}{L_1} a^3 b = 4.66 \text{ l/s.}$$

The gas flow passing through the first gateway is equal to

$$Q_1 = U_1 \left(p_0 - p_1 \right) = 2633 \text{ l·Torr/s.}$$

The effective pumping speed of the mechanical pump in the first gateway at $p_1 = 195$ Torr is

$$S_1 = Q_1 / p_1 = 13.5 \text{ l/s.}$$

Similarly let us perform the calculation of the conductivity of the second and third gateway channels:

$$p_{av2} = 101 \text{ Torr, } U_2 = 2 \text{ l/s, } Q_2 = 376 \text{ l·Torr/s, } S_2 = 54 \text{ l/s,}$$
$$p_{av3} = 3.5 \text{ Torr, } U_3 = 0.14 \text{ l/s, } Q_3 = 0.97 \text{ l·Torr/s, } S_3 = 10 \text{ l/s.}$$

If there are two identical gateways (input and output) S_3 must be doubled. In this example, for pumping the first and second gateways, and also the vacuum chamber, it is possible to use mechanical pumps. At least in the first channel of the gateway can be used the pump of type NVR-16D, and in the second channel – AVR-50. However, given the additional (not included in the calculations) leakages and releases of gas from the chamber walls, walls of the gateways, fittings, metal surface (especially in the cleaning process), etc. it is needed to increase the performance of the vacuum pumps. This primarily relates to the pump in the vacuum chamber itself. Based on the characteristics of the vacuum chamber evacuation of the specific unit VU-2MBS as a pump for pumping the chamber, it is advisable to use a pump of type AVR-150.

Studies have shown that the process of controlling the cathode spots for the uniform cleaning of the entire surface during the tape-cathode movement is subject to the same laws as in the vacuum arc evaporators [8, 9] with end type cathodes or long cathodes.

Calculation of the shape and geometry of the electrodes (especially the anode) was performed taking into account the balance of power, released at the electrodes. The desire to increase the cleaning rate by increasing the arc discharge current may lead to the occurrence of an anode spot, and as a result, to the installation failure. Therefore, the shape and dimensions of the anode are determined by the technological parameters of the installation.

For the cathode while ensuring uniform cleaning and a predetermined pulling speed, the most important is the temperature of the surface and change of its surface geometry.

References

[1] Kuznetsov V G 2009 *Vacuum technique and technology* **2** 81–4

[2] Kuznetsov V G, Kurbanov T A, Krymov D V and Ponimatkin V P 2014 *Vacuum technique and technology* **1** 78–81
[3] Kuznetsov V G, Kurbanov T A, Ponimatkin V P and Rizakhanov R N 2013 *Proceedings of the 11-th International Conference "Films and coatings-2013" (Saint-Petersburg)* pp 15–20
[4] Kuznetsov V G, Kurbanov T A, Krymov D V and Ponimatkin V P 2014 *Vacuum technique and technology* **1** 70–2
[5] Kuznetsov V G, Kurbanov T A and Ponimatkin V P 2015 *Patent № 2560112 RU. Method of defectoscopy of metal products during their surface treatment*
[6] Steshenkova N A, Shumilov V P, Kuznetsov V G, Shalimov A G and Sokolov O G 2005 *Herald of shipbuilding technology* **13** 43–50
[7] Alekseev G A, Nesvizhskiy M B and Shelyubskiy M L 1970 *Reviews on electronic engineering. Microelectronics series* **6** 94

IOP Conf. Series: Journal of Physics: Conf. Series **872** (2017) 012007 doi:10.1088/1742-6596/872/1/012007

Equilibrium gas pressure in various operating modes of ion-plasma accelerators

C V Korolev, Yu B Movsesyants and P M Tyuryukanov

All-Russian Electrotechnical Institute, Krasnokazarmennaya str. 12, 111250, Moscow, Russia

E-mail: tyuryukanov@vei.ru, svulm@list.ru

Abstract. The results of measurements gas equilibrium pressure and elemental composition for various operating modes of two-stage ion-plasma accelerators using a discharge in a transverse strongly inhomogeneous magnetic field are presented. It is established that the gas pressure in the chamber is most strongly changed at small ion currents to the collector. The main process of gas separation is the desorption from the surface of the electrodes under the action of low-energy ions. In steady state, with an accelerating voltage more than 1 kV, the gas pressure changes slightly. The main process is spraying.

1. Introduction

Ion-plasma accelerators without differential pumping systems and separation by ion masses are used in many technological processes. These devices are installed in the immediate vicinity of the surface to be treated. The method of carrying out the process should ensure the implementation of a continuous technological cycle for the formation coatings. The cycle includes cleaning, etching, activation of substrates, the formation of sub-layers and ion implantation on substrates of complex geometry. Vacuum conditions in such process must be highly-pure and stationary. The most effective cleaning of the surfaces of samples and electrodes from adsorbed gas and contamination is carried out by a flow of strongly nonequilibrium plasma. The ion energy in the stream should not exceed the sputtering threshold ~30 eV. The temperature and electron concentration should be sufficiently large and provide a high probability of dissociation of polluting molecules [1, 2]. The binding energy molecules of the adsorbed gas with the surface varies from 1 eV for the lowest monolayer to 0.01 eV for the upper monolayer. This layer can be desorbed already under the action of a random molecular flow from the volume of the vacuum chamber ~$NV/4$. The desorbed gas creates a flux of ~0.1 A/cm^2. When a pure working gas is injected into the chamber volume at a pressure of ~0.01 Pa, the polluting layers on the surface of the sample are completely replaced by the working gas molecules. This is realized at an ion current density of 0.1 A/cm^2 in less than 1 s. The surface of the sample is etched by spraying the crystal lattice. The energy of the ion flux (as a rule, ions of inert gases are used) should be above the threshold. But it should not exceed 1 keV so that defects of the crystal lattice do not arise. The etching process takes considerably longer time and requires a high uniformity of the ion flux along the sample surface. For deposition of a thin film coating, a stream of neutral molecules or ions with energy below the sputtering threshold of the crystal lattice of the substance is introduced into the plasma composition, which must be introduced into the film. By exciting their own electronic and molecular levels of the molecules and crystal lattices of the target deposited on the surface, the plasma activates the coating process. For the implantation of semiconductor materials, the ion energy should not exceed 1 keV, while for doping the surface of instruments with the purpose of their hardening, ions in the stream must have an energy of more than 30 keV. The energy dispersion should not exceed 10 % at an ion current density of up to 10 mA/cm^2. Two-stage ion-plasma accelerators based on a discharge in a transverse strongly inhomogeneous magnetic field have been developed in All-Russian Electrotechnical Institute [3, 4]. They are able to work in all the modes considered. Accelerators are easily reconfigured in a wide range of parameters and create flows of various configurations - annular, ribbon, radially converging and divergent. This allows the processing of extended parts of arbitrary curvature for one entry into the vacuum chamber.

2. Experimental results and their discussion

The parameters of ion sources using cathodes from various materials are measured. Figure 1 shows the current-voltage characteristics of the discharge in argon at the same pressure of 0.03 Pa in the chamber. The use of various cathode materials shows that the burning voltage of a high-current discharge with a graphite cathode is much larger than when using a molybdenum cathode. The discharge characteristics are significantly affected by cathode sputtering. Surfaces of electrodes in a stationary discharge are quickly cleaned of adsorbed gas. Therefore, the composition of the ion flux will mainly include the ions of the gas of the cathode material being pumped into the ion source and ionized atoms. The ion flux to the cathode surface is determined by the coefficient of ion-electron emission γ and the discharge current density: $J_i = J_p(1 + \gamma)^{-1}$. The flux of atoms atomized from the surface of the cathode is determined by the sputtering coefficients for various types of ions.

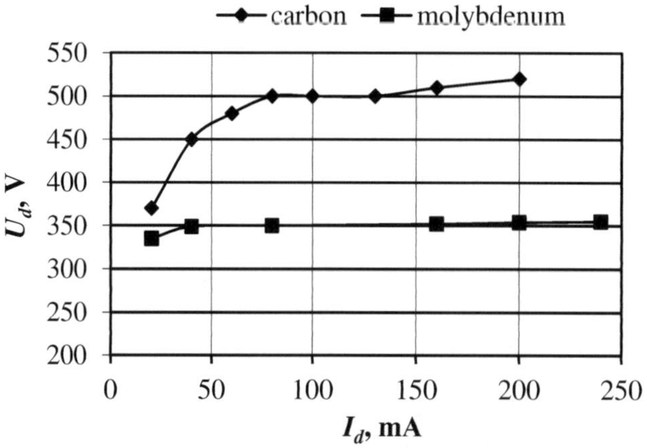

Figure 1. Current-voltage characteristics of a high-current discharge.

Let us consider how the discharge characteristics change when the effect of sputtering on the elemental composition of the ion beam is taken into account. The change in the concentration of sputtered N_p atoms and N_g gas in the discharge gap is described by the equations:

$$\frac{dN_g}{dt} = -\nu^0_{ir} N_g N_e;$$

$$\frac{dN_p}{dt} = -\nu^0_{ip} N_p N_e - \frac{v_a N_p}{4} \frac{\nu_p d}{v_p} \frac{S}{V_p} - \nu_p N_p \chi_p \left(1 - \frac{\nu_p d}{v_p}\right) \frac{S_p}{V_p} + K_z \nu^0_{ig} N_g N_e \frac{d_k S_p}{V_p} + K_p \nu^0_{ip} N_p N_e \frac{d_k S_p}{V_p},$$

where $\nu_{ip}{}^0$, $\nu_{ig}{}^0$ are the frequencies of ionization of sputtered atoms and gas per particle; ν_p is the frequency of elastic collisions of atoms; χ_p is the probability of adsorption of sputtered atoms; S is the surface area of the discharge through which the gas flows; S_p is the cathode area; V_p is the discharge volume; K_r and K_p are the cathodes sputtering coefficients under the action of gas ions and ions formed in the discharge gap as a result of the ionization of sputtered atoms at an average ion energy.

IOP Conf. Series: Journal of Physics: Conf. Series **872** (2017) 012007 doi:10.1088/1742-6596/872/1/012007

Stationary conditions are realized when the concentration gradient of the atomized substance is small and the cathode can be neglected by gas ions.

Assuming $\nu_p = \nu_p^0 N_p$, the average concentration of sputtered atoms in the discharge gap can be estimated from the formula

$$N_p = \frac{K_p \frac{d_k}{d} - 1 - \chi_p \frac{v_p}{\nu_{ip}^0 N_e d}}{\chi_p + \frac{S_b}{S_p} \frac{v_a}{4 v_p}} \frac{\nu_{ip}^0}{\nu_p^0} N_e.$$

It follows that discharge in crossed electric and magnetic fields with a preferential cathode sputtering process can be considered stationary if the following condition is satisfied:

$$K_p > \frac{d}{d_k}\left(1 + \chi_p \frac{v_p}{\nu_{ip}^0 N_e d}\right).$$

Using the known expressions for the sputtering coefficient $K_p \approx K_0(eU_p/E_b)^{1/2}$ and the width of the active zone of the discharge $d_k \approx (2U_{pm}\nu^0/eB^2\nu_i)^{1/2}$ from this ratio, it is possible to estimate the value of the minimum voltage necessary for burning the discharge in the atomized substance

$$U_p^{\min} \approx \left(\frac{\nu_i E_b}{2\nu_0 m}\right)^{1/2} \frac{Bd}{K_0}\left(1 + \chi_p \frac{v_p}{\nu_{ip}^0 N_e d}\right),$$

where $N_e = \varepsilon_0 B^2 \nu_i / m\nu_0$. The parameters K_0, E_b and χ_p depending on the cathode material. This explains the experimentally observed strong dependence of the magnitude of the burning voltage of the discharge on the cathode material.

For the realization of various technological processes, the pressure and composition of the gas in the process chamber under various operating conditions of the ion accelerator are of great importance. Figure 2 shows the dependence of the gas pressure on the current density of the ion beam at different discharge current values in argon at a flow rate of ~0.5 A. At each point of the graph, the gas pressure and the parameters of the ion accelerator are steady. Measurements of the pressure in the chamber were carried out using a PMI-27 ionization manometer, and the elemental composition of the neutral gas and ion beam was monitored by a magnetic mass spectrometer.

The pressure changes most strongly at low values of the accelerating voltage at the second stage of the accelerator. If no voltage is applied to the second stage of the accelerator between the cathode and the collector, the ion current to the collector is small, and the cathode takes a negative floating potential. The main fraction of ions passing through the gap in the cathode from the high-current discharge is returned to the cathode. The average energy of the ions that fly to the collector in the retarding field becomes below the spraying threshold. The main process of gas evolution is desorption from the surface of the cathode and the collector. With the ion flow formed, the pressure remains sufficiently low and depends little on the ion current density. In this mode, the mass spectrometer records a significant fraction of ~30 % of the atomized cathode atoms made of molybdenum.

Another important characteristic of the ion source is the dependence of the ion current density on the collector on the magnitude of the accelerator cathode potential relative to the collector. These characteristics are presented in figure 3 and for a complete understanding of the physical processes in the accelerator should be compared with the graph in figure 2.

It is seen that under negative voltages the ions accelerated in the anode layer of the high-current discharge of the first stage of the accelerator decelerate. Acceleration of ions in the second stage is most effective at voltages greater than 1 keV. In the region of the $J_i(U_k)$ dependence plateau, the efficiency of the second stage of the accelerator is rearranged, by changing the position of the emitter.

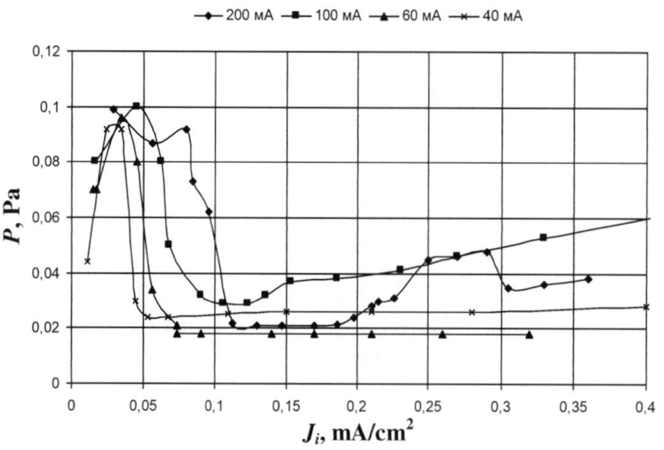

Figure 2. Dependence of pressure in the discharge chamber on the current density of the ion source at different values of the discharge current.

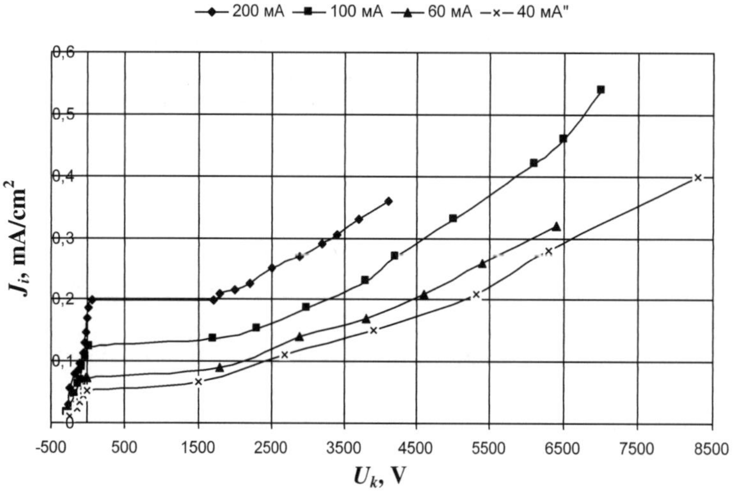

Figure 3. Dependence of the ion current density on the collector on the value of the cathode potential of the ion source relative to the collector.

3. Conclusion

Thus, it is established that the gas pressure in the chamber of the ion-plasma accelerator is most strongly changed at small ion currents to the collector. The main process of gas separation is desorption from the surface of the electrodes under the action of low-energy ions. In steady state, with an accelerating voltage of more than 1 kV, the gas pressure varies little, and the main process is spraying.

References

[1] Pleshivcev N and Semashko N 1989 *Results of science and technology. Series the Physical basis of laser and beam technology (VINITI)* 555
[2] Guseva M 1989 *Results of science and technology. Series the Physical basis of laser and beam technology (VINITI)* 55
[3] Tyuryukanov P, Nikolsky A, Fetisov I and Thick I 1982 *Plasma physics* **8**(6) 1220–7
[4] Movsesyants Yu and Tyuryukanov P 2011 *Engineering physics* **3** 21–5

Some questions of designing installations for X-ray inspection

N N Potrakhov[1], A Yu Gryaznov[1], K K Zhamova[1], V B Bessonov[1], A V Obodovskiy[1], N E Staroverov[1], E D Khlopova[1] and M A Nikolskiy[2]

[1] Department of electronic instruments and devices, Saint Petersburg Electrotechnical University "LETI", 197376, Saint Petersburg, Russia
[2] FSBSI Anapa Zonal Experimental Station of Viticulture and Winemaking North Caucasian Regional Research Institute of Horticulture and Viticulture, 353456, Anapa, Russia

E-mail: kzhamova@gmail.com

Abstract. The problems of choosing the source of X-ray radiation for use as part of technological X-ray control installations are considered. The design and main characteristics of a domestic X-ray source based on a microfocus X-ray tube are described.

1. Introduction

Improvement of the electronic component base in terms of increasing the degree of integration and reducing the size of microchips, as well as passive *RLC* components, led to the creation of new technologies for their assembling on printed circuit boards. To ensure quality control of the assembling of printed circuit boards directly in production conditions, in the early 2000s, specialized X-ray units were developed [1].

At present, such an installation generally includes a radiation source based on an X-ray tube with a focal spot of micron sizes [2] and an image receiver based on a flat panel X-ray detector, which are located in a special chamber for radiographic work. During the control of printed circuit boards, two methods of obtaining an X-ray image are commonly used: contact and projection. To provide appropriate X-ray optical survey schemes, the camera is equipped with automated mechanisms that allow the X-ray tube, radiation detector and controlled object to move relative to each other along a given trajectory within the required limits. An important role in ensuring the claimed functional capabilities of the installation is carried out by specialized software for computer processing of acquired images.

2. Materials and methods

Leading manufacturers of X-ray inspection systems, for example, "Nordson", currently design and manufacture installations exclusively on the basis of microfocus X-ray tubes with constant pumping. The advantage of this design, the foundations of which were proposed back in 1982 by German engineer A. Reingold, is the possibility of repeated replacement of the cathode and anode target. This makes it possible to operate the tube at the maximum loads – emission to the cathode and thermal to the target, but it causes a number of shortcomings in comparison with the vacuum tubes:

– large dimensions, weight and complexity of the construction, for example, the use of a metal-ceramic insulator assembly designed for the total working voltage of the tube, as well as vacuum-tight mechanical collapsible connections of a metal balloon of a tube with replaceable units;

– presence in the tube of an inalienable constructive element – a specialized pumping system;

– need for high-voltage tube training after the replacement of individual units, which imposes additional requirements on the generator device of the power source of the tube, and subsequent alignment of its electron-optical system;

– high price.

Simpler in design and convenient to operate, and also cheaper, were three-electrode metal-glass tubes with a thin-film target of the BS series, taken out on a long anode pipe from a vacuum balloon, which were developed and mass produced in the Design Bureau of the LOEP "Svetlana" starting from the mid-70 of the last century.

In 2016 specialists of JSC "Svetlana-X-ray", the current name of the Design Bureau of the LOEP "Svetlana", mastered the serial production of the X-ray tube BS-16(III) of the third design, which, in its basic characteristics, is closer to collapsible tubes with a constant pumping out [3]. The maximum voltage is 150 kV, the maximum current is 150 μA, the diameter of the focal spot measured by the grid method (according to GOST 22091.9-86), at a voltage of 150 kV is 5 μm (figure 1).

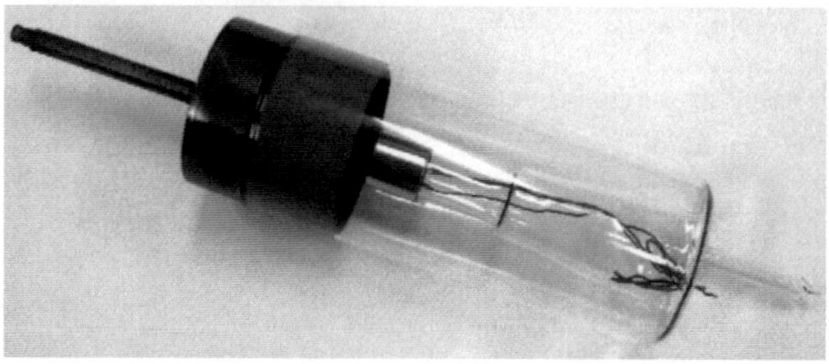

Figure 1. X-ray tube BS-16(III).

The diameter and length of the glass balloon of the tube are 73 mm and 215 mm, the diameter and length of the copper anode tube are 10 mm and 100 mm. The tungsten target is deposited on a beryllium substrate (exit window of an X-ray tube) 0.2 mm thick. The minimum focal distance is 0.5 mm. During the development of the tube, the possibility of using an artificial diamond as an exit window material was investigated. An anode assembly based on a plate made of artificial diamond of a large area and 0.2 mm in thickness was constructed.

However, due to the higher cost of such a unit, the issue of serial production of a tube with a diamond exit window is under consideration. To ensure the specified diameter and position of the electron beam on the target, internal electrostatic (based on three-electrode EOS) and external magnetic (based on a permanent ring magnet) focusing systems are used.

The guaranteed service life of the tube in the nominal mode of operation is 300 hours. Practical service life can be up to 1000 h or more. The main reason for the final tube failure is the breakage (burnout) of the cathode due to the thinning and embrittlement of the filament due to the evaporation of tungsten from its surface.

Based on the BS16(III) tube, an X-ray source (IRI) of the monoblock type RAP150M-0.15N.3 was developed. The appearance of the IRI monoblock is shown in figure 2. It implements a high-voltage power circuit for an X-ray tube with a grounded anode; an accelerating voltage of negative polarity is applied to the cathode-grid tube assembly from the IRI generator device. In the generator set, in addition to the high-voltage source of accelerating voltage, there are power sources for the cathode and control (grid voltage) heating. The modern circuit technologies are applied in the IRI: direct rectification of the mains voltage, conversion of the rectified voltage to a frequency of several tens of kHz, the frequency-pulse operation of the converter (modulation of the rectified voltage), and multistage multiplication of the modulated voltage.

In the monoblock used a combined insulation is from solid component and oil. All elements of high voltage are filled with epoxy composition, and the gap between the walls of the balloon tube and the high-voltage socket of the monoblock is filled with transformer oil.

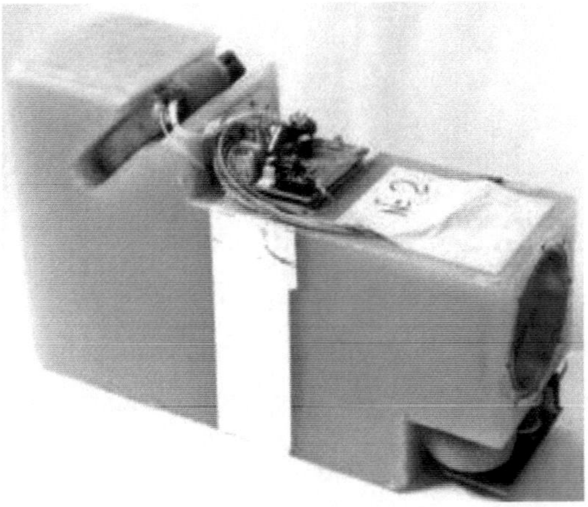

Figure 2. The monoblock RAP150M-0.15N.3 (without case and without a tube).

Control of the operating modes of the X-ray tube: voltage, heat current and current of the anode, exposure time, diameter and position of the focal spot on the target is carried out using a microprocessor device. The power consumed by RAP150M-0.15N.3, in the nominal operating mode does not exceed 200 W. Dimensions monoblock RAP150M-0.15.1 (without case and without a tube) are 110×220×260 mm, weight 6 kg.

3. Conclusion

In general, the experience gained in the work with RAP150M-0.15N.3, including in the microfocus X-ray computer tomograph MRKT-01, shows that, from the point of view of the "price/quality" ratio, it can compete with the most modern IRI designs based collapsible X-ray tubes with constant pumping in the development of industrial X-ray machines for X-ray quality control of PCB assembly.

Acknowledgment

The work was carried out with the financial support of the Presidential Grant for the state support of young Russian Ph.D candidates in the framework of the project "Development of a technique for digital X-ray diagnostics of contact (spot) welding of structures and products of the aerospace industry" Project No. 14.Y30.17.1582-MK.

References

[1] Potrakhov N N 2015 *Biomedical Engineering* **48** 237–40
[2] Blinov N N, Vasilyev A Yu, Bessonov V B, Gryaznov A Yu, Zhamova K K, Potrakhov E N and Potrakhov N N 2014 *Biomedical Engineering* **48** 58–61
[3] Podimsky A A and Potrakhov N N 2017 *Journal of Physics: Conference Series* **808** 012011

IOP Conf. Series: Journal of Physics: Conf. Series **872** (2017) 012009 doi:10.1088/1742-6596/872/1/012009

Linear motion feed through with thin wall rubber sealing element

V P Mikhailov and E A Deulin

Bauman Moscow State Technical University, 105005, Moscow, Russian Federation

E-mail: mikhailov@bmstu.ru, deulin@bmstu.ru

Abstract. The patented linear motion feedthrough is based on elastic thin rubber walls usage being reinforced with analeptic string fixed in the middle part of the walls. The pneumatic or hydro actuators create linear movement of stock. The length of this movement is two times more the rubber wall length. This flexible wall is a sealing element of feedthrough. The main advantage of device is negligible resistance force that is less then mentioned one in sealing bellows that leads to positioning error decreasing. Nevertheless, the thin wall rubber sealing element (TRE) of the feedthrough is the main unreliable element that was the reason of this element longevity research. The theory and experimental results help to create equation for TRE longevity calculation under vacuum or extra high pressure difference action. The equation was used for TRE longevity determination for hydraulic or vacuum equipment realization also as it helps for gas flow being leaking through the cracks in thin walls of rubber sealing element of linear motion feedthrough calculation.

1. Introduction

The main purpose of the feedthroughs design was long travel three-coordinate drive creation with unique parameters, which are: travel length about 300–500 mm; high pressure till $3\cdot10^5$ Pa or vacuum till 10^{-6} Pa; angles of motion stock inclination around Y, Z axis about 3–5°; gas flow penetration (dissolution) through thin rubber wall less $Q = 10^{-9}$ m$^3\cdot$Pa/s; longevity more $N = 10^6$ cycles. The base element of the feedthroughs is the elastic thin-wall rubber element (TRE) reinforced with analeptic strings being fixed into the middle part or the rubber walls.

2. Feedthrough design

The feedthroughs may be used as in high vacuum equipment, also as in pneumatic and in hydraulic equipment with pressure till two atmosphere [1, 2]. The main advantage of the feedthroughs is the negligible value of the resistance force, which is many times less the resistance forces in metal bellows, membranes, traditional rubber collar sealing elements, being used in pneumatic, hydraulic or in vacuum technology practice. From another side the limited longevity of feedthroughs (TRE) element is the main disadvantage of the drive based on TRE. The patented linear motion feedthrough [3] is based on elastic thin rubber walls usage being reinforced with analeptic string fixed in the middle part of the walls. The scheme of the linear motion vacuum feedthrough is shown in figure 1(a). The elastic rubber wall thickness is about 0.3–0.4 mm and stock diameter is about D = 20 mm, as it is shown in figure 1(b).

The TRE longevity research was done on experimental installation, the scheme of which is shown in figure 2.

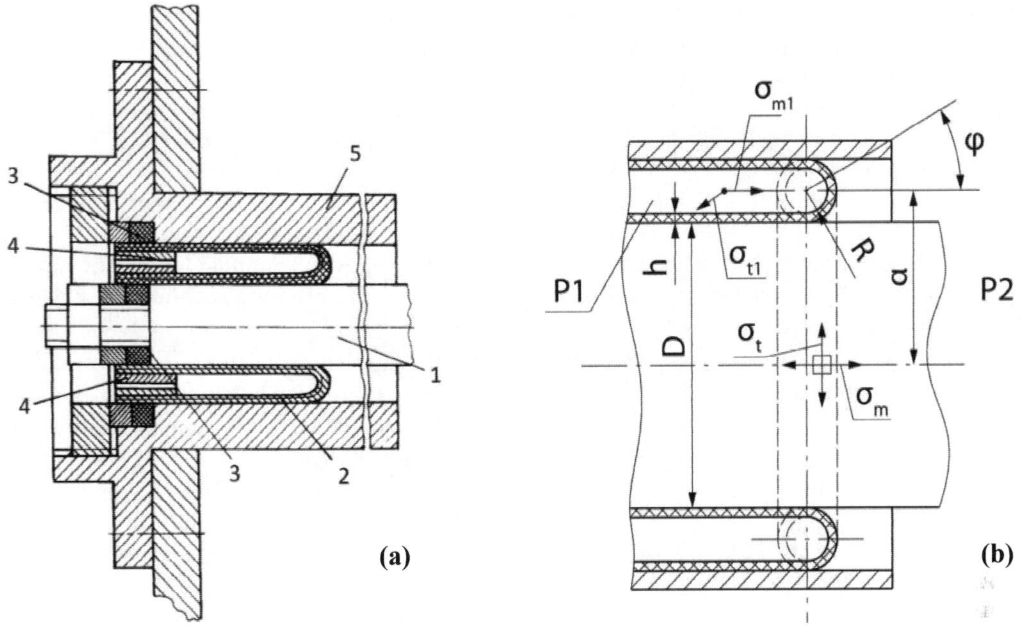

Figure 1. Scheme of linear motion feedthrough: design elements **(a)**: *1* – linear moving stock; *2* – thin wall rubber elastic element; *3* – sealing collar; *4* – clamping rings; *5* – body and parameters of TRE **(b)**.

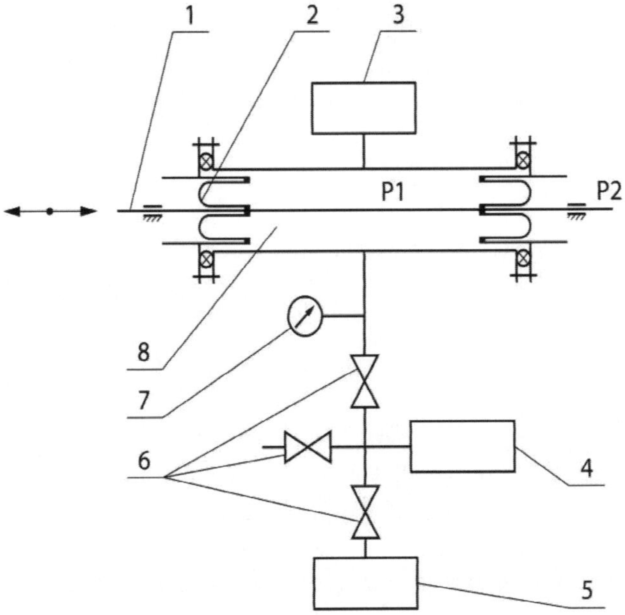

Figure 2. Scheme of experimental installation for linear motion feedthrough parameters research: *1* – stock; *2* – TRE; *3* – vacuum chamber; *4*, *5* – receiver; *6* – valves; *7* – manometer; *8* – work volume.

3. Theoretical base

The theory [1, 2] and experimental results help to create equation for TRE longevity N calculation under pressure difference action. The equation for TRE means longevity calculation in working cycles:

$$N = 10^{\frac{\sigma_0 - \sigma_a}{4.55}}, \qquad (1)$$

where σ_a – TRE amplitude stress, MPa; σ_0 – TRE ultimate strength, MPa (σ_0 = 30 MPa).

The equation (1) was used for TRE longevity N determination in equipment for variable media technology realization also as it helps for gas flow being leaking through the cracks in thin walls of rubber sealing element of linear motion feedthrough calculation.

The pneumatic or hydro actuators create linear movement of stock. The linear motion feedthrough contains linear moving stock *1*, elastic rubber flexible sealing elements *2*, and body.

The experiments show, that under condition:

$$\sigma_a = \sigma_0, \tag{2}$$

where σ_0 – ultimate strength under stretch load, the TRE longevity is equal $N = 1$, i.e. the TRE expectable disruption is after first cycle of load. The experimental result show, that for rubber IRP-1345 (RF standard) $\sigma_0 = 30$ MPa.

The TRE longevity research was done on experimental installation (figure 1), where additional pressure in work volume *8* was about $p_1 = 0.1$ MPa. The mean TRE longevity figure was about $N = (2.18 \pm 1.28)\cdot 10^6$ cycles at mean stress amplitude value $\sigma = 30$ MPa. The scheme and directions of the stresses amplitude σ_a appearing in TRE we can see in figure 1(b).

The distance W between inner surface of body *5* hole and stock *1* outer surface (figure 1) plays the main role in circumferential stress forming. Circumferential stress may be calculated

$$\sigma_{t1} = \frac{pD}{2h}, \tag{3}$$

where p – additional (differential) pressure on TRE wall ($p = 0.1$ MPa); D – stock diameter ($D = 20$ mm); h – wall TRE thickness ($h = 0.4$ mm).

Maximal values of circumferential stress σ_{t1} also as pressure difference $p_1 - p_2 = 0.046$ MPa are realized at $2R = W = 3$ mm. Axial stress in TRE wall may be calculated

$$\sigma_{m1} = \frac{p\left[(D+2W)^2 - D^2\right]}{8Dh}, \tag{4}$$

where

$$W = \frac{pD^2}{2Eh}\left(1 - \frac{\mu}{2}\right), \tag{4.1}$$

where E – modulus of elasticity of rubber ($E = 6$ MPa); μ – Poisson's ratio for rubber ($\mu = 0.5$).

Axial stress in TRE wall may be increased till $\sigma_{m1} = 0.43$ MPa at pressure difference $p_1 - p_2 = 10^5$ Pa and may be calculated

$$\begin{aligned} \sigma_t &= \frac{pR}{2h}; \\ \sigma_m &= \frac{pR}{2h}\frac{2a + R\sin\varphi}{a + R\sin\varphi}, \end{aligned} \tag{5}$$

where R – radius of TRE being bended; φ – measured angle of TRE being bended.

The calculation shows, that $\sigma_t = 0.21$ MPa and $\sigma_m = 0.43$ MPa at $\varphi = 90°$. In our case, maximum stress values were $\sigma_a = \sigma_{t1} = 1.15$ MPa and for rubber IRP-1345 being used in experimental TRE the stress-longevity equation:

$$1.15 + m\lg(2.16\cdot 10^6) = 30, \tag{6}$$

where constant $m = 4.55$ MPa and

$$\sigma_a + 4.55\lg N = 30. \tag{7}$$

In this case data of equation (7) may be used in the equation (1) for TRE longevity N determination in equipment under variable pressure, also as it helps us for gas flow leaking through the thin walls of rubber sealing element calculation.

4. Experimental results

The results of residual pressure variation in sealed chamber after crack in TRE appearing are shown in figure 3. The graphs point $t = 0$ corresponds to experiment beginning after 2000 cycles of TRE service period, and after crack in TRE appearing.

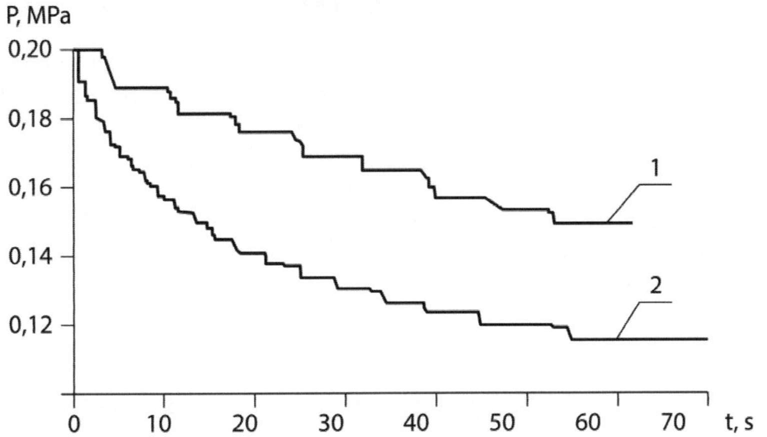

Figure 3. Graphs of residual pressure variation in sealed chamber after crack in TRE appearing, where: *1* – cycle time 8 s; *2* – crack in TRE "open" position.

The experiments show, that in stock position with the crack placed in the bended part of the TRE, the crack is open for leaking gas flow, also as it is closed when the TRE wall is pressed to stock or to body surface. The result of this crack behavior is step-form pressure variation at reciprocal stock linear motion. The gas flow leaking through the crack after periodically opening–closing crack narrow may be calculated:

$$Q_n = \frac{G_1 - G_t}{t}, \tag{8}$$

where $G_{1,t}$ – gas amount in the vacuum chamber at the initial moment; t – experimental time (in our case $t = 65$ s). Gas amount in the vacuum chamber at the final moment may be calculated:

$$G_{1,t} = P_{1,t} V, \tag{9}$$

where $P_{1,t}$ – gas pressure in chamber at final moment; V – chamber volume (in our case $V = 3.31 \cdot 10^{-3}$ m^{-3}). The crack conductivity:

$$U = \frac{Q_n}{P_{av}} = \frac{G_1 - G_t}{t \cdot P_{av}} = \frac{V(P_1 - P_t)}{tP_{av}}, \tag{10}$$

where P_{av} – gas mean pressure in chamber at time t, when the crack is open.

The crack conductivity value may be used for crack conditional diameter calculation:

$$d = \sqrt[4]{\frac{Uh}{1360P_{av}}}, \tag{11}$$

where h – TRE wall thickness ($h = 3.5 \cdot 10^{-4}$ m). So, taking into account equation (10) and pressure variation according figure 3 graphs, we may use these data for cracks size parameters estimation. The results of mentioned experiments presentation we can see in table 1.

Table 1. The results of the TRE wall crack's conditional diameter estimation.

The cycle number after crack appearing	P_t, MPa	Q_n, m^3·Pa/s	U, m^3/s	d, m
0	0.127	3.72	$2.84 \cdot 10^{-5}$	$8.64 \cdot 10^{-5}$
2000	0.108	4.68	$3.69 \cdot 10^{-5}$	$9.3 \cdot 10^{-5}$

5. Conclusions

The results show, that fatigue crack of the TRE appearing and increasing is being correlated with pressure variation as a result of leaking flow intensity value.

The demanded work pressure may be created, when the TRE crack is pressed to the shaft or to body's surfaces the gas flow through the crack is equal to zero, because the crack's tunnel is covered with the wall's surface in the marked shaft position.

The process of the crack spontaneous "closing" is the advantage of this feedthroughs type because it helps to ensure demanded work vacuum at any moment of the working period with help of shaft stopping the marked shaft position.

Acknowledgements

This study was performed within the framework of implementing the design part of the State project No. 9.8503.2017/BP of the Russian Ministry of Education and Science in the field of research. The authors thank the student of BMSTU V S Yakovlev for assistance in the design and preparation of this article.

References

[1] Deulin E A, Mikhailov V P, Panfilov Yu V and Nevshupa R A 2010 *Mechanics and Physics of Precise Vacuum Mechanisms. Fluid Mechanics and Its Applications* **91** (Springer)
[2] Deulin E A et al. 2010 *Fluid Mechanics and its Applications* **91** 137–66
[3] *Patent RF 2037702, dated 19.06.1995*

Research and development complex with remote access

Y V Panfilov, L L Kolesnik, V T Ryabov and S V Sidorova

Department of Electronic technologies in mechanical engineering, Bauman Moscow State Technical University, "BMSTU", 105005 Moscow, Russia

E-mail: panfilov@bmstu.ru, sidorova@bmstu.ru

Abstract. Vacuum modular tool for thin films vacuum deposition was considered. Deposition tool automatic control system built-in to the remote access. Methodological support for the preparation of laboratory works and virtual simulator for mastering the ability to work on the vacuum depositiontoolwere developed.

1. Introduction

In the framework of the State contract were designed vacuum modular deposition tool for deposition of thin films modular automatic controlsystem, technical support for the preparation for laboratory worksperformance, system of remote access to the installation and a virtual simulator performing the trainerrole for mastering the skills [1].

However, the wide distribution in the educational process of other universities, developed an interactive educational-scientific complex have not received. That for several reasons are: the absence of industry regulations for its use, the need for additional funding works to improve the property and introduce it into the educational process and limited options for laboratory work because of the use of while only one method of applying thin films of resistive thermal evaporation.

2. Vacuum modular deposition tool

Vacuum modular deposition tool "UVN-1M" (figure 1(a)) is a part of research and development (RND) complex and equipped with oil-free pumping system [2], a vacuum chamber with a volume of 2 l, quick-detachable flanges, one of which is a resistive evaporator, and another substrate holder with a dielectric substrate (figure 1(b)). Formed on the substrate contact padsto which through a sealing current lead is connected measuring device: picoammeter, LCR-station or other device that measures one of the characteristics of a thin film [3]. The automatic control system connected to the local network with access to the Internet. External view of the vacuum modular deposition tool shown in figure 1.

This configuration allows measurement of key characteristics of coating films during their formation. As an example, in figure 2 shows a graph of the current flowing between the pads depending on the time of the first island formation and then a continuous film [4].

In the second portion of the graph the non-linear dependence in the current increase indicates the thin film discontinuous (islet) growth with the coalescence of islands and the change of the distance between them. On this site it is possible to study the initial stages of thin films growth depending on the deposited material, deposition rate and energy of film-forming particles, which depends on the application a thin film method.

IOP Conf. Series: Journal of Physics: Conf. Series **872** (2017) 012010 doi:10.1088/1742-6596/872/1/012010

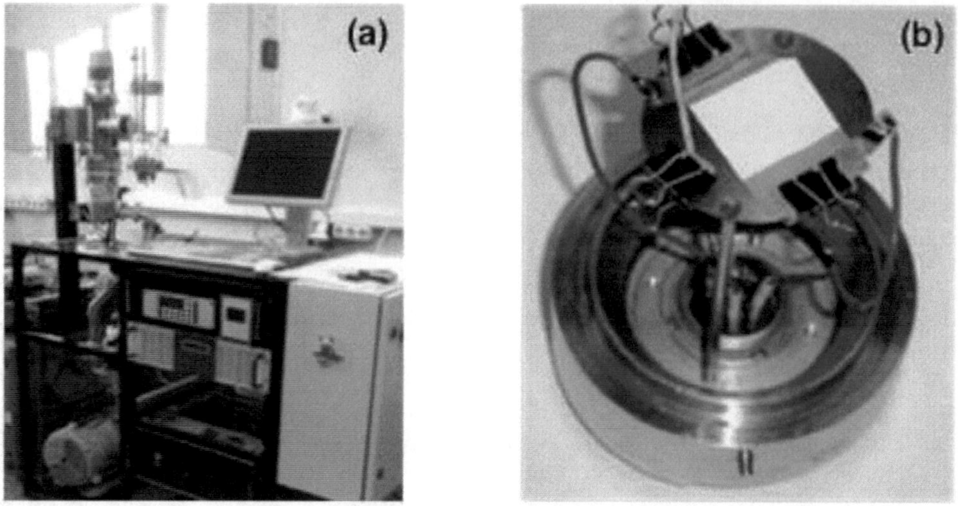

Figure 1. External view of the vacuum modular deposition tool "UVN-1M" (**a**) and the upper flange with the substrate based on the substrate holder (**b**).

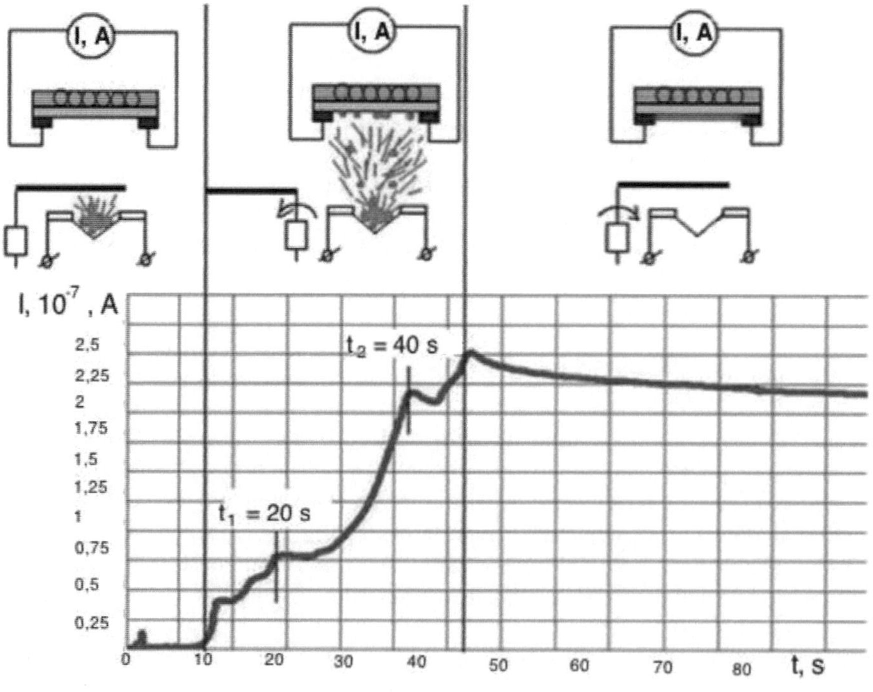

Figure 2. Current flowing between the pads depending on the thin film formationstage

The laboratory work methodology includes the theoretical foundations study of the thin films technology and mastering the ability to control the deposition tool using the built-in simulator. Passing the test and getting admission to the laboratory work, students in the computer class, to commence laboratory work in remote access mode.

3. Remote access system

A special program – remote access server – to provide remote access to equipment has developed [5]. General scheme for the remote access to the equipment is presented in figure 3.

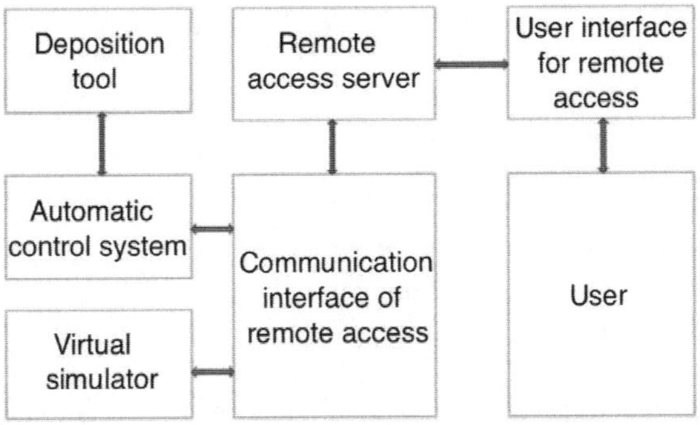

Figure 3. General scheme of remote accessorganization to equipment.

The vacuum tool automatic control system is connected via the driver pairing to the system for remote access. The driver pairing is an intermediate link between the equipment and the remote access server. It has a clear specification of the controlled format descriptions and control signals, through which it interacts with the system. Implementation of interaction with the real automatic control system is hidden inside the driver. This solution allows you to connect virtually any process equipment equipped with a control system by writing to it the driver corresponding to the specifications of the remote access systems without changing the functionality of the software part of the complex.

As a real vacuum deposition tool can be a virtual simulator equipment, which also connects via the same driver. A virtual simulator allows you to fully simulate a job, giving the user the feeling that he is working with real equipment. This option requires the presence of an operator, allows you to simultaneously conduct multiple virtual experiments and can be used for training purposes.

When working with equipment in the remote access mode in real time and in virtual simulator implemented techniques and algorithms with the transactions exception that require manual action on the equipment side, for example, the installation of the substrate. For such operations the simulator just gives a signal about readiness of the equipment to work.

You can connect the required number of virtual simulators or facilities to the remote access server. Their number is determined by the power computing means and the data channel. The remote access server on the one hand communicates with the hardware through drivers pairing, and on the other hand accepts client connections from the external network.

In this project adopted the concept of a two-tiered approach to creating a system of collective access to the laboratoryresources. The first level access is the equipment access the virtual laboratory through a local network, the second – via the Internet.

Educational-scientific complex can operate in three modes: in the mode of directive management, remote access in real time and in virtual simulation. When working in the mode of directivecontrol, the user is immediately around the vacuum tool of independently producing all of actions. In this case, the algorithm and methodology for end-user interactions with the educational-scientific complex is no different from working with any other technological equipment.

In the remote access mode in real time, the user controls the installation via a LAN or the Internet. In this case around the installation requires the operator to perform manual operations. In case of emergencies or loss of communication with the user, the operator can translate the installation to the legislative management regime.

If the mode user directive control (if it necessary) has the opportunity to conduct a technological process, using the controls located directly on the power supply, the remote control all the actions happen only through a personal computer with installed software. Any command received remotely, is processed in the control system, checks for the correctness of the action and passed to the appropriate subsystem installation.

In the virtual simulator, as mentioned earlier, the user interacts with the simulator installation. The presence of the operator is not required. You can connect to conduct virtual experiment for multiple users. Accept connections when working in remote access mode without using a web server, connecting directly to the remote access server. To connect, you need to use a special program – a thin client.

This connectivity option allows you to reduce the server response time and provides process control equipment closest to the real control mode.

Software product interacts with the user, implements the concept of thin client. To work with complex sufficient to have only the special client software.

4. Conclusions

The developed method is suitable to almost allmethods of thin films vacuum depositionsuch as electron beam evaporation, magnetron sputtering, molecular beam epitaxy, atomic layer deposition etc. It can be implemented in any laboratory vacuum tool for thin filmdeposition with automatic control system.

For example, there are small vacuum systems series developed in JSC NIITM (Zelenograd, Moscow, Russia), which could be included fortraining of thin film deposition in vacuum according to the manufacturer intended (figure 4) [6].

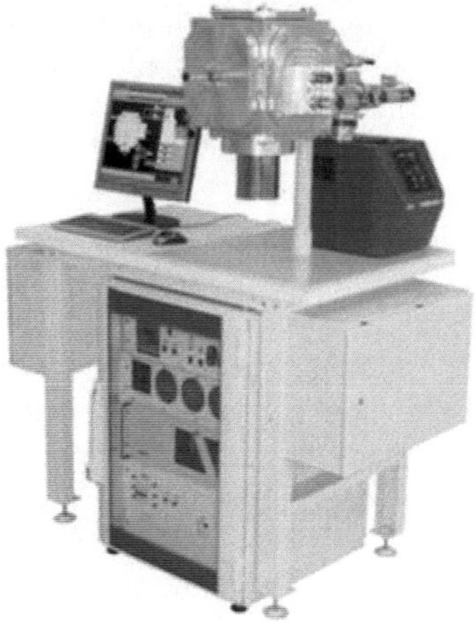

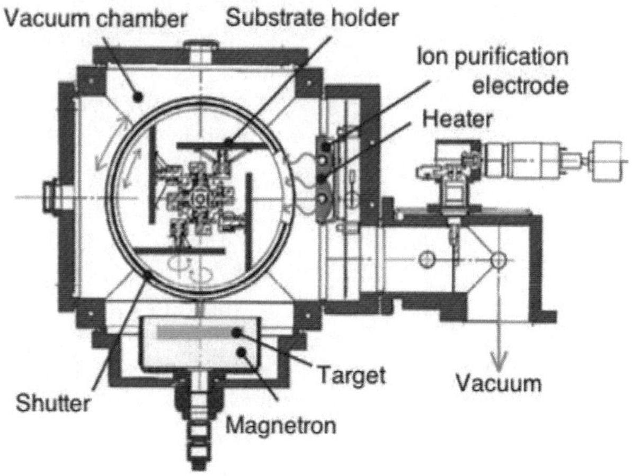

Figure 4. Vacuum tool for thin films vacuum deposition ("MVU TM", JSC NIITM).

It can be notedthat these tools are not equipped with training manuals for practical classes with students, which makes such implementation complicated. Equipment settings system remote access with a virtual simulator and supplying in conjunction with the teaching materials can significantly expand the circle of potential consumers.

References

[1] Panfilov Y V, Kolesnik L L, Ryabov V T, Moiseev K M and Sidorova S V 2012 *Interaktivnye uchebno-nauchnye kompleksy dlya vypolneniya rabot v rezhime udalennogo dostupa*
[2] Kapustin E N, Burmistrov A V and Salikeev S I 2015 *Vakuumnaya tekhnika i tekhnologiya* **25** 76–180
[3] Panfilov Y V, Ryabov V T and Sidorova S V 2014 *Nanoinzheneriya* **8** 14–8
[4] Panfilov Y V and Sidorova S V 2012 *Nanoinzheneriya* **9** 8–11
[5] Kolesnik L L 2013 *Nanoinzheneriya* **12** 31–7
[6] Odinokov V V and Panfilov Y V 2013 *Nanoinzheneriya* **11** 7–18

IOP Conf. Series: Journal of Physics: Conf. Series **872** (2017) 012011 doi:10.1088/1742-6596/872/1/012011

New developments of JSC "Svetlana-X-Ray"

A S Baklin[1], A A Podimskiy[1] and N N Potrakhov[2]

[1] JSC "Svetlana-X-Ray", 198095, Saint Petersburg, Russia
[2] Saint Petersburg Electrotechnical University "LETI", 197022, Saint Petersburg, Russia

E-mail: nn@eltech-med.com

Abstract. The designs and basic electrical, X-ray optical and mass-dimensional parameters of X-ray tubes developed by JSC "Svetlana-X-Ray" during 2016–2017 are presented.

1. Introduction

The company "Svetlana-X-Ray" was actively involved in the solution of the tasks set by the Government of the Russian Federation on import substitution. In the period from 2016 to 2017 years developed and mastered in mass production about a dozen new X-ray tubes [1, 2] – analogs of products produced by the world's leading manufacturers. Among the developed tubes, it should be noted, at least, three: 4RHV8; 0.09BPK30-90; 1.3BPK32-320(II).

2. Research results

X-ray tube 4RHV8 (figure 1) is intended for spectral analysis. The construction of the tube is a metal-glass with two electrodes. Straight tungsten cathode of annular shape. The anode is massive copper, located inside the vacuum balloon of the tube. The material of the anode target is rhodium. The exit window is beryllium. The output window thickness is 0.1 mm; the radiation output directed along the longitudinal axis of the X-ray tube. The tube supplied together with a protective casing as part of an X-ray emitter. High-voltage insulation of the radiator – transformer oil. The design of the radiator provides complete protection against unused radiation, as well as cooling of the tube anode and the radiator itself. Electrical power supply system for a tube with a grounded cathode. Cooling is by a water flow system.

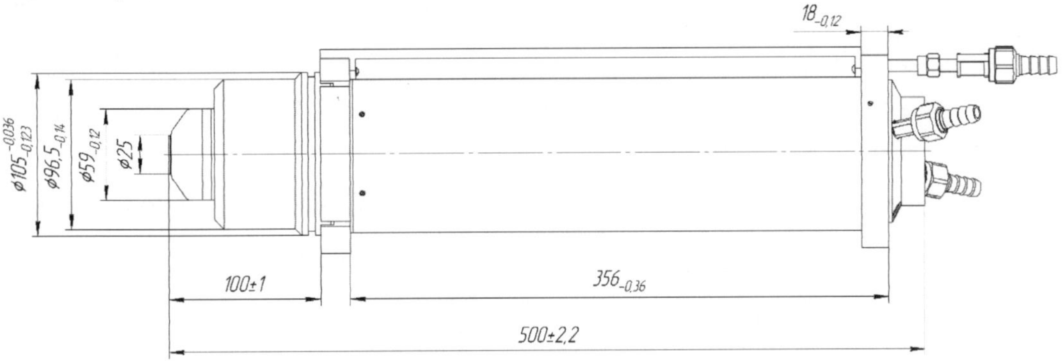

Figure 1. X-ray tube 4RHV8 in the casing.

IOP Conf. Series: Journal of Physics: Conf. Series **872** (2017) 012011 doi:10.1088/1742-6596/872/1/012011

The source based on the described X-ray tube used as part of a multichannel X-ray spectrometer CPM-25/35 (produced by the NJSC "Nauchpribor").

X-ray tube 0.09BPK30-90 (figure 2) is designed for inspection equipment. The construction of the tube is a three-electrode and metal-glass. Straight-heat tungsten cathode in the form of a flat spiral. The anode removed from the vacuum balloon of the tube.

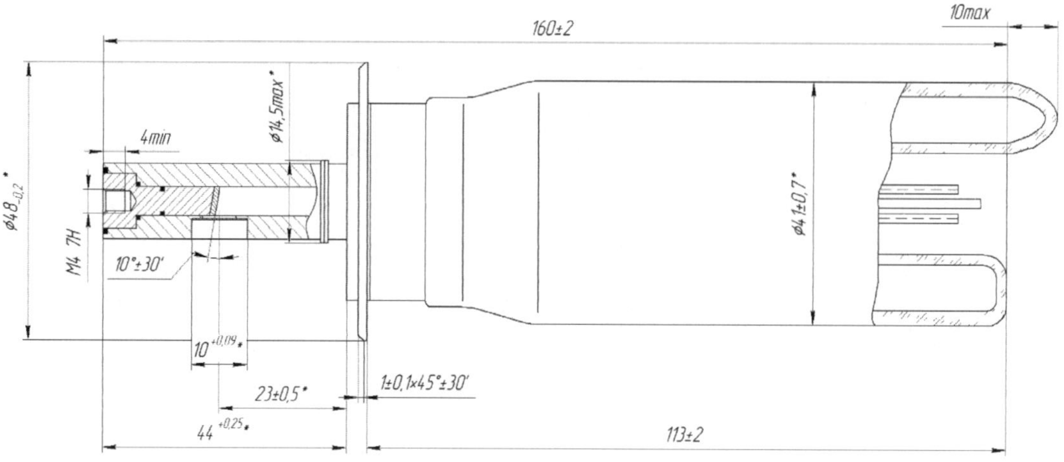

Figure 2. X-ray tube 0.09BPK30-90.

The X-ray tube 1.3BPK32-320(II) (figure 3) intended for industrial X-ray radioscopy. The tube is an analog of a tube from Lohmann model 300/5 NRF (ICM C3005/A; ICM 3003/5). The construction of the tube is a two-electrode and metal-glass. Straight-heat tungsten cathode in the form of a flat spiral. The anode is located inside the vacuum balloon of the tube. The target is a massive cone. The material of the target is tungsten. The radiation output is panoramic, perpendicular to the longitudinal axis of the tube. Electrical circuit power supply tube with a grounded anode. Cooling of the tube is forced air. The tube used for radioscopy thick-walled pipes of gas and oil pipelines.

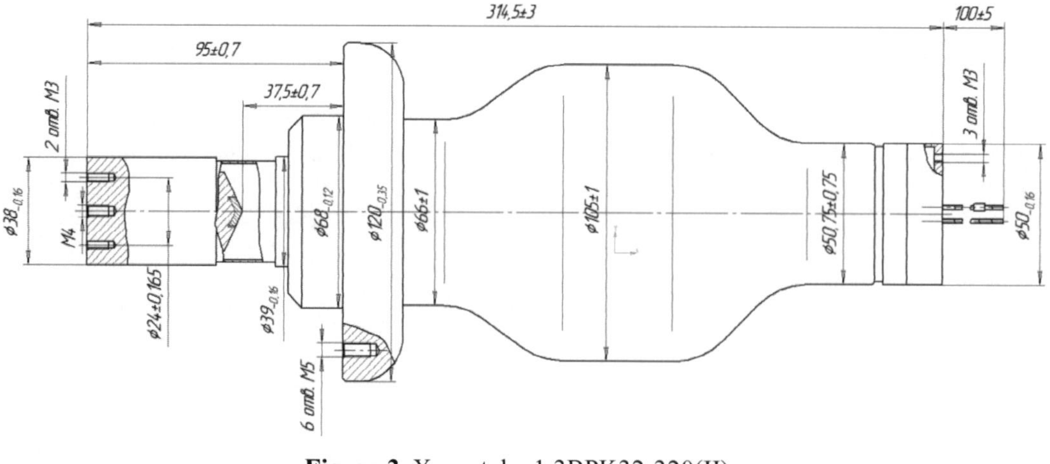

Figure 3. X-ray tube 1.3BPK32-320(II).

The main parameters of the tubes are presented in table 1.

3. Conclusion

On the basic parameters, all the developed X-ray tubes are not inferior to foreign analogs, and in terms of price/quality ratio, they significantly exceed them.

Table 1. The main parameters of X-ray tubes.

Parameter	4RHV8	0.09BPK30-90	1.3BPK24-320(II)
Voltage, kV	20–75	90	150–320
Max current, mA	150	2	5
Power, kW	4	0.09	1.3
Size of effective focal spot, mm	Ø13.5×3.5	0.5–1.2	5×2
Shape of effective focal spot	Ring	Oval	Trapezoid
Angle of the divergence of the effective radiation beam, deg		6×80	40×360
Max heat voltage, V	13	2.2	3.7
Max heat current, A	8.5	3.6	2.1

Acknowledgments
The work was carried out with the financial support of the Russian Science Foundation in the framework of the project “Robotic system of multi-angle microfocus X-ray radiography of aviation parts and assemblies of polymer composite materials under complex impact”. The project number is 16-19-00155.

References
[1] Podymsky A A and Potrakhov N N 2014 *Biomedical Engineering* **47** 78–80
[2] Podimsky A A and Potrakhov N N 2017 *Journal of Physics: Conference Series* **808** 012011

IOP Conf. Series: Journal of Physics: Conf. Series **872** (2017) 012012 doi:10.1088/1742-6596/872/1/012012

Regarding the radio-technical operation mode calculation of the generator lamps

E D Pryalukhin[1], M A Senin[2], M I Pikus[2], S A Martsinukov[2], V P Valuev[3] and A N Ramazanov[2]

[1] JSC "S.E.D.-SPb", St. Petersburg, Russia
[2] Saint-Petersburg State Electrotechnical University "LETI", St. Petersburg, Russia
[3] Peter the Great St. Petersburg Polytechnic University, St. Petersburg, Russia

Abstract. The variant of calculating parameters of high-power generator lamps in a dynamic mode was offered in this work. The output power of the lamp, the load on the electrodes and the efficiency of the anode circuit are determined. The calculation of experimental mode was confirmed with engineering accuracy (~10 %).

Powerful generator lamps (PGL) are one of the main types of vacuum devices [1]. It currently used in television, radio broadcasting, etc. (figure 1). By electronically lamps group refers electronic devices in which in the working volume a high vacuum is created, and the operation principle is based on the use of an electron stream, whose motion is controlled by electric fields [2, 3]. Structurally, a lamp consists of a cathode, an anode and grid electrodes to control the electron beam motion.

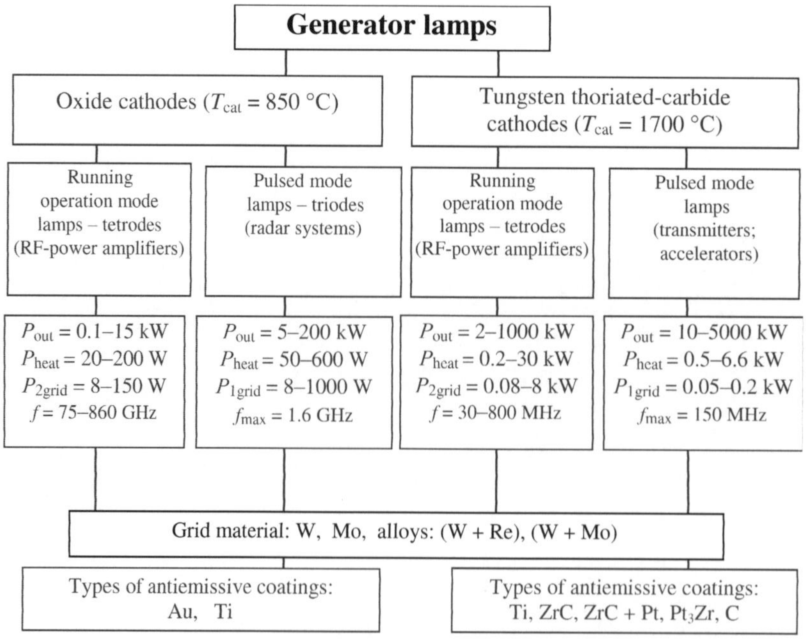

Figure 1. Technical characteristics of powerful generator lamps.

The generator lamps' output power level varies from kilowatt to 2.5 MW in running mode and up to 5 MW in pulsed mode, and the anode voltages range varies from units to hundreds kilovolts [4]. The main areas of generator lamps improvement are an increase in the specific power per unit volume and an increase in durability with a high level of reliability. The increase in specific power is associated with an increase in the operating temperature of the grid nodes and anodes, which leads to a decrease in reliability and lifetime.

Let's consider a variant of a dynamic mode operation of a generator lamp. First, the according parameters chosen with the maximum given permissible parameters:

- anode supply voltage U_a;
- efficiency of the anode chain η;
- residual voltage of the anode $U_{a\,\mathrm{res}}$;
- screen voltage U_s.

The calculations begin with the determination of the oscillatory voltage at the anode

$$U_{ma} = U_a - U_{a\mathrm{res}}.$$

The anode voltage usage factor is determined by the formula

$$\xi = \frac{U_{ma}}{U_a}. \tag{1}$$

Next step is to define the lamp operation class for the bias voltage. The concept of the operating mode is determined by the ratio of the anodic quiescent current to the magnitude of the signal current. Thus, the anode current usage factor

$$\gamma = \frac{2\eta}{\xi}. \tag{2}$$

The radio engineers use the terms angular pulse width and cut-off angle to characterize the duration of part of the half-cycle during which the anode current flows.

Finally, using the tables of the expansion of the cosine-wave pulse in a Fourier series by the Berg method, the cutoff angle of the anode current is found. The operation class for calculated lamp is C. In C operating mode the time of the anode current flow is less than the time of the positive half-wave of the input signal. The mode is characterized by much higher values of efficiency and distortion level in comparison with the used in amplifiers B operating mode [5].

The voltage on the grid 1 is determined from the graph of the anode-grid characteristic

$$I_{a\max} = \frac{I_{a1}}{\alpha}. \tag{3}$$

The operating bias voltage providing the required cutoff angle of the anode current is found by graphical construction. The coordinates of the anode current pulse $I_{a\,\max}$ are used to determine the locking point with the residual voltage at the anode $U_{a\,\mathrm{res}}$ and half the pulse at the supply voltage of the anode $0.5I_{a\,\max}$. The locking voltage is a result of intersecting two points with an axis with a straight line.

First the points $0.5I_{a\,\max}$ and $0.5U_{\mathrm{ma}}$ are found then a straight line through the points of the anode current pulse amplitude and the voltage on the grid 1 before crossing with the anode voltage line is drawn, a locking voltage is obtained. The bias voltage is found by the formula

$$E_{s1} = \frac{U_{s1} - U_{s1}^{+}\cos\Theta_a}{1-\cos\Theta_a}. \tag{4}$$

The amplitude value of the excitation voltage

$$U_{mc} = U_{s1}^{+} - E_{s1}. \tag{5}$$

The cutoff angle of the grid 1 current is found by formula

$$\Theta_{s1} = \arccos\frac{-E_{s1}}{U_{ms}}. \quad (6)$$

The pulse currents according to the characteristics (for grids) are determined. Then the constant component of grid current 1 is found

$$I_{s10} = \alpha'_{0s1} I_{s1\,\max}. \quad (7)$$

The first harmonic of grid 1 current is determined

$$I_{s11} = \alpha'_{1s1} I_{s1\,\max}. \quad (8)$$

The constant component of grid 2 current is found by the formula

$$I_{s20} = 0.8\alpha_{0a} I_{s2\,\max}. \quad (9)$$

To calculate the power the currents values obtained above are used. The grid 2 dissipated power is

$$P_{s2} = U_{s2} I_{s20}. \quad (10)$$

The grid 1 dissipated power is

$$P_{s1} = 0.5U_{ms} I_{s11} + E_{s1} I_{s10}. \quad (11)$$

The anode dissipated power is

$$P_a = U_a I_{a0} - P_{\text{out}}. \quad (12)$$

The grids and anode loading of the lamp do not exceed 87 % of the maximum permissible powers is checking

$$K_{s2\,\text{res}} = \frac{P_{s2}}{P_{s2\,\text{add}}}. \quad (13)$$

Next the equivalent resistance of the anode circuit is found

$$R_{oe} = \frac{U_{am}}{I_{a1}}. \quad (14)$$

Finally, the lamp power amplification factor is determined

$$K_P = \frac{P_{\text{out}}}{0.5U_{ms} I_{s11}}. \quad (15)$$

The results of the calculations are then compared with the experimentally obtained data and the deviations are determined (%):

- output power, kW;
- anode circuit efficiency factor, %;
- grid 2 loading, W;
- grid 1 loading, W;
- anode dissipated power, W.

The level of power released on the tetrode's screen grid is determined by its operating mode and determines the output of electrons from its surface into the discharge gap. The Richard–Dashman formula [6, 7] for thermionic emission:

$$j_{TEE} = AD_0 T^2 \exp(-e\varphi / kT), \quad (16)$$

where $A = 4\pi m_e k^2/h^3 = 1.2 \cdot 10^6$ A/(m^{2}K^2) – Richardson constant; D_0 – quantum-mechanical coefficient of electron reflection from the metal-vacuum boundary; $k = 1.38 \cdot 10^{-23}$ J/K – Boltzmann constant; φ – work function, W.

Real conditions work requires from the developers of devices to include the development of the process of applying antiemission coatings in their technological map [8–10]. The most effective coatings used for oxide cathodes are Au-, Ti-based coatings, for tungsten cathodes these are zirconium and platinum based compounds [11, 12] are often used as antiemission materials, which have a higher work function of the grid material.

The calculations of the dynamic mode of a powerful generator lamp were confirmed with engineering accuracy (~10 %).

References

[1] Vetrov N Z, Kostrin D K, Lisenkov A A and Popova M S 2015 *Journal of Physics: Conference Series* **652** 012032
[2] Tsarev B M 1967 *Calculation and design of electron tubes* (Moscow: Energia)
[3] Bystrov Yu A, Kostrin D K, Lisenkov A A and Vetrov N Z 2015 *Vakuum in Forschung und Praxis* **2** 22–5
[4] Perez Vasquez N O, Kostrin D K and Uhov A A 2017 *Journal of Physics: Conference Series* **808** 012007
[5] Fridrikhov S A and Movnin S M 1982 *Physical basis of electronic engineering* (Moscow: Higher School)
[6] Kleiner E Yu 1979 *Fundamentals of the theory of electron tubes* (Moscow: Higher School)
[7] Bystrov Yu A et al. 2009 *Modern technological processes in the production of powerful generator lamps* (St. Petersburg: SPbETU "LETI")
[8] Kostrin D K and Lisenkov A A 2016 *Materials Science Forum* **843** 278–83
[9] Kostrin D K and Lisenkov A A 2017 *Vakuum in Forschung und Praxis* **3** 35–9
[10] Kostrin D K, Pikus M I, Smirnov E A and Lisenkov A A 2017 *Journal of Physics: Conference Series* **789** 012030
[11] Kostrin D K and Lisenkov A A 2016 *Materials Science Forum* **870** 371–6
[12] Lisenkov A A, Vetrov N Z and Kostrin D K 2017 *Technical Physics Letters* **4** 390–2

Modelling of a high-current magnetron discharge in a plasma electron emitter

S Yu Udovichenko[1], D K Kostrin[2] and A A Lisenkov[3]

[1] Department of experimental physics and nanotechnology, research and education center "Nanotechnology", Tyumen State University, 625003, Tyumen, Russia
[2] Department of electronic instruments and devices, Saint Petersburg Electrotechnical University "LETI", 197376, Saint Petersburg, Russia
[3] Institute of Problems of Mechanical Engineering, Russian Academy of Sciences, 199178, Saint Petersburg, Russia

E-mail: udotgu@mail.ru

Abstract. An analytical model of a high-current form of a low-pressure glow discharge in an inverted cylindrical magnetron, which performs the function of plasma electron emitter, is shown. Were found conditions of the discharge self-sustaining, allowing to estimate the voltage of the discharge and determine the critical value of the magnetic field and residual gas pressure below which the existence of this type of discharge is impossible. A comparison of the calculated discharge characteristics with experimental data obtained on the setup for studying the emission properties of the magnetron discharge was carried out.

1. Introduction

For generation of high-current electron beams with pulse duration of microsecond range, it is reasonable to use the plasma emitters with cold cathodes. An efficient source of high-current electron beam was created on the basis of a plasma discharge in crossed electric and magnetic fields excited in the electrode system of an inverted cylindrical magnetron type [1–3]. Due to the electron oscillations the discharge in crossed fields is easily ignited at low pressures, providing a high degree of plasma gas ionization and stability of the plasma parameters. In [1] is shown a possibility of using discharge system of an inverted magnetron type for receiving a tubular beam of charged particles. In [3] the results of experimental research of a high-current magnetron discharge in conditions of generating tubular electron and ion beams in pulsed and in continuous modes are shown.

Created plasma emitter on the basis of the magnetron discharge system is a part of an electron accelerator designed for the relativistic microwave electronics. At an accelerating voltage of 200 kV and pulse duration of 15 μs electron beam current is 500 A.

For further upgrade of such a source of electrons, optimization of its geometry and parameters of plasma discharge, and also for definition of the discharge self-sustaining conditions in the present work modelling of the discharge cell is carried out.

2. Experimental setup

Figure 1 shows a diagram of the experimental setup with electrode system of an inverted magnetron type [2, 3]. The discharge is ignited between the cathode *1* and the anode *2*. The diameter of the cathode is changed from 80 to 140 mm, the interelectrode gap varies in the range of 10–20 mm, and the length of the electrodes is 60–100 mm. The end electrode *5* is electrically connected with the cathode. The electron emission from the discharge is carried out from the annular surface of plasma through a slit in the electrode *5* with a width of 2 mm. Particle acceleration is carried out when voltage is applied between the electrodes *5* and *8*. The magnetic field in the discharge and accelerating gaps is created using solenoids *3* and *4*, respectively. The magnitude of the magnetic field varies from 100 to 500 G, pressure of the plasma gas (xenon, nitrogen, argon) in the discharge chamber is 10^{-3}–$6\cdot10^{-3}$ Torr.

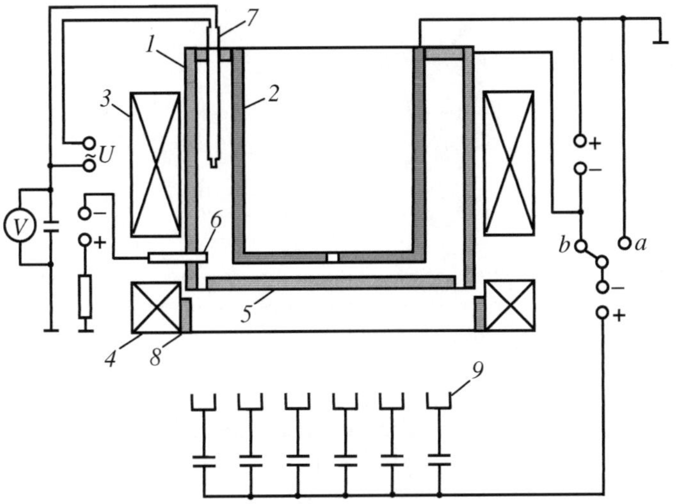

Figure 1. The experimental setup for studying the emission properties of the magnetron discharge: *1* – cathode; *2* – anode; *3*, *4* – solenoids; *5* – end electrode; *6* – Langmuir probe; *7* – emission probe; *8* – accelerating electrode; *9* – collector.

When the gas pressure is below 10^{-3} Torr, discharge exists in a low-current form with current less than 1 A and discharge voltage of several kV. With growing pressure, increases the intensity of ionization and occurs an abrupt transition to a high-current form of the discharge, which is accompanied by a reduction of discharge voltage to several hundred volts, raise of the current by hundreds of times and a significant increase in the brightness of the discharge.

3. Model of a discharge cell and comparison of the theoretical and experimental results

A discharge cell of an inverted magnetron is a hollow cylinder with radius R, on the axis of which is situated the anode in a form of a rod with radius r_0. The cylindrical surface acts as the cathode. Uniform magnetic field B is directed along the cylinder axis. We investigate only stationary high-current mode when the voltage drop U is concentrated in a thin near-cathode layer, and all the interelectrode space is occupied by the positively charged plasma. Cathode layer is formed by a stream of fast electrons with the energy corresponding to the voltage of the discharge U. These electrons with velocity $V_f(R) = V_{fR} = (2eU/m)^{1/2}$ come into the weakly ionized quasi-neutral plasma and maintain the necessary level of the gas ionization. Slow plasma electrons are not able to produce the ionization of atoms and provide a discharge current to the anode.

Let us consider the one-dimensional problem in the cylindrical coordinate system and analyze the conditions of discharge maintenance, considering that the transfer of fast electrons along the radius of the cylinder is in the diffusive regime, and the influence on the movement of fast particles of weak electric field in quasi-neutral plasma can be neglected. With the help of the stationary equation of continuity and the equation for the flow of fast electrons [4, 5] we can get the following equation for radial distribution of particle concentration n_f:

$$\frac{1}{\rho}\frac{d}{d\rho}\left(\rho\frac{dn_f}{d\rho}\right)-n_f=-\frac{\gamma n_{ic}V_s\tau_r}{2(R-r_0)}=-n_{f0}, \tag{1}$$

where $\rho = r/\lambda_f$; $\lambda_f = (D_{fr}\tau_r)^{1/2}$; $D_{fr} = eU\nu_{f0}/3m(\omega_{Be}^2 + \nu_{f0}^2)$ – diffusion coefficient; ν_{f0} – frequency of elastic collisions with gas atoms; $\omega_{Be} = eB/m$ – Larmor frequency; γ – effective coefficient of ion-electron emission; $\tau_f = eU/\nu_i W$ – characteristic time of relaxation of fast electrons; ν_i – ionization frequency; W – energy spent by fast electrons on ionization of the gas atoms.

The term on the right side of equation (1) describes the appearance of fast electrons in the volume due to the γ-processes on the cathode under the action of incident plasma ions. Here n_{ic} and $V_s = (T_e/M)^{1/2}$ – concentration and the rate at which ions leave the region of quasi-neutral plasma to the cathode; T_e – temperature of the plasma electrons; M – mass of the ions. The solution of the equation (1) is

$$n_f(\rho) = n_{f0}\left(1 - AI_0(\rho) - BK_0(\rho)\right),$$

where A, B – constants; I_0, K_0 – modified Bessel functions of zero order of the first and second kind respectively. This solution satisfies the boundary conditions at the anode and cathode:

$$n_f(r = r_0) = 0; \quad \left(n_f V_f\right)_R = D_{fr}\left(\frac{dn_f}{dr}\right)_R. \tag{2}$$

In the right part of the second condition of (2) there is no minus sign, because as the positive direction for the flow of fast and slow electrons we consider the direction from the cathode to the anode, i.e. against the positive direction of the cylindrical coordinate r. Taking into account the boundary conditions we can get the following expression for the concentration of fast electrons:

$$\frac{n_f(\rho)}{n_{f0}} = 1 - \frac{I_0(\rho)K_1(\rho_R) + I_1(\rho_R)K_0(\rho) + 2(\rho_R - \rho_{r0})\left(I_0(\rho_{r0})K_0(\rho) - I_0(\rho)K_0(\rho_{r0})\right)}{I_0(\rho_{r0})K_1(\rho_R) + I_1(\rho_R)K_0(\rho_{r0})}, \tag{3}$$

where $\rho_R = R/\lambda_f$; $\rho_{r0} = r_0/\lambda_f$.

The condition of the discharge self-sustaining in a cylindrical magnetron can be found in the following way. As follows from the work [4], the multiplication factor of the plasma electrons $\alpha = eU/W$ is connected with the ratio of the fast electrons current at the anode to the complete discharge current $\psi = I_{fa}/I$:

$$\alpha \approx \frac{1}{\gamma - \psi}. \tag{4}$$

Current of electrons at the anode is transferred mainly by plasma electrons $I = I_{fa} + I_{ea} \approx I_{ea}$, that is formula (4) is valid when $\psi << 1$. Determining the flow of fast electrons from the equation that was used for the second boundary condition (2), and the flow of plasma electrons to the anode

$$j_{ea} = \frac{1}{r_0}\int_{r_0}^{R} e_i \nu_i n_f r dr, \tag{5}$$

using (4) we can find the condition of the discharge self-sustaining in the magnetron:

$$u = 1 + \frac{1}{\frac{\rho_{r0}N}{2}\left(5R^2/r_0^2 - 4R/r_0 - 1\right) - 1}, \tag{6}$$

where

$$N = \frac{I_0(\rho_{r0})K_1(\rho_R) + I_1(\rho_R)K_0(\rho_{r0})}{I_1(\rho_R)K_1(\rho_{r0}) - I_1(\rho_{r0})K_1(\rho_R) + 2(\rho_R - \rho_{r0})\left(I_0(\rho_{r0})K_1(\rho_{r0}) + I_1(\rho_{r0})K_0(\rho_{r0})\right)}.$$

In the formula (6) are used for convenience the dimensionless variables $u = U/U_0 = \gamma\nu_i\tau_f$ and $b = B/B_0$, where $U_0 = W/e\gamma$; $B_0 = 1.5(mW\nu_{f0}/\nu_i)^{1/2}/e\gamma(R - r_0)$, besides $\nu_{f0} \approx 2\nu_i$; $\rho_R = (2.61b/u)R/(R - r_0)$ and $\rho_{r0} = \rho_R r_0/R$ in the case of magnetized electrons, when $\omega_{Be}^2 >> \nu_{f0}^2$.

Figure 2 presents the dependence of the discharge sustainment voltage from an induction of the magnetic field obtained using equation (6).

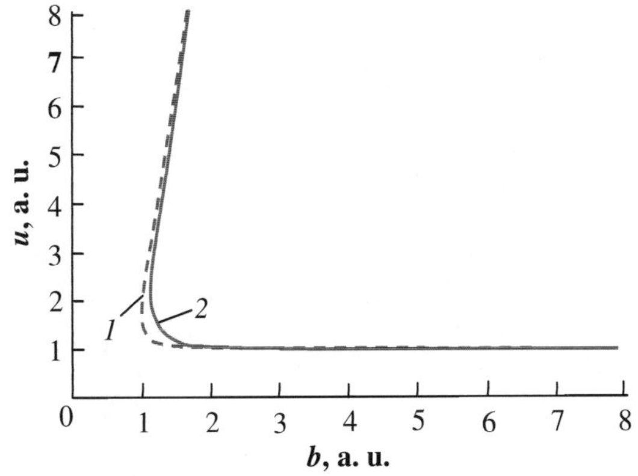

Figure 2. Dependence of the discharge sustainment voltage from the magnetic field induction obtained using equation (6): *1* – for flat electrodes with $d = R - r_0 = 2$ cm [9]; *2* – $r_0/R = 0.86$, $R = 7$ cm, $r_0 = 6$ cm.

It can be seen that in the inverted magnetron the minimum value of the magnetic field for which the existence of the discharge is still possible is more than in planar magnetron [6]. The magnitude of the magnetic field B_0, shown in the formula (6), and curve *2* in figure 2 are in a good agreement with experimental results [3], shown in figure 3.

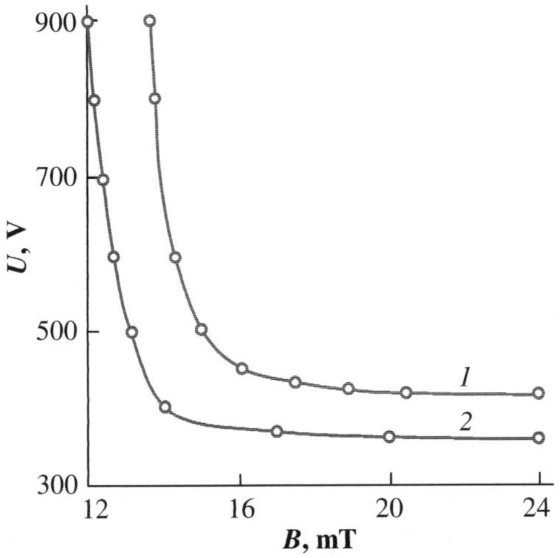

Figure 3. Dependence of the ignition voltage from the induction of the magnetic field at nitrogen pressures: *1* – $0.3 \cdot 10^{-2}$ Torr; *2* – $0.6 \cdot 10^{-2}$ Torr.

From these results it also follows that an increase in gas pressure leads to a decrease in the discharge voltage and shifts the voltage curve to smaller values of the magnetic field.

This model does not describe the dependence of the voltage of a high-current discharge from the residual gas pressure. Influence of the gas pressure on the conditions of discharge self-sustaining can be considered if in the equation (4) for the current density of plasma electrons to the anode across the magnetic field instead of (5) use the following expression [4]:

$$j_{ea} = \frac{en_0 V_{Te}}{4\sqrt{1+\beta_e^2}},$$

where n_0 – plasma density in the interelectrode gap; $V_{Te} = (T_e/m_e)^{1/2}$, $\beta_e = 1/r_{Te} n_g \sigma_{e0}$, $r_{Te} = mV_{Te}/eB$ – thermal velocity, Hall parameter and cyclotron radius of the plasma electrons, respectively; n_g – density of the gas atoms; σ_{e0} – transport cross section of electron-atom collisions. Then in the assumption of the constant concentration of plasma in the interelectrode gap, i.e. $n_0 \approx n_{ic}$ (see formula (1)) instead of (6) we have:

$$u\left(1-\frac{\psi}{\gamma}\right)=1; \qquad (7)$$

where $\frac{\psi}{\gamma}=\frac{1}{(\rho_R-\rho_{r0})}\frac{T}{p}$; p – gas pressure;

$$T=\frac{I_1(\rho_R)K_1(\rho_{r0})-I_1(\rho_{r0})K_1(\rho_R)+(\rho_R-\rho_{r0})[I_1(\rho_{r0})K_0(\rho_{r0})+I_0(\rho_{r0})K_1(\rho_{r0})]}{I_0(\rho_{r0})K_1(\rho_R)+I_1(\rho_R)K_0(\rho_{r0})}.$$

Figure 4 shows the dependence of the discharge voltage from the xenon pressure at different values of the magnetic field, plotted according to the formula (7). Here b = 1 corresponds to the magnetic field $B = B_0$ = 140 G; γ = 0.1; W = 30 eV; $\sigma_{e0} = 1.5\cdot10^{-15}$ cm^2. The calculated dependence of the discharge voltage from the pressure confirms the results of the experiment – when the gas pressure is lower than 1 mTorr magnetron discharge exists in a high voltage form and has a voltage more than 1 kV.

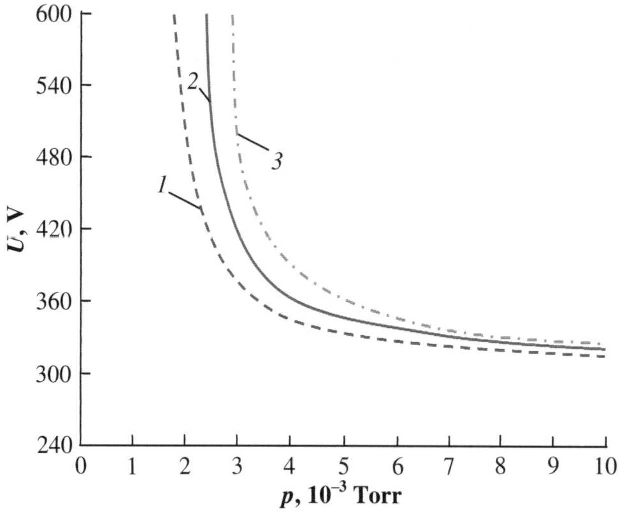

Figure 4. Dependence of the discharge voltage from the xenon pressure: *1* – b = 1.2; *2* – b = 1; *3* – b = 0.9.

4. Conclusions

In this work was built a hydrodynamic model of a high-current plasma discharge in the crossed electric and magnetic fields excited in the electrode system of an inverted cylindrical magnetron type.

Were found the self-maintaining conditions for a high-current discharge in a cylindrical magnetron cell. The minimum value of the magnetic field induction for which in such geometry the existence of a discharge is still possible is somewhat larger than in the planar magnetron. As well as in direct magnetron, with a decrease in the magnitude of the magnetic field increases the minimum gas pressure that is necessary to maintain a high-current discharge form in an inverted magnetron.

Minimum gas pressure for which may exist a high-current magnetron discharge is on the order of magnitude smaller than the corresponding pressure of $2\cdot10^{-2}$ Torr in the combined reflective discharge with a hollow cathode [4].

Built in this model dependences of the discharge voltage of a high-current plasma discharge from the pressure and magnitude of the magnetic field are in a good agreement with the corresponding experimental curves.

References

[1] Oks E M and Chagin A A 1988 *Journal of technical physics* **6** 1191–3
[2] Oks E M and Chagin A A 1991 *Journal of technical physics* **6** 204–6

[3] Oks E M, Chagin A A and Schanin P M 1991 *Proceedings of the I All-Union conference on plasma emission electronics (Ulan-Ude)* pp 18–23
[4] Kozyrev A V 2009 *Proceedings of the III International Kreyndelevskiy seminar "Plasma emission electronics" (Ulan-Ude)* pp 5–9
[5] Zulkova L A, Kozyrev AV and Proskurovskiy D I 2005 *Journal of technical physics* **11** 59–64
[6] Nikulin S P 1998 *Journal of technical physics* **7** 56–63

IOP Conf. Series: Journal of Physics: Conf. Series **872** (2017) 012014 doi:10.1088/1742-6596/872/1/012014

Combined vacuum transducers for measuring pressure in the range from 10^5 to 10^{-7} Pa

L N Rozanov[1], A N Sokolov[2], A Y Skryabnev[1] and D E Abrosimov[1]

[1] Peter the Great Saint Petersburg Polytechnic University, Institute of Metallurgy, Mechanical Engineering and Transport, 195251, Saint Petersburg, Russia
[2] Scientific and Design Center for Optoelectronic Observation Systems "Corporation "Cometa", 194021, Saint Petersburg, Russia

E-mail: scarts@mail.ru

Abstract. A combined vacuum-measuring transducer with a measurement range from 10^5 to 10^{-7} Pa is developed. The device combines heat and magnet-discharge transducers.

In modern vacuum systems, the most frequent work is with pressures in a wide range from 10^5 to 10^{-7} Pa. As for pressure measurements, it is impossible to cover the entire range with a vacuum transducer of the one type, therefore vacuum gauges with transducers operating on different principles (thermal, magnetic-discharge, ionization, etc.) are used [1]. Therefore, it is of interest to develop combined transducers (figure 1) that combine sensors of different types [2].

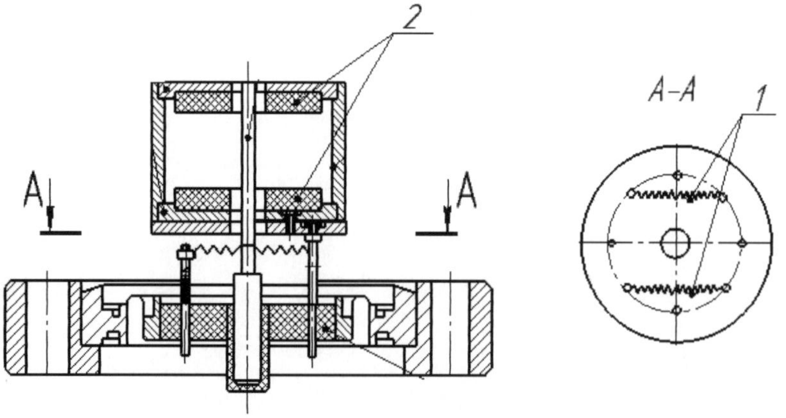

Figure 1. General view of the combined transducer: *1* – thermal transducer; *2* – magnetic transducer.

Manufacturers are currently producing wide-range vacuum gauges, for example Edwards WRG-S or Televac CC-10, with a measurement range of 10^5–10^{-7} Pa. To cover a wide range of pressures, a combination of different types of converters is used in them. However, the above-mentioned vacuum gauges have one significant drawback. Despite the fact that they are designed to measure pressure in high vacuum systems, rubber vacuum seals are used in their designs. This results to a decrease in the efficiency of the high-vacuum pumping system due to increased gassing of the rubber in comparison with the metal seals [3]. Thus the accuracy of measurement at pressures close to the lower limit is deteriorates.

To eliminate these drawbacks, has been designed combined a vacuum-metric transducer that does not include parts made of polymer materials. In this device, two types of transducers are combined: thermal and magnetic-discharge (figure 2).

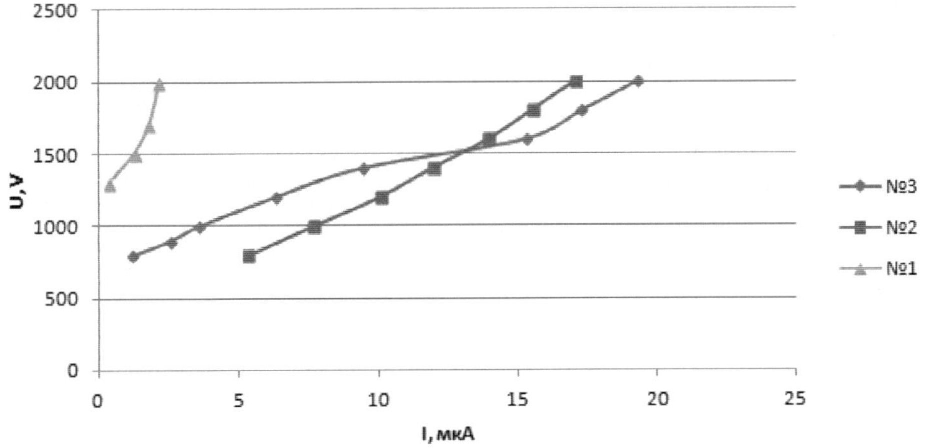

Figure 2. Volt-ampere characteristics of various versions of the magnetic part of the combined transducer at a pressure $p = 10^{-3}$ Pa.

The principle of the action of thermal converters is based on the dependence of the heat conductivity of the gas on pressure. In this case, a tungsten filament is used, which is heated by passing an electric current through it. The principle of the action of magnet-discharge transducers is based on the dependence on the pressure of the current of an independent gas discharge in a crossed electric and magnetic fields. In this case, a scheme of the inverse-magnetron type is used. In this scheme, the central rod acts as an anode, to which a high voltage is applied (figure 3). In this case, both the anode and the cathode are in a constantly-acting magnetic field.

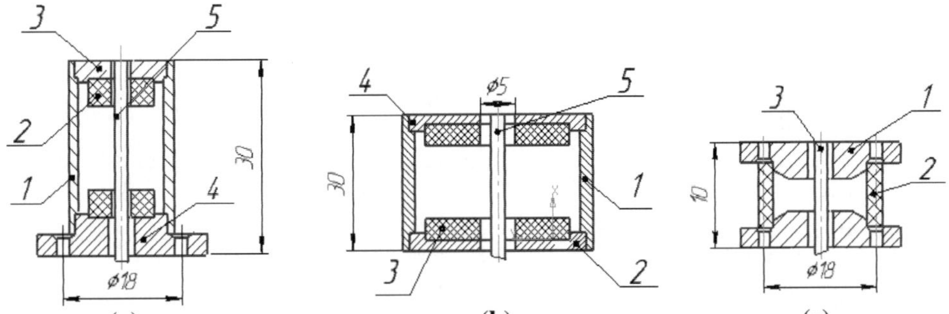

Figure 3. Various versions of the magnetic part of the transducer: **(a)**: *1* – cylindrical sleeve; *2* – magnet; *3* – upper washer; *4* – lower washer; *5* – anode; **(b)**: *1* – magnet; *2* – cylindrical sleeve; *3*, *4* – washer; *5* – anode; **(c)**: *1* – washer; *2*, *3* – magnet.

The thermal converter makes it possible to measure pressure in the range of 10^5–10^0 Pa (figure 3(a)). In terms of construction, it is made in the form of two identical tungsten filaments *1* witch thick 50 μm. The filament has a spiral shape. This allows you to get more resistance than a direct straight filament at the same length. When measuring one filament is involved, and the second is entered into the construction as a backup. Thermal vacuum transducers can operate in two modes – a constant filament temperature mode and a constant filament current mode [4]. This transducer operates in a constant temperature mode. The working temperature of the filament is 50 °C.

To measure lower pressures in the range of 10^0–10^{-7} Pa, a magnetic transducer is used. In total, three versions of the magnetic system were considered during the development (figure 3). In two versions (figures 3(a), (b)), two magnets were used to create the magnetic field, one at the top and one at the bottom. These constructions, in fact, differed only in the geometric dimensions of the chamber. In the third version (figure 3(c)), one magnet was used and acted as a cylindrical wall of the ionization chamber. A comparative analysis of the volt-ampere characteristics of various variants at a fixed pressure (figure 2) showed that the third variant has the best sensitivity, which was chosen as the main (figures 4, 5).

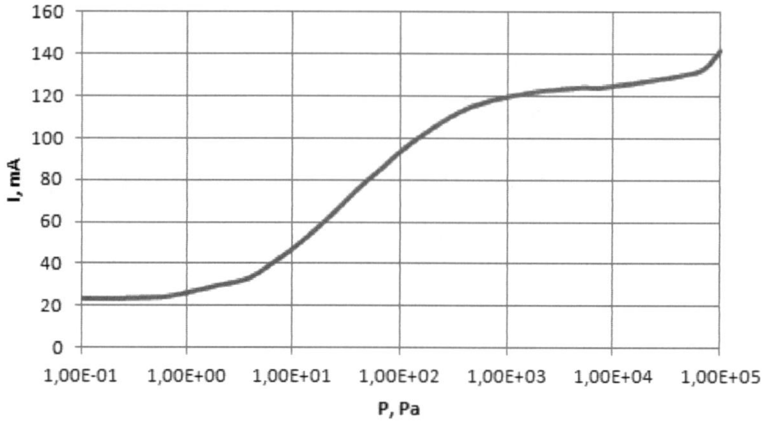

Figure 4. Calibration characteristics of the thermal part of the transducer.

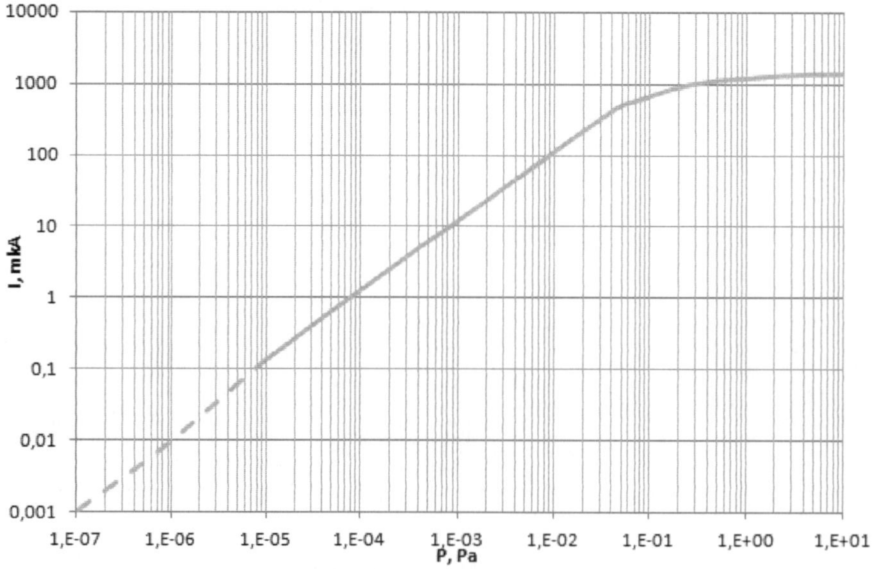

Figure 5. Calibration characteristics of the magnetic part of the transducer.

To connect the transducer to the electronic unit, a ceramic electrical input *6* with nine connectors welded to the CF40 rotary flange is used. Such a constructive solution allowed us to refuse the use of rubber seals. The electronic block of the vacuum gauge is designed for self-reading and control of vacuum sensors with optional leads. The output is carried out on the computer monitor and on its own LED panel. It is also possible to control the relay, depending on the pressure value.

As a result of the work done, a vacuum gauge model was developed, operating in a pressure range of 10^5–10^{-7} Pa. The design of the converter does not use rubber seals, which significantly reduces its gas emission in comparison with the analogs.

IOP Conf. Series: Journal of Physics: Conf. Series **872** (2017) 012014 doi:10.1088/1742-6596/872/1/012014

References

[1] Rozanov L N 2007 *Vacuum technology* (Moscow: Higher School)
[2] Rozanov L N and Skryabnev A Y 2014 *Vacuum technology and technology* **24** 67–70
[3] Pliskovsky V Y 1976 *Structural materials and elements of vacuum systems* (Moscow: Mechanical Engineering)
[4] Frolov E C, Minaichev V E, Aleksandrova A T et al. 1992 *Vacuum technology: Reference book* (Moscow: Mechanical Engineering)

Investigation of low-temperature ion nitriding technology of titanium alloy Ti-6Al-4V

R D Agzamov, K N Ramazanov, A F Tagirov, E L Vardanyan and K K Shulakov

Ufa State Aviation Technical University, Ufa, 450000, Russia

E-mail: ainurtagirov@mail.ru

Abstract. The paper presents the results of low-temperature nitriding in non-self-maintained high-current arc discharge and in glow discharge. Measurement results of microhardness across the depth of a nitrated layer are given. It is found that the depth of a nitrated layer depends on treatment temperature. The temperature increase from 450 to 600 °C during nitriding in non-self-maintained high-current arc discharge leads to the increase in the depth of a nitrated layer from 10 to 24 µm, and from 10 to 22 µm when treated in a glow discharge. The paper also describes data on residual stress in a surface layer of samples after low-temperature ion nitriding.

1. Introduction

Compressor blades of a gas-turbine engine (GTE) are classified as critical parts of an engine and determine its lifetime and operational integrity. Nowadays, the aircraft engine industry faces a problem of increasing the strength properties of compressor blades [1].

Reliability and durability of blades significantly depend on physical-chemical and structural-phase composition, microgeometry and residual surface stress. The residual stress in a surface layer has a significant influence on fatigue resistance within parts operating under static and fluctuating loads. The compressive residual stress positively influences the fatigue resistance, while the compressive residual stress of a surface layer is caused by formation and development of fatigue crack growth and, hence, the decrease in lifetime of blades [2, 3].

The formation of compressive residual stress is solved through ion implantation. However, due to its characteristics ion implantation cannot be applied for treatment of monowheels since it may cause interblade area overlap (only treatment of single blades is possible).

Gas nitriding does not seem possible since treatment is carried out at 800–900 °C which leads to the growth of grains and deterioration of material properties. The authors suggest low-temperature ion nitriding having the following advantages: monitoring of treatment parameters in a variety of operating modes (temperature, pressure, ion current density, particle energy); high saturation rate, high degree of fineness; high performance of parts made from hard titanium alloys [4, 5].

2. Methodology

The research was carried out using a two-phase titanium alloy Ti-6Al-4V widely applied in the aviation industry. Two types of samples were used in the experiment: disc-shaped 80×10×1 mm plates of 20 mm in diameter and 5 mm thickness.

A series of experiments were performed at various modes on low-temperature ion nitriding of Ti-6Al-4V titanium alloy in non-self-maintained high-current arc discharge generated by plasma source and in a glow discharge. Different modes of low-temperature ion nitriding are shown in table 1.

Table 1. Modes of low-temperature ion nitriding.

№ modes	Temperature, °C	Duration *t*, h	Operation environment composition
Nitriding modes in non-self-sustaining high-current arc discharge			
1	450 ± 10	1	100 % N_2
2	600 ± 10	1	100 % N_2
Nitriding modes in glow discharge			
1	450 ± 10	1	85 % Ar + 15 % N_2
2	600 ± 10	1	85 % Ar + 15 % N_2

The inclined microsections with an angle of 6° to the monitored surface were prepared to measure the microhardness along the depth of the diffusion layer of nitrided samples. Microhardness measurements were carried out applying the method of reconstructed imprint in accordance with the Russian State Standard GOST 9450-76 using the Struers Duramin-1/-2 microhardness tester. The static load applied to the diamond tip for the period of 10 s was 490 [MN] (50 g). The depth of a hardened layer was determined from the hardness distribution curve to the value of hardness of the basic material.

The microstructure of a hardened layer was studied using optical microscope Olympus GX-51. To determine the influence of the nitriding mode on the residual stresses, a method for precise measurement of changes of the interplanar distances determined from the displacement of the diffraction line was used. The measurement was carried out using the DRON-4 X-ray diffractometer.

3. Results and discussion

Measurements of surface microhardness showed that after nitriding there is an increase in surface microhardness which is caused by the formation of a golden nitride film and the presence of a diffusion layer [6]. Figure 1 shows variation in microhardness with respect to the depth of the diffusion layer.

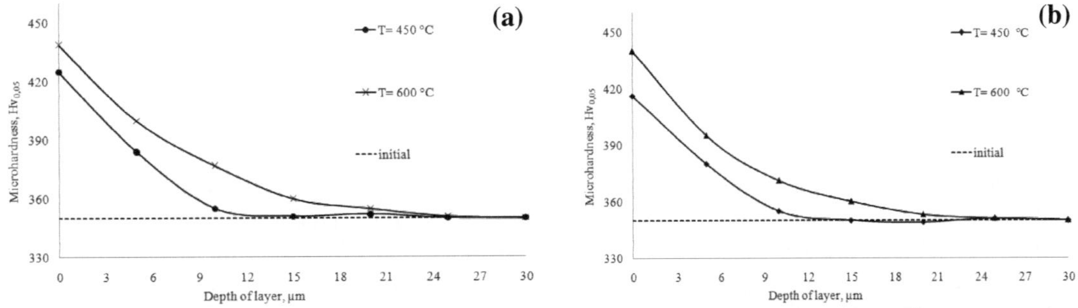

Figure 1. Dependences of microhardness variation of the nitrided layer depth of Ti-6Al-4V alloy: **(a)** – basic and modified by non-self-sustaining high-current discharge; **(b)** – initial and modified in glow discharge.

The diagrams show that as the distance between the surface increases, the microhardness on the samples decreases smoothly, which indicates the presence of an extended layer with increased hardness.

Dependences of microhardness of the depth of the nitrided layer show that the temperature of ion nitriding, both in non-self-sustained high-current arc discharge and glow discharge, has a significant effect on the nature of nitrogen diffusion into the interior of titanium alloy during low-temperature nitriding. For example, in case of nitriding in non-self-sustaining high-current arc discharge, an increase in temperature from 450 to 600 °C leads to an increase of the nitrided layer depth from 10 to 24 μm. In case of nitriding in glow discharge, an increase in temperature from 450 to 600 °C also leads to an increase of the nitrided layer depth from 10 to 22 μm. Figure 2 shows the structure of Ti-6Al-4V titanium alloy samples after low-temperature ion nitriding.

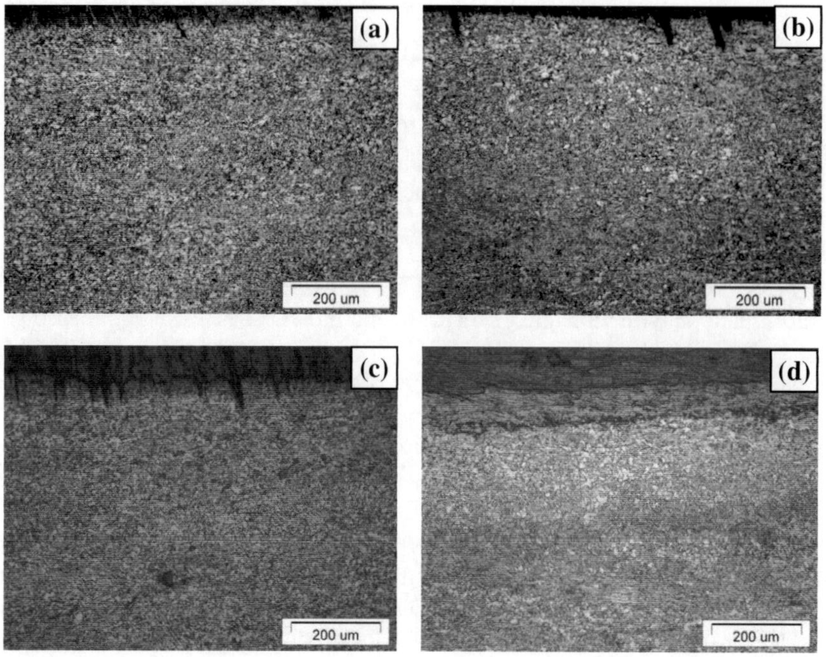

Figure 2. Microstructure of Ti-6Al-4V titanium alloy samples after nitriding: **(a)**, **(b)** – after nitriding in non-self-maintained high-current arc discharge; **(c)**, **(d)** – after nitriding in glow discharge.

The microstructure of samples after various surface treatment modes is mainly presented by equiaxial α-phase grains with boundary separation of drop-, acicular, and prolate shape β-phase (figure 2). At the same time the average size of α-and β-phase grains in their initial state makes 7.5 and 2.1 μm correspondingly.

Treatment at 450 °C does not change microstructural parameters of samples and grain phases remain equiaxial. At 600 °C there is a growth of separate crystallites which illustrates the beginning of secondary recrystallization. Microstructural parameters of a boundary layer are similar to the parameters of a material.

Residual stress in a surface layer after low-temperature ion nitriding was determined for disc-shaped samples. The values of the residual stresses after ion modification are presented in table 2. The presence of compressive stresses on the surface of the initial sample of 7.1 kgf/mm^2 could be bound to preliminary machining (turning, grinding, etc.), which usually leads to the residual stress in a thin surface layer of materials.

After nitriding in non-self-sustained high-current arc discharge, tensile stresses arise on the surface of samples, which can be the cause of accelerated generation and development of cracks. The results presented in table 2 show that the temperature affects the magnitude of the residual stress. Long-time holding at a temperature of 600 °C leads to an increase of tensile stresses from 5.7 to 12.4 kgf/mm^2, which is unacceptable for key workpieces. High temperature and prolonged exposure, apparently, leads to the degradation of structure and appearance of an α-layer. The decrease in temperature leads to the decrease in surface tensile stress.

Table 2. Residual stress values on the samples.

№ modes	Residual stress values σ, kgf/mm^2
Initial	+7.1 ± 1
After treatment in non-self-sustained high-current arc discharge	
1	+0.5 ± 1
2	+12.4 ± 1
After treatment in glow discharge	
1	–23.7 ± 1
2	–8.3 ± 1

Compressive stresses appear on the surface of samples treated in glow discharge, possibly due to formation of phases with an increased specific volume. Moreover, a decrease in temperature and exposure time of treatment leads to an increase in the compressive residual stresses and amounts to –23.7 ± 1 kgf/mm^2.

4. Conclusion

Increase in microhardness of Ti-6Al-4V titanium alloy samples is observed after low-temperature ion nitriding both in non-self-maintained high-current arc discharge and in glow discharge.

It is found that the depth of a nitrated layer significantly depends on treatment temperature. The temperature increase from 450 to 600 °C during nitriding in non-self-maintained high-current arc discharge leads to the increase in the depth of a nitrated layer from 10 to 24 μm, and from 10 to 22 μm when treated in a glow discharge.

The study revealed that nitriding at 450 °C does not result in the change of sample structure and at 600 °C the growth of separate crystallites is observed

It was defined that the temperature affects the sign and magnitude of residual stresses. Thus, in case of ion nitriding in non-self-sustained high-current arc discharge at a temperature of 600 °C, the residual stress was presented at the range of +12.4 ± 10 kgf/mm^2. When nitriding is performed in glow discharge at 450 °C, the residual stress amounted to –23.7 ± 10 kgf/mm^2.

References

[1] Smyslov A M, Smyslova M K and Dubin A I 2012 *Vestnik USATU* **46** 24–9
[2] Smyslov A M, Smyslova M K, Dubin A I, Sazonov V P and Pavlov V F 2016 *News of higher educational institutions* **37** 121–30
[3] Smyslov A M, Mingazhev A D, Smyslova M K, Selivanov K S and Mingazheva A A 2011 *Vestnik USATU* **41** 109–12
[4] Panayoti I A and Soloviev G V 1996 *Metal Science and Heat Treatment* **5** 28–31
[5] Smolyakova M Yu and Vershinin D S 2011 *Journal of Advanced Materials* **1** 478–82
[6] Lopatin I V and Akhmadeev Yu H 2011 *Scientific News of Belgorod State University* **22** 180–6

Study of the thermal mode of a silicon carbide field emission cathode

A S Ivanov, V A Ilyin and V N Titov

School of Electronics, Saint Petersburg Electrotechnical University "LETI", 197376, Saint Petersburg, Russia

E-mail: asivanov@etu.ru

Abstract. The dependency of the thermal mode of the silicon carbide (6H-SiC) field emission arrays on the integral emission current density is considered. Experimental temperature estimates reaching values of 1100 °C are presented. Computer simulation of the thermal mode accounting for the emission of electrons from a limited area of the apex of the tip, the Joule heating at the tips, and the nonlinear dependence of the thermal conductivity of the material is performed. The possibility of increasing the integral density of the emission current above 10 $A{\cdot}cm^{-2}$ is demonstrated.

1. Introduction

Field emission cathodes have found wide application in information display systems. They are promising sources of electrons for microwave vacuum electronic devices, requiring values of emission current exceeding hundreds of mA, usually provided by thermal cathodes. In recent years, thanks to the use of silicon carbide, it has been possible to create field emission arrays (FEA) with an integral emission current density exceeding 6 $A{\cdot}cm^{-2}$ [1].

According to the results of the studies [2–5], a high obstacle to the increase in the density of the emission current is the Joule heating of the cathode tips. Particularly, at a temperature of 1400 °C, instability of the emission current can arise due to the transition to the intrinsic electrical conductivity of the material. The temperature of the field emission cathode depends on many factors, primarily on its configuration, material features, and effective heat dissipation. On the basis of experimental study and computer simulation, the article gives estimates of the dependence of the cathode temperature on the emission current density for the developed silicon carbide field emission arrays.

2. Experimental data

The silicon carbide (6H-SiC) field emission arrays (figure 1) is fabricated on a single-crystal *n*-type substrate by the method of two-stage reactive ion etching (RIE) in a fluorine-containing medium [6, 7]. At the first stage, the periodic structure of rectangular pedestals is formed. At the second stage, the upper horizontal faces of the pedestals are transformed into arrays of nanoscale tips with a surface density of $5{\cdot}10^8$ cm^{-2}. Dimensions of a single tip are the following: the radius of the apex is 30 nm, the height is 2 μm and the base diameter is 0.3 μm. The two-tier configuration of the field emission arrays provides an amplification of the electric field near tips and, thus, increases the emission current. The field emission arrays is placed in a specially assembled vacuum diode with a fixed electrode gap (40–60 μm) and an anode area of 0.5 mm^2.

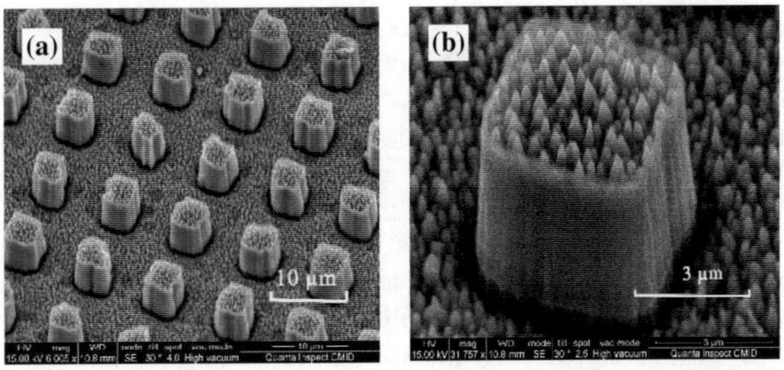

Figure 1. SEM micrographs: **(a)** – 6H-SiC array of pedestals; **(b)** – 6H-SiC pedestal with nanoscale tips.

Measurements of the current-voltage characteristics of the diode have been carried out in a high-vacuum chamber. Figure 2 shows the typical characteristics for the phenomenon of field electron emission.

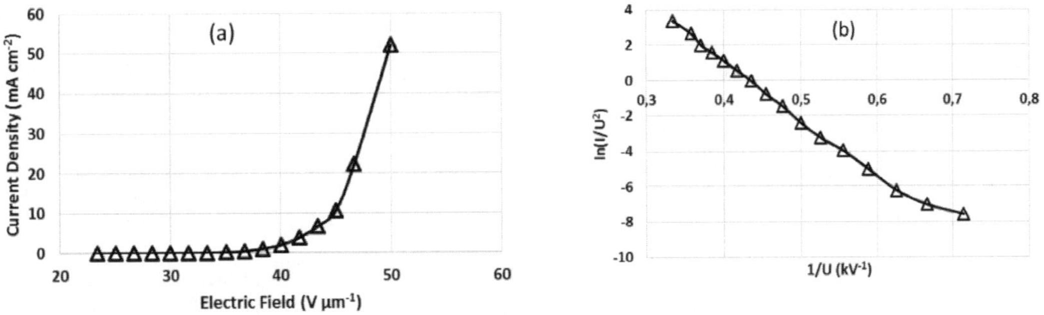

Figure 2. Experimental characteristics of 6H-SiC FEA: **(a)** – current density vs electric field; **(b)** – Fowler–Nordheim plot.

At a current density level near 50 mA·cm^{-2}, a soft cathode glow was observed. When the voltage reached 2.6 kV, the current spontaneously increased, and the temperature of the cathode raised up to 1100 °C. The photograph of the cathode at this temperature is shown in figure 3.

Figure 3. A photograph of cathode glow at a temperature of 1100 °C.

3. Numerical setup

The tiny scale of the objects under investigation and the experimental difficulty of achieving the peak current density make it necessary to use computer simulation to estimate the cathode's capabilities.

The 6H-SiC substrate has high electrical conductivity, and most of the Joule heat is released in the volume of the tips and pedestals. The 6H-SiC has a significant non-linear temperature dependence of the thermal conductivity. We propose and assess the hypothesis that the change in the thermal conductivity of a part of the substrate under the pedestals of the FEA, when heated by a heat flux from the pedestals, can affect the thermal mode.

To simplify the solution of the thermal problem, we consider the case of a sufficiently dense arrangement of the pedestals, and in the first approximation simulate the thermal mode of a single column on the substrate having a cross-sectional area equal to that of one pedestal. This approach allows us to further reduce the system under consideration to a single tip located on a parallelepiped with a cross-section equal to the area of one element of the array of tips. The height of the parallelepiped is a sum of the height of the pedestal and the thickness of the substrate plate below it. The results of the solution of the simplified thermal problem can be regarded as an upper estimate of the temperature for a real system.

Thus, the simulated simplified system (thermal element) consists of a tip on a parallelepiped, which we divide into two parts: the pedestal and the substrate. Since the substrate is much thicker than the height of the pedestal, we normalize the thickness of the substrate to the height of the pedestal, adjusting the thermal characteristics of the model substrate material accordingly.

The geometric configuration of the simulated thermal element is shown in figure 4. The following physical processes are taken into account in the construction of the model: a) an increase of Joule heating at tips with the rise of the emission current; b) the nonlinear dependence of the thermal conductivity of the material [8], which leads to an increase in the temperature of the tips due to the decrease of the heat sink efficiency; c) heat transfer through the substrate under cathode. The heat transfer by radiation and the Nottingham effect are not taken into consideration.

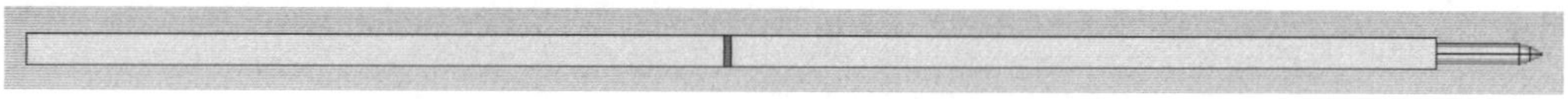

Figure 4. A geometric configuration of the simulated element of the cathode: from left to right are the substrate, the pedestal, and the tip.

The general procedure is as follows: given the value of the integral current density from the surface of the FEA and the cathode configuration, the current emitted from one pedestal, as well as the current emitted from one tip can be found. From the solution of the field emission problem we obtain the effective emission area of one tip. Setting the current emitted from one tip and the effective area, we find the current density in the bulk of the element. After the calculation of the volume density of Joule heating in the bulk of the element the thermal problem is solved. The boundary conditions are as follows: on the lower (left) face is an isothermal condition of ambient room temperature; on the lateral faces of the computational domain, and over the tip, the adiabatic conditions are applied.

The maximal density of Joule heating is in the region near the apex of the tip. The difference in temperature of the tip and the pedestal is slight, being on the order of several tens of degrees. In the bulk of the substrate the temperature decreases gradually.

4. Simulation results

The general results of computer simulation in the form of the dependence of the maximum temperature of the cathode tips on the emission current density are shown in figure 5. The maximum temperature of the cathode tips starts to raise at emission current density about 2.5 $A \cdot cm^{-2}$. The temperature increases linearly to about 6 $A \cdot cm^{-2}$ and reaches a value of 250 °C. Then a nonlinear rise in temperature begins and a current with density about 10.8 $A \cdot cm^{-2}$ heats the cathode up to 1400 °C.

5. Conclusions

Numerical simulation, within the limits of the assumptions made, shows that in the developed configuration of the silicon carbide (6H-SiC) field emission arrays, it is possible to obtain an integral emission current density of no less than 10 $A \cdot cm^{-2}$. In such case, the maximum temperature of the cathode does not exceed a value of 1400 °C, at which a transition to the intrinsic electrical conductivity of the material occurs and instability of the emission current can arise.

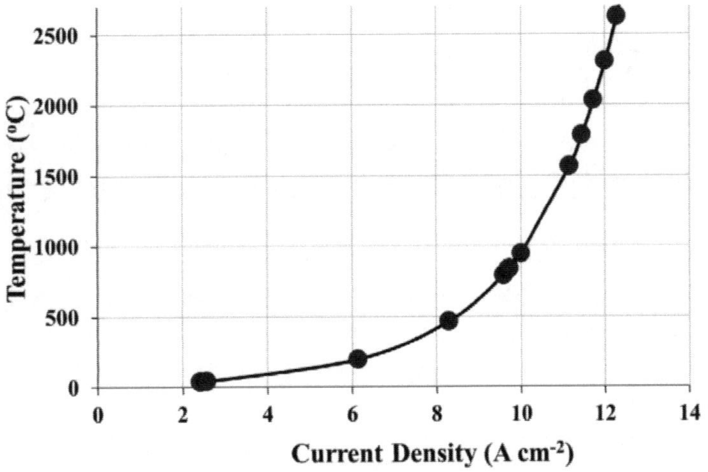

Figure 5. Dependence of the maximum temperature of cathode tips on the emission current density.

Acknowledgments
This work was supported by Russian Foundation for Basic Research under Grant 15-19-30022.

References

[1] Kang M G, Lezec H J and Sharifi F 2013 *Nanotechnology* **24** 065201
[2] Martin E E, Trolan J K and Dyke W P 1960 *J. Appl. Phys.* **31** 782
[3] Vincent P, Purcell S T, Journet C and Binh V T 2002 *Physical Review B* **66** 075406
[4] Chen S, Ying P, Wang L, Wei G and Yang W 2014 *Applied Physics Letters* **105** 133106
[5] Wang L, Wei G, Gao F, Li C and Yang W 2015 *Nanoscale*
[6] Afanasyev A V, Ivanov B V, Ilyin V A , Kardo-Sysoev A F, Kuznetsova M A and Luchinin V V 2013 *Mat. Sci. Forum* **1010** 740
[7] Golubkov V A, Ivanov A S, Ilyin V A, Luchinin V V, Bogdanov S A and Vikharev A L 2016 *Journal of Vacuum Science & Technology B* **34** 062202
[8] Nilsson O, Mehling H, Horn R, Fricke J, Hofmann R, Müller S G, Eckstein R and Hofmann D 1997 *High Temperatures-High Pressures* **29** 73

Influence of hydrogen content in working gas on diffusion processes at ion nitriding of martensitic and austenitic steels

U G Khusainov, K N Ramazanov, R D Agzamov, E L Vardanyan and R S Esipov

Ufa State Aviation Technical University, 450000, Ufa, Russia

E-mail: uldash990@mail.ru, ramazanovkn@gmail.com, esromles@mail.ru

Abstract. The paper describes the influence of hydrogen content in a working gas on diffusion processes at ion nitriding of martensitic and austenitic steels. It is found that the presence of hydrogen in a working space of a vacuum chamber contributes to the increase in nitrogen diffusion depthward the material. Dependence of the hardened layer thickness and depth distribution of microhardness on gaseous composition was obtained. The structure of steel nitrated layer with various content of hydrogen in a vacuum chamber was studied.

1. Introduction

At present the emphasis is placed on the increase in wear resistance of workpiece surface operating under friction. One of the efficient surface hardening methods applied for machine components and mechanisms is glow-discharge nitriding (ion nitriding) [1, 2]. This method gained widespread use due to some advantages: high growth rate of the nitrated layer; diffusive layers with set phase composition and structure; ecological compatibility of the process [3]. Despite all mentioned advantages ion nitriding is time consuming (25–30 h).

It is known [4, 5] that in case with glow-discharge nitriding the composition of gaseous medium has significant impact on structure, properties and growth rate of a hardened layer. Hence, the use of working fluids containing argon with small amount of hydrogen allows to increase the speed of nitriding (up to 5 %). Being good reducing agent, hydrogen prevents the formation of iron oxides on a workpiece surface, which in their turn make nitriding more complicated [4]. When subject to bombardment, argon with its high weight forms numerous defects contributing to acceleration of nitrogen diffusion in metals [5, 6]. Thus, in order to enhance the efficiency of ion nitriding it is critical to control the composition of a working gas. However, at present there is no data on the influence of multicomponent saturating fluids on the diffusion speed of a hardened layer.

2. Experiment

Ion nitriding was carried out on installation ELU-5M. Prior to ion nitriding, samples of martensitic and austenitic steels were treated by heat to obtain a homogeneously grained microstructure. The chemical composition of steels and the parameters of heat treatment are given in table 1. Prior to nitriding samples were subject to sputter cleaning at temperature of 250 °C under pressure of 10 Pa at argon atmosphere. Ion nitriding was performed in flowing N_2, Ar, H_2 at 550 °C for 6 h under 150 Pa pressure. Microindentation was conducted on the Struers Duramin 1/-2 microhardness testing machine with a load of 100 g (980.7 mN) for 10 s. At least four measurements were taken at each given depth. The microstructure of the specimen was observed under optical microscope Olympus GX51.

Table 1. Chemical composition of steels and parameters of heat treatment.

Type of steel	Chemical composition, %	Heat treatment
Martensitic	0.10–0.16 C; 10.5–12.0 Cr; 1.5–1.8 Ni; 0.18–0.30 V; 0.35–0.50 Mo; 1.60–2.0 W	Quench 1050 °C + annealing 800 °C
Austenitic	≤0.12 C; 17–19 Cr; 9–11.0 Ni; ≤0.8 Ti	Quench 1050 °C

3. Results and discussion

It is known that the surface temperature at glow discharge nitriding without additional heating of a vacuum chamber working space (cold wall nitriding) is determined by energy and density of charged particle flux. The density of ion flux regarding any component of a gaseous mixture depends on its percentage in a vacuum chamber and pressure of the environment, while the energy depends on the mass of particles and voltage between electrodes. Thus, at a different equivalent ratio of working gas components there is a change in surface temperature and workcycle parameters. In order to determine technological parameters of ion nitriding the dependence of voltage between electrodes on working gas composition in a vacuum chamber at a temperature $T = 550$ °C and pressure $P = 150$ Pa (figure 1) was obtained.

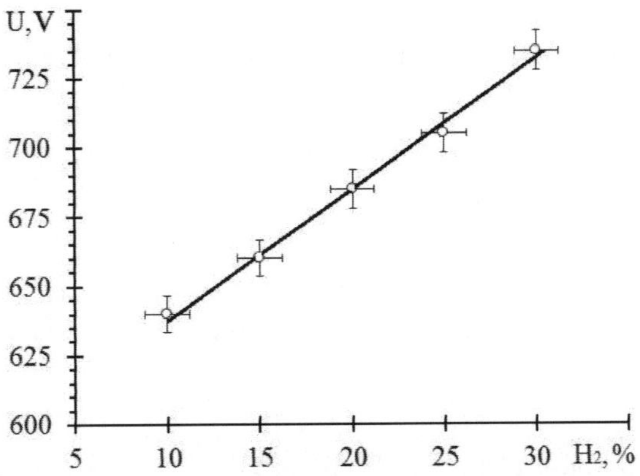

Figure 1. Dependence of the potential difference between the electrodes on the hydrogen content in the vacuum chamber, at which the temperature of the sample is $T = 550$ °C.

The analysis of the given dependence (figure 1) showed that when hydrogen content in a working gas varies from 10 to 30 % there is a need to increase the energy of particle flux due to the increase in charge voltage from 640 to 730 V to ensure the desired treatment temperature (550 °C). This is caused by the fact that at the decrease in argon concentration in gaseous mixture there is a reduction of density of argon ion flux.

Heat treatment of martensitic and austenitic steels is carried out to ensure stabilization of structure and to obtain homogeneity of material bulk properties. The structure presented as diversely focused martensitic grains is formed as a result of heat treatment of martensitic steel. Upon quenching and annealing the hardness of a sample made 365 ± 10 HV_{100}. The microstructure of austenitic steel after quenching is characterized by uniform structure of austenite grains with twins and a small amount of TiC carbides on crystallite boundaries. Sample hardness after heat treatment made 190 ± 10 HV_{100}.

In order to study diffusion processes the following distributions of microhardness on depth were obtained (figure 2).

The analysis of the measured data showed that the distribution of microhardness on depth is characterized by area (I) where the hardness steadily decreases with the increase of distance from the

IOP Conf. Series: Journal of Physics: Conf. Series **872** (2017) 012017 doi:10.1088/1742-6596/872/1/012017

surface, which is followed by its sharp decrease (transition zone II) with subsequent smooth alignment to the hardness of the basic material (area III). When comparing dependences of microhardness distribution on depth it was identified that the diffusion zone of martensitic steel has smoother decrease of hardness without sharp transition area (II). Sufficient hardening of a steel surface happens when the content of hydrogen in a vacuum chamber varies from 15 %. With further increase in hydrogen content up to 30 % the hardness of martensitic steel surface increases slightly and falls within the limits of 1650 ± 30 HV_{100}. The hardness of austenitic steel surface increases from 820 to 1100 ± 10 HV_{100} in case hydrogen content ranges from 15 to 20 % and slightly changes (1100 ± 30 HV_{100}) with its further increase.

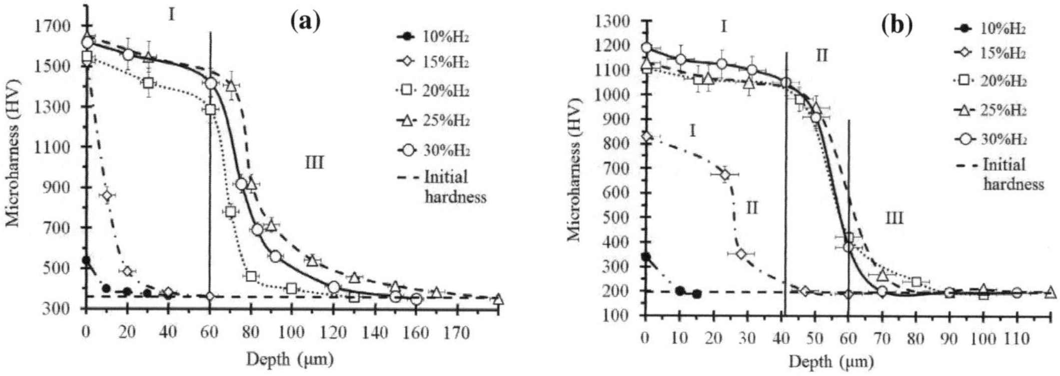

Figure 2. Distribution of microhardness along depth of a hardened layer of martensitic **(a)** and austenitic **(b)** steels depending on hydrogen content in a vacuum chamber.

When using gaseous media one shall consider the phenomenon of hydrogen embrittlement of a hardened layer. Bitmap images of sample sections were obtained to study the structure of a nitrated layer (figure 3).

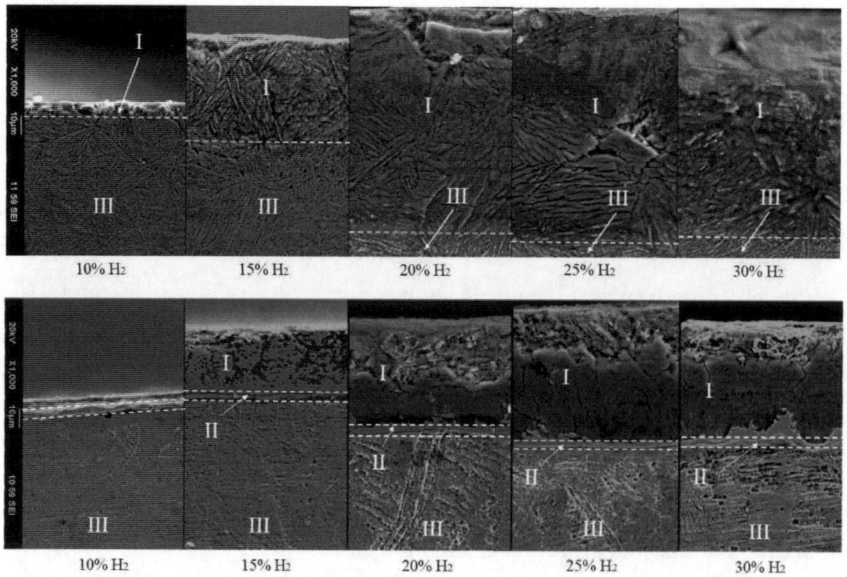

Figure 3. Bitmap images of a surface layer of martensitic **(a)** and austenitic **(b)** steels after nitriding with various content of hydrogen in a working gas.

The study of a hardened layer structure showed correlation with distribution of microhardness on depth of all samples. The structure of steels after ion nitriding is characterized by the presence of three zones: I – nitriding layer; II – diffusion zone; III – basis. It was found that depending on hydrogen content in a working gas and a category of steel the morphology of a hardened layer differs essentially. Intense crack formation is observed on martensitic steel (figure 3(a)) when hydrogen content in gas medium is from 25 %, whereas chips of a near-surface area (I) are local and there are some cracks along the boundaries of martensitic grains. In case of austenitic steel (figure 3(b)) when hydrogen content in a vacuum chamber ranges from 20 to 30 % in near-surface sections of a zone of nitriding layer (I) there are cracks and chips along the whole cross-section, which indicates brittle nature of a hardened layer. The above data confirm diffusion saturation of steel samples with hydrogen when its content in a vacuum chamber falls within the range from 20 to 30 %. The absence of cracks and chips in a surface layer of samples treated with hydrogen content of 10 and 15 % indicates low hydrogen diffusion on a surface due to low concentration.

4. Conclusion

As a result study of the influence of hydrogen content in a working gas on diffusion processes at ion nitriding of martensitic and austenitic steels the following was identified:

1. Efficient hardening of a surface of martensitic and austenitic steels happens when hydrogen content in a vacuum chamber is from 15 %. With further increase the hardness of a steel surface does not considerably change.
2. Change of hydrogen content from 10 to 25 % leads to the increase in nitrogen diffusion depthward the material. As a result, the growth rate of the nitrated layer increases. When hydrogen content exceeds 25 % due to predominance of surface diffusion coating with hydrogen there is the reduction of nitrogen diffusion and the thickness of a hardened layer decreases.
3. Presence of hydrogen in a working chamber in the amount of over 20 % leads to steel saturation with hydrogen which causes hydrogen embrittlement of a hardened layer. Therefore, there is near-surface damage of a nitrated layer.

Acknowledgments

The study was supported by the grant from the Russian Science Foundation (#15-19-00144).

References

[1] Park M J, Leyland A and Matthews A 1990 *Journal of Surf. and Coat. Technol.* **43** 481

[2] Michel H, Czerwiec T, Gantois M, Ablitzer D and Ricard A 1995 *Journal of Surf. and Coat. Technol.* **72** 103–11

[3] Gary G T 1974 *Journal of Applied Physics* **45** 5072–3

[4] Mehmet K, Akgün A and Ayhan Ç 2002 *Journal Materials Characterization* **49** 241–6

[5] Budilov V V, Ramazanov K N, Khusainov Yu G and Zolotov I V 2014 *Reinforcing technol. and coatings.* **120** 27–30

[6] Shinichiro A and Nobuhiro U 2012 *Journal of Thin Solid Films* **523** 11–4

Processing of metallization technology aluminum oxide ceramics for electro-vacuum devices elements and power electronics devices

L L Kolesnik, T S Zhuleva, P O Predtechenskiy, Myo Kyaw Hlaing and Zaw Phyo Aung

Bauman Moscow State Technical University, Department of "Electronic engineering technologies", 2nd Baumanskaya str. 5, bld. 1, 105005, Moscow, Russia

E-mail: kolesnik@bmstu.ru

Abstract. In this journal are considered the results of processing titanium film deposition technology on the aluminum oxide substrate. The results of measuring the characteristics of adhesives are given.

Electro vacuum devices producing by the domestic industry are contained in the structure of metal-ceramic compound elements. These compounds are required for strength, vacuum tightness, sustainability to temperature changes. At the moment, these connections are performed on the basic of thick film technology.

Thin films technology ensures generally the absence of the several types of thick film technologies and on the other indicator surpasses them. Transition to thin film technology for creating the basis soldering elements of vacuum devices and power modules will be reduced the percentages of rejects and will be improved the quality of these connections.

Since 2016 at Bauman Moscow State Technological University has been taken modernization of laboratory apparatus for processing thin film coating technology into vacuum. Modernization was made in laboratory's stand with the magnetron spraying method for thin film deposition on simples of aluminum oxide ceramics. The modernization process consisted of the mounting of gas and magnetron system, which provided the technology base for processing the development of thin film coating.

Special flange was designed and manufactured for apparatus of magnetron system. Cooling water continuously supplied to magnetron system along a close circuit and we have been used that cooling system for prevention of overheating and its demagnetization. However, the cooling system is realized so that was cooled magnets, but the target of magnetron by itself was not cooled and the process was heated temperature about 900 °C. Figure 1 shows an external view of the apparatus after completion of the modernization.

After completion the apparatus of installation work we were made the series of experiments for film deposition. We have been used titanium as the target material (cathode), and the corundum ceramics was used as the substrate material.

During the experiments of coatings deposition, the target was trained for 5 min before starting of the deposition. It was the sign, after the training time for target (cathode) 5 min was sharply increased the current with its subsequent stabilization. At the same time with the cleaning target (cathode) was heated temperature to 800 °C, and then after this condition deposition to substrate have been made from the hot cathode.

Figure 1. An external view of the laboratory apparatus after modernization.

We have been conducted the preliminary research and defined the range of gas flow and the range of power. From this research we have got thin films without visual defects. For titanium substrate these values are: gas flow 30 sccm, the range of power from 350 to 420 W, the range of current are from 520 to 600 mA. In these regimes, a series of experiments was conducted to deposit titanium films on an aluminum oxide substrate. In the following table 1 show the regimes of films deposition for power stabilization.

Table 1. The regimes of films deposition for power stabilization.

No	Gas flow Q, sccm	Power P, W	Chamber pressure p, mbar	Voltage U, V	Current I, mA	Deposition time t, min
1	30	350	$4.3 \cdot 10^{-2}$	630–673	550–510	5
2	30	385	$4.4 \cdot 10^{-2}$	690–750	570–530	5
3	30	420	$4.3 \cdot 10^{-2}$	745–770	580–510	5

There were 9 experimental researches and each mode was reproduced 3 times. Before the deposition the polished substrate was treated with isopropyl alcohol in an ultrasonic bath. The deposition time was 300 s for each simple film. The film thickness was measured with the scanning probe microscope Solver NEXT after the etching steps with hydrofluoric acid (figure 2).

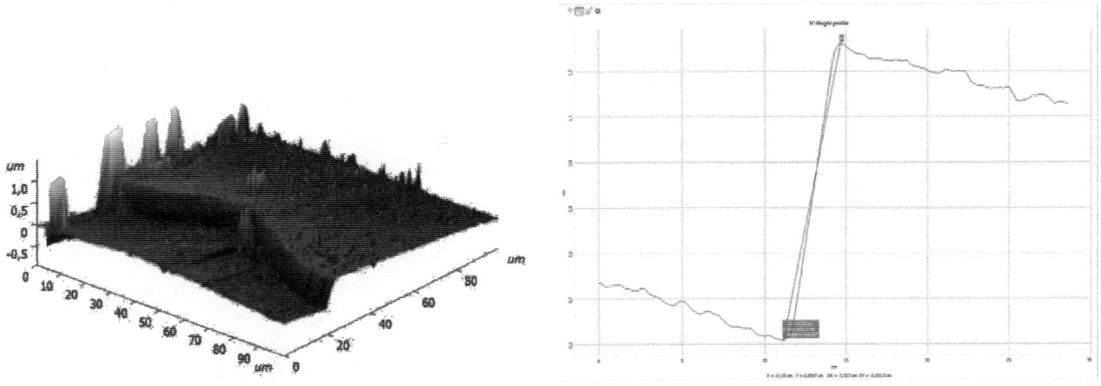

Figure 2. Scanning surface and measuring thickness of film.

By the thickness of the film and the time of deposition by an indirect way, the average film deposition rate was calculated. The calculating results of deposition rate are shown in figure 3.

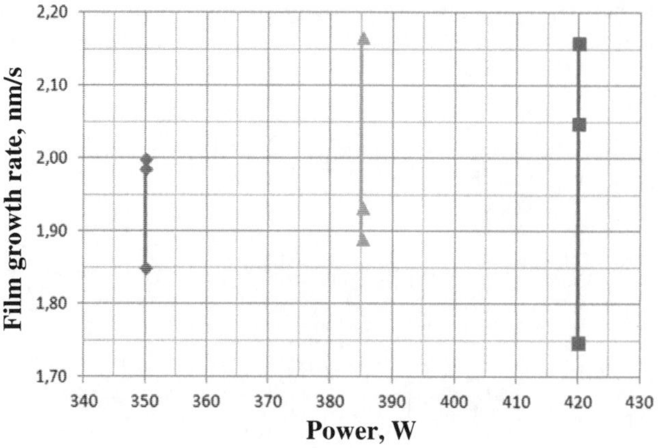

Figure 3. Results of calculating the rate of titanium film deposition.

The samples of films were tested the values of adhesion with the net cut incision method and the separation method. In the first case, samples were scratched two opposite directions, and in the second, a glued fungus broke off from the film. The devices for measuring the values of adhesion by each of methods are shown in figure 4.

Figure 4. Adhesive measuring devices (to the left side is the net cut incision method and the right side is force separating method).

All samples of films was shows the highest values of adhesion. The measuring of the net cut incision method is 5 B, and the measuring of force separating method is from 5.5 to 8 MPa, at that the destruction occurred on the glue, which the fungus was glued to the surface coating. Consequently the real values of adhesion not less than the values of following table 2. For further work, measurements of adhesion will be used the method of force separation with soldered funguses, in order to avoid the destruction of the fungus-coating connection.

Table 2. Results of measurement adhesion for titanium films by two methods.

No	Power P, W	Values of adhesion by net cut incision method	Values of adhesion by force separating method	Note
1	350	5 B	5.5 N/mm^2	Destruction has occurs at the joint of glue onto surface films
2	385	5 B	8 N/mm^2	
3	420	5 B	6 N/mm^2	

These results which describes in the upper table suggest that the possibility of using the selected regimes for deposition sublayer of titanium under the materials which uses for soldering components (copper, tin-gold, etc.).

IOP Conf. Series: Journal of Physics: Conf. Series **872** (2017) 012018 doi:10.1088/1742-6596/872/1/012018

References

[1] Kolesnik L L, Deev V A and Seleznev A V 2013 *Vacuum science and technology: proceedings of the XXth scientific-technical conference with participation of foreign experts* pp 165–7

[2] Martin P M 2010 *Handbook of deposition technologies for films and coatings: science, applications and technology* (Oxford: Elsevier) 902–12

[3] *US Patent 5571332 Electron Jet Vapor Deposition System 1996*

[4] Danilin B S and Syrchin V K 1982 *Magnetron sputtering systems* (Moscow: Radio and communication)

[5] Marakhtanov M K 1990 *Magnetron systems of ion sputtering* (Moscow: Izd-vo MGTU)

Influence of technological parameters on the mechanical properties of titanium nitride films deposited by hot target reactive sputtering

A A Kozin[1], V I Shapovalov[1], V V Smirnov[1], A S Useinov[2], K S Kravchuk[2], E V Gladkikh[2], A V Zavyalov[1] and A A Morozova[1]

[1] Saint-Petersburg Electrotechnical University "LETI", Saint-Petersburg, Russia
[2] Technological Institute of Superhard and Novel Carbon Materials, Troitsk, Moscow

E-mail: vishapovalov@mail.ru

Abstract. The impact of the discharge current density and nitrogen flowrate on the mechanical properties of titanium nitride films deposited by reactive magnetron sputtering of a hot target was studied. It was found that the films deposited at the highest discharge current density have the highest hardness and modulus of elasticity. These films have a texture with a significant predominance of the peak corresponding to the [111] direction in a face-centered cubic lattice of TiN.

Titanium nitride (TiN) keeps leading positions among nitrides of transition metals synthesized as films or coatings [1–3]. Having high electrical conductivity TiN films possess high adhesion, hardness and chemical inertness, and they are used as decorative, wear-resistant and corrosion-resistant coatings to modify the metal surface against fretting corrosion. These films are also applied in the silicon technology as the diffusion barriers for aluminum and copper metallization in production of MOS transistors, ohmic and rectifying contacts. Recently, high-power reactive sputtering including hot target sputtering [4–7] has become the most popular method for the synthesis of TiN films and coatings.

This paper presents a study on the influence of the discharge current density and nitrogen flowrate on the mechanical properties of the titanium nitride films deposited by reactive magnetron sputtering of a hot target.

The films are made in a vacuum chamber with a volume of $7.8 \cdot 10^{-2}$ m^3 equipped with a flat dc magnetron with a titanium target of 130 mm in diameter. The hot target mode is provided by fixing a 1 mm thick titanium disk with a gap of 1 mm on a 4 mm thick copper plate cooled by running water. The residual pressure in the chamber did not exceed 10^{-2} mTorr. Sputtering was carried out in an argon and nitrogen environment at a pressure of p_{Ar} = 1 mTorr, a nitrogen flowrate of 2–6 cm^3/min and a discharge current density of 14 and 33 mA/cm^2. The measurement of the mechanical properties (nanohardness and modulus of elasticity) was performed by the nano-hardness tester "NanoScan-4D" using the nanoindentation method. X-ray phase studies of the films were carried out on the X-ray powder diffractometer D8-Advance "Bruker" (CuK$_\alpha$ radiation, 40 kV operating voltage at a current of 40 mA).

Analysis of the measurements of nanohardness and the Young's modulus of films revealed that the nitrogen flowrate has an impact on the above properties, and this impact cannot be expressed functionally. The relationship is shown by scattering the dependences *H*(*d*) and *E*(*d*) using the highly simplified qualitative form of the dispersion analysis [8].

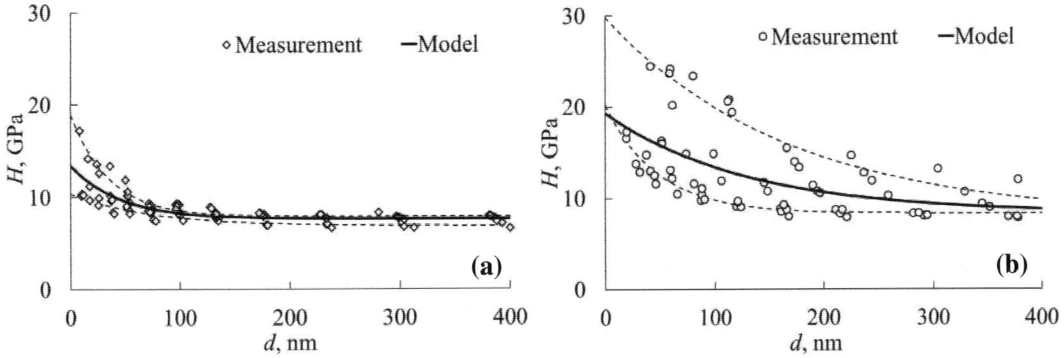

Figure 1. Nanohardness of TiN films deposited at a nitrogen flowrate in the range 2–6 cm^3/min and a current density (mA/cm^2): **(a)** – 14; **(b)** – 33. Points show the measurement results when the indenter is immersed in depth *d*, solid lines are approximated

The samples were combined into two groups deposited at different current densities. The results for each of them, shown in figure 1 by the points, were considered as measurements performed on one sample. The solid lines in figure 1 reflect the approximation of the experimental data using the formula

$$H(d) \approx a_H + b_H \exp(-d/c_H), \tag{1}$$

with a_H, b_H and c_H model parameters calculated by the least square method.

The dashed lines in figure 1 outline the ranges of nanohardness variation in each group of samples. They are described by the expressions of the type (1) but with other sets of parameters. They were calculated based on the measurements of individual samples having the largest and smallest nanohardness in a given batch. The results for the modulus of elasticity, which were obtained similarly using expression (1) are shown in figure 2.

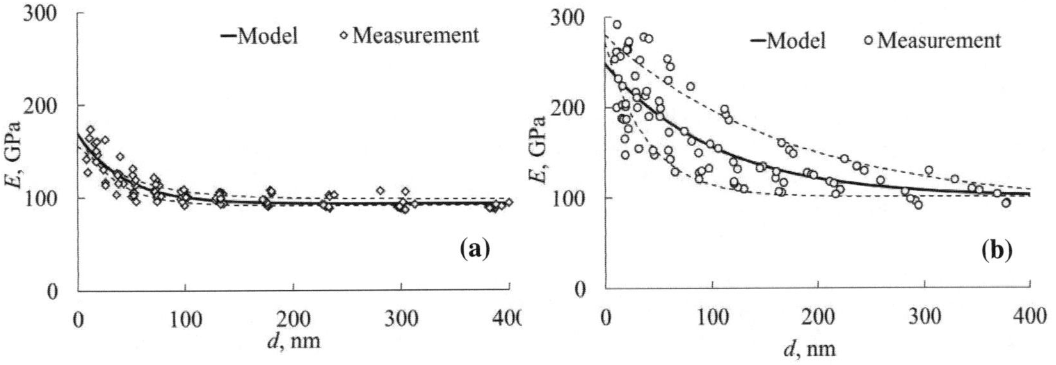

Figure 2. The Young's modulus of TiN films deposited at a nitrogen flowrate in the range 2–6 cm^3/min and a current density (mA/cm^2): **(a)** – 14; **(b)** – 33. Points show the measurement results when the indenter is immersed in depth *d*, solid lines are approximated

Thus, figure 1 and 2 reflect the effect of nitrogen flowrate on the mechanical properties of titanium nitride films. This influence appears in the scattering of experimental points. It is insignificant under the lowest current density (figure 1(a) and 2(a)) that indicates an insignificant impact of nitrogen flowrate on the properties of the films. The width of each scattering region increases under a larger current density (figure 1(b) and 2(b)) substantially. This effect is related to the increasing influence of the flow rate of the reactive gas on the parameters of the films.

The existence of the above influence is supported by the results of X-ray phase analysis. Two series of X-ray patterns corresponding to two groups of films deposited at different current densities are shown in figure 3.

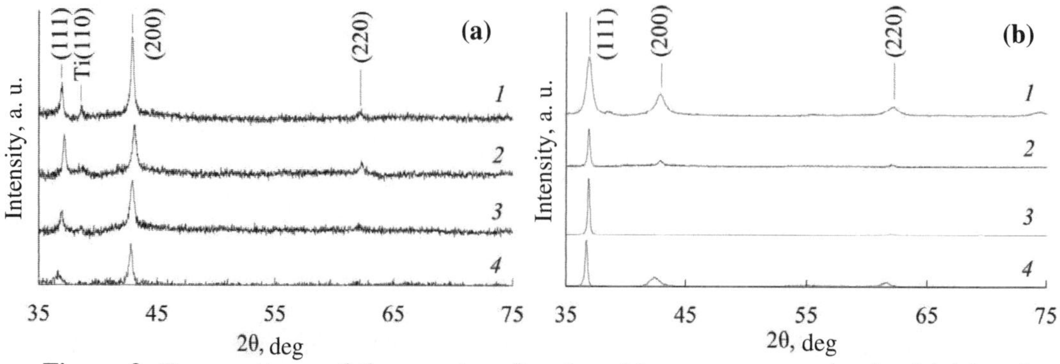

Figure 3. X-ray patterns of the samples after deposition at a current density 14 **(a)** and 33 **(b)** mA/cm^2. Nitrogen flowrate (cm^3/min): *1* – 2; *2* – 3; *3* – 4; *4* – 6.

All studied films had a crystalline phase. Strongly pronounced peaks were observed on the X-ray patterns at 2θ ~ 36.7°, ~ 42.6° and ~ 61.8°, corresponding to the face-centered cubic lattice TiN (space group of symmetry Fm-3m (225)), with the preferred orientation of the planes (111), (200) and (220). Moreover, the intensity ratio of the peaks on the X-ray patterns *I*(111)/*I*(200) < 1 at a current density of 14 mA/cm^2, which corresponds to the X-ray powder diffraction pattern (card 01-087-0630). In addition, they contain metallic titanium at a flowrate of nitrogen less than 4 cm^3/min. This is indicated by the peak at 2θ ~ 38.5° corresponding to the cubic lattice Ti with the predominant orientation of the plane (110), referring to the transformation group Im-3m (229) (card 00-044-1288). The films are textured ((*I*(111)/*I*(200) > 1) in the [111] direction and there is no metal phase a current density of 33 mA/cm^2. At the same time, an increase in nitrogen flowrate leads to a decrease in the width of the main peak (111), which indicates an increase of the crystallite size.

Therefore, based on the analysis of X-ray patterns it can be concluded that the nitrogen flowrate has stronger impact on the film properties when the current density is higher.

Acknowledgments
In the part of manufacturing samples, implementation of XRF, analysis of measurement results and preparation of the article, the research is supported by the Russian Science Foundation (grant 15-19-00076). Measurements of the mechanical properties of the films were made at the Technological Institute of Superhard and Novel Carbon Materials.

References

[1] Othman M F, Bushroa A R and Abdullah W N R 2015 *J. Adhes. Sci. Technol.* **29** 569–91
[2] Mishra S K, Kumar R, Soni S M et al. 2015 *J. Mater. Engineer. Performan.* **24** 5013–21
[3] Bavadi R and Valedbagi S 2012 *Mater. Phys. Mechan.* **15** 167–72
[4] Jiang F, Zhang T F, Wu B H et al. 2016 *Sur. Coat. Technol.* **292** 54–62
[5] Lapshin A E, Levitskii V S, Shapovalov V I et al. 2016 *Glass Phys. Chem.* **42** 359–62
[6] Bleykher G A, Krivobokov V P and Yuryeva A V 2015 *Techn. Phys.* **60** 1790–5
[7] Shapovalov V I, Karzin V V and Bondarenko A S 2017 *Phys. Lett. A* **381** 472–5
[8] Sheffe H 1959 *The Analysis of Variance* (New York: Willy)

Thermal radiation shielding by nanoporous membranes based on anodic alumina

E N Muratova[1], L B Matyushkin[1], V A Moshnikov[1], K V Chernyakova[2] and I A Vrublevsky[2]

[1] Saint Petersburg Electrotechnical University "LETI", ul. Prof. Popova 5, Saint Petersburg, 197376, Russia
[2] Belarusian State University of Informatics and Radioelectronics, P. Brovki Street 6, Minsk, 220013, Belarus

E-mail: SokolovaEkNik@yandex.ru

Abstract. The paper is devoted to infrared thermography studies of nanoporous alumina membranes with various geometric parameters of the porous layer: its thickness and average pore diameter. Thermal radiation shielding by anodic alumina membranes is presented. The result obtained showed that nanoporous alumina membranes can be used as heat shields to smooth contrast of thermal radiation of the object and the surrounding background.

1. Introduction

Investigation of the optical properties of nanomaterials is an important task in modern solid state physics. This is because nanocomposite media are the basis on which to base development and create new materials with specified physical and optical properties, the parameters of which are determined by the size, shape, and density of nanoscale objects.

The properties of porous dielectrics are important in research. For example, the structure of nanoporous anodic aluminum oxide is formed by creating a uniformly distributed network of nano-sized channels in a bulk material. Therefore, the physical and optical properties of such nanoporous materials can be very different from those of the whole material. The procedure for obtaining nanoporous anodic aluminum oxide is simple. The properties of the porous structure can be controlled by changing the formation regimes [1, 2]. Technology features make alumina a very convenient object for studying physical phenomena in nanostructures. It is possible to note such important properties of porous anodic aluminum oxide as lower than in bulk aluminum oxide, the refractive index and the dielectric constant, as well as the ordered structure of nanosized pores.

In our previous works [3–5] we studied transmission spectra of thin alumina membranes in a wide range of wavelengths and analyzed spectral regions that characterized certain membrane properties. It was shown that porous alumina membranes significantly blocked infrared radiation in the range of 8–14 μm. This funding is a good agreement with [6]. Within the above limits, there is a maximum of thermal radiation from bio-objects at a wavelength of ≈10 μm (figure 1).

2. Experiment

In this paper, the processes of shielding the heat flux by porous anodic aluminum oxide with a nanosized structure using thermal imaging measurements were studied.

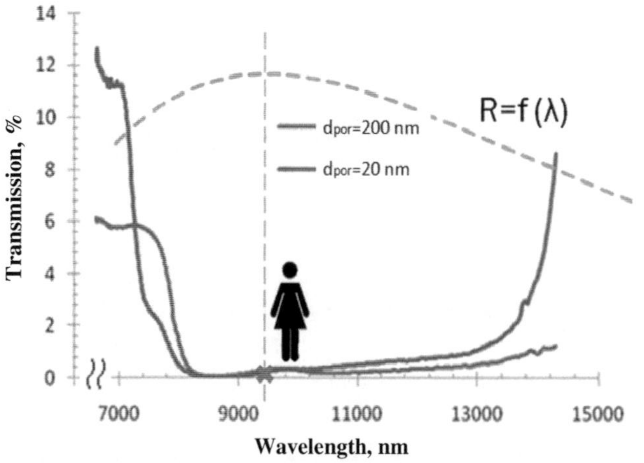

Figure 1. The IR transmission spectra of porous alumina membranes thickness of 10 μm and spectral distribution of blackbody radiation with temperature of 36.6 °C ($R = f(\lambda)$) [7].

The investigations were carried out using two thermal imagers: RGK TL-160 (temperature range –20…+350 °C) and uncooled infrared imaging camera system MobIR M4 (–20…+250 °C). The aluminum foil of 99.99 % purity, 100 μm thickness was used in the experiments. The specimens were etched in 1 M solution of NaOH and then in acetone. Then they were air-dried and thoroughly rinsed with distilled water and air-dried.

To prepare nanoporous anodic alumina with the highly ordered nanoporous structure the two-step anodizing process was used. At the first stage aluminum was anodized in 0.3 M aqueous solution of oxalic acid in the two-electrode cell equipped with Pt cathode at (15.0 ± 0.1) °C. The anodizing voltage was linearly increased from 0 to 40 V with the rate of 0.5 $V \cdot s^{-1}$, and then the anodizing process was carried out at a potentiostatic mode. The oxide layer was subsequently treated with chemical etching in the solutions of phosphoric and chromic acids. At the second stage the aluminum was iteratively anodized potentiostatically at 40 V. To form nanoporous anodic alumina films of different thicknesses (2–60 μm) the process was carried out during different periods of time. After anodizing the samples were rinsed and dried in hot air. Then from the opposite site to the anodic oxide the remaining aluminum layer was selectively etching. After complete metal dissolution the impermeable membrane was formed.

Anodic alumina membranes had ordered porous structure with a pore size of about 21 nm and the distance between them was 84 nm. The pores were oriented parallel to each other and arranged perpendicular to the growth surface, and possessed the same cylindrical shape.

Transmission of thermal energy to the object due to thermal radiation, accompanied by the processes of reflection or absorption of heat in the object, leads to the fact that its temperature varies on the environment. The main contribution to the observed thermal contrast is made by the object and background radiation. To visualize the thermal image of objects, two types of thermal imagers are used: cooled thermal imagers operating in the short-wave range (3–5 μm), and uncooled thermal imagers in the medium-wave range (8–14 μm). To detect and identify thermal objects, such as a person whose spectral wavelength of thermal radiation is 9.3 μm, uncooled thermal imagers are used.

As a source of heat radiation, the human palm was chosen. A Teflon plate (5 mm thickness) was applied as a heat shield. The plate had two 14 mm-diameter holes: the first one was used to arrange the samples and study the characteristics of the thermal shielding, and through the second one the thermal radiation passed without obstacles. Teflon plate (heat shield) was located above human palm (heat source), blocking the main heat radiation and passing it only through the described holes. The heat shield was located at the distance of 4 mm from the heat source.

3. Results and discussion

As a result of the studies, data on the temperature change of the object for various membranes were obtained (figure 2) and the temperature profile was constructed (figure 3).

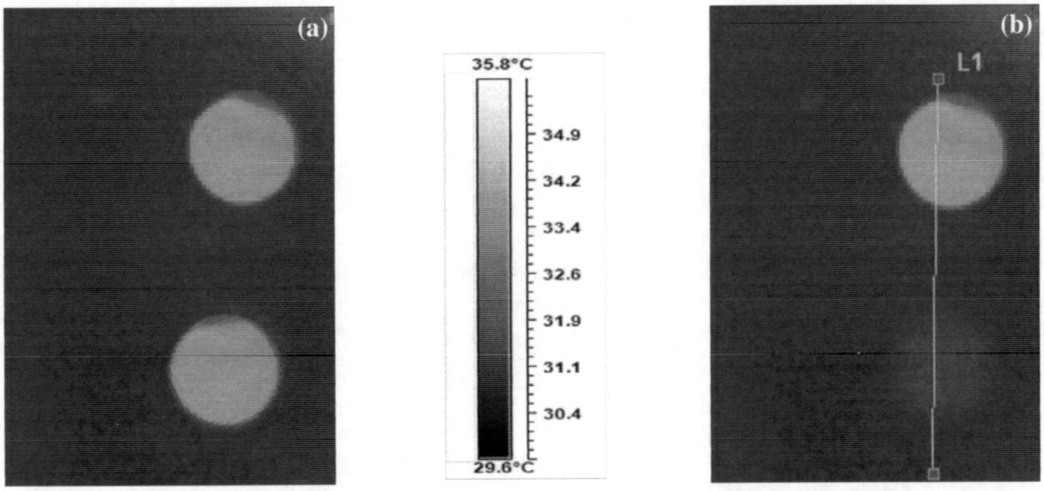

Figure 2. The results of thermal studies: an appearance of the bio-object through the lens of the thermal imaging camera for the case of **(a)** the passage of heat flow through the control and operation hole and **(b)** masking the operation hole with nanoporous anodic alumina.

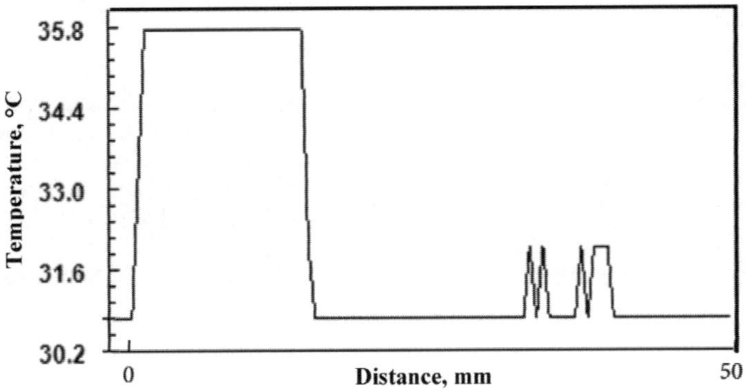

Figure 3. The temperature profile for the line passing through the control and operation holes.

In addition, the dependence of the change in the thermal temperature of the object on the thickness of the nanoporous alumina film was revealed (figure 4).

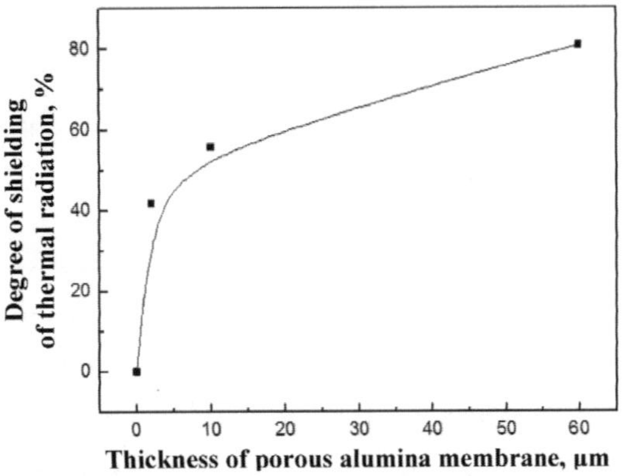

Figure 4. The curve of the degree of shielding vs. thickness of alumina membrane.

The degree of heat radiation shielding (D) for the anodic alumina membrane with thickness 10 μm has reached 56 %. To compute the degree of shielding the following equation was used

$$D = \frac{T_s - T_m}{T_s - T_{sh}} \cdot 100\ \%,$$

where T_s – temperature of heat source; T_m – temperature of membrane; T_{sh} – temperature of shield. This can be explained by the diffusive scattering of the optically inhomogeneous media [1], where the sources of inhomogeneity are small diameter pores.

The results showed that the use of porous anodic alumina films for shielding significantly reduced the passage of thermal radiation from heat sources and reduced the temperature of the heat spot emitted by the bioobject (human palm) from 37.1 to 32.0 °C (figure 2).

In addition, the studies of shielding radiation using a membrane thickness of 100 μm from a thermal source with a temperature of 68 °C were carried out (figure 5).

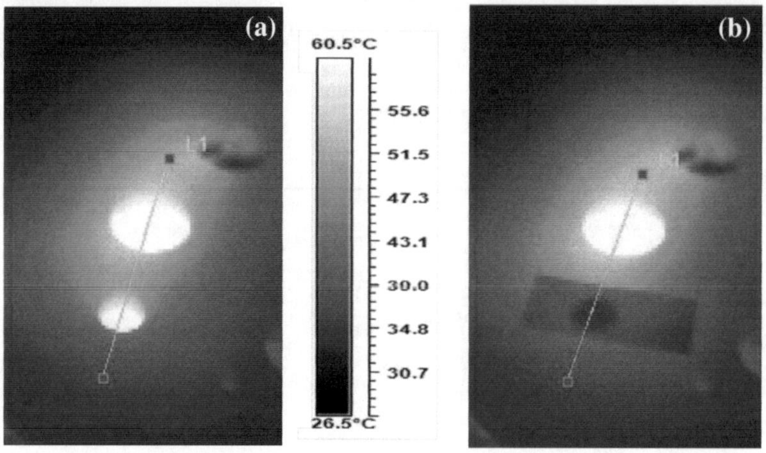

Figure 5. The results of thermal studies: an appearance of a thermal source through the lens of the thermal imaging camera for the case of **(a)** the passage of heat flow through the control and operation hole and **(b)** masking the operation hole with nanoporous anodic alumina thickness of 100 μm.

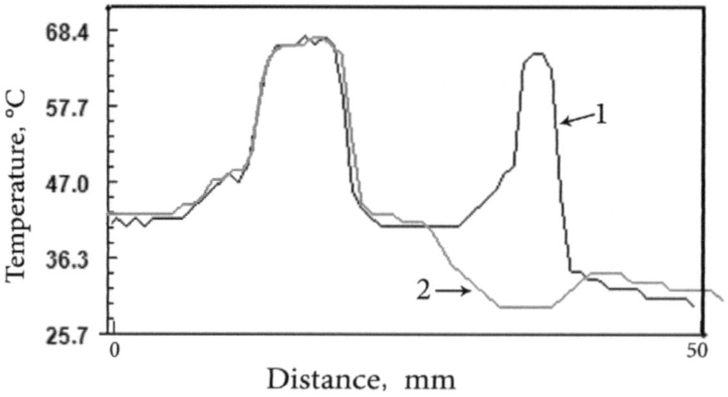

Figure 6. The temperature profiles for the line passing through the control and operation holes: 1 – without a membrane; 2 – with a membrane.

As it can be seen from the results obtained (figure 6), the use of a porous anodic alumina membrane with a thickness of 100 μm allows completely blocking thermal radiation and smoothing out the thermal profile for the radiation source relative to the background.

4. Conclusions

Based on the conducted studies, the temperature profiles of the bio-object in the holes of the Teflon plate for the case without a mask and using a mask from a porous anodic alumina film were obtained. It was found that the porous anodic alumina did not absorb the heat radiation in the wavelength region 8–14 μm and its properties could be controlled by choice of the appropriate geometric parameters. Therefore, such membranes can be used as heat shields to smooth contrast in the infrared range thermal radiation of the object and the surrounding background.

Acknowledgments

This work was supported by RFBR research project № 16-38-60110 mol_a_dk.

References

[1] Luchinin V V, Moshnikov V A, Muratova E N and Samigullin R Sh 2015 *Journal of Physics: Conference Series* **586** 012008
[2] Vrublevskiy I A, Dik S K, Terekh A S, Smirnov A V and Chernyakova K V 2012 *Problemy fiziki, matematiki i tekhniki* **12(3)** 101–5
[3] Matyushkin L B, Muratova E N and Panov M F 2016 *Micro & Nano Letters* **12** 100
[4] Matyushkin L B, Muratova E N, Spivak Yu M, Shimanova V V, Korlyakova S A and Moshnikov V A 2014 *Journal of Physics: Conference Series* **572** 012031
[5] Vrublevsky I, Chernyakova K, Ispas A, Bund A and Gaponik N 2011 *J. Lumin* **131** 938
[6] Wackelgard E 1996 *J. Phys.: Condens. Matter* **8** 4289
[7] Muratova E N, Luchinin V V, Moshnikov V A, Lifshits V A, Matyushkin L B, Panov M F, Potrakhov N N, Galunin S A, Ishin V V and Shemukhin A A 2017 *Glass Physics and Chemistry* **43(2)** 163

Cleaning of the dielectric surfaces using a controlled gas-discharge source of fast neutral particles

A S Fadeev[1], A S Talanov[1], D K Kostrin[1], S A Trifonov[1], A A Lisenkov[2] and V D Goncharov[1]

[1] Saint Petersburg Electrotechnical University "LETI", 197376, St. Petersburg, Russia
[2] Institute of Problems of Mechanical Engineering, Russian Academy of Sciences, 199178, St. Petersburg, Russia

E-mail: lisran@yandex.ru

Abstract. In the present work is proposed the design of a controlled gas-discharge source of fast neutral particles, allowing to clean surfaces of dielectric and semiconductor materials. Also are shown the results of the research of the source main working characteristics.

Development of the gas-discharge sources using fast neutral particles is very relevant. These sources are used in various technological processes at one of the main preparatory stages for cleaning the surface of the substrate before depositing a coating on a dielectric surface [1, 2]. Quality cleaning of a dielectric substrate before spraying of a coating on it is a key to achieving a good adhesion.

Depending on the type of the processed surface and its conductivity, for cleaning prior to coating deposition, are used gas-discharge sources of charged and neutral particles. Cleaning of the dielectric surfaces using the charged particles is impossible due to the charge accumulation on the surface of a workpiece. Collected charge significantly reduces the speed of impact on the surface, up to the complete cessation of the process, and leaves the defects on the substrate surface.

While separating from the gas-discharge plasma the neutral particles and processing the surface by them there are no defects associated with charge accumulation on the surface of a substrate [3]. Therefore, the aim of this work is the design of a controlled gas-discharge source of fast neutral particles. Design of a developed source is shown in figure 1.

The source scheme includes the following components: *1* – system for the supply and pre-heating of a working gas; *2* – solenoid; *3* – anode (source housing); *4* – ceramic insulators; *5* – gas-distributing disk; *6* – massive neutralizer (main cathode); *7* – screen; *8* – treated workpiece; *9* – additional cathode.

In the developed scheme as a working gas is used an inert gas argon. The gas feed system consists of two input tubes, which are heated by the filament coils, and a gas-distributing disk, providing uniform gas supply to the zone of ionization and resonant charge exchange, also it forms the pressure difference between the source volume *A* and the vacuum chamber *B*.

The source electrical scheme includes the main cathode *6*, the anode *3* and an additional electrode *9*. On the main cathode *6* is a negative discharge voltage U_{dis}, while the additional cathode *9* is under a floating potential. The anode *4* (source housing) is grounded. The electrode arrangement shown in figure 1 provides sustaining of the discharge inside the source volume.

As the magnetic field source is used an external solenoid *2*, which provides a directional movement of the charged particles produced in the gas discharge to the neutralizer *6*.

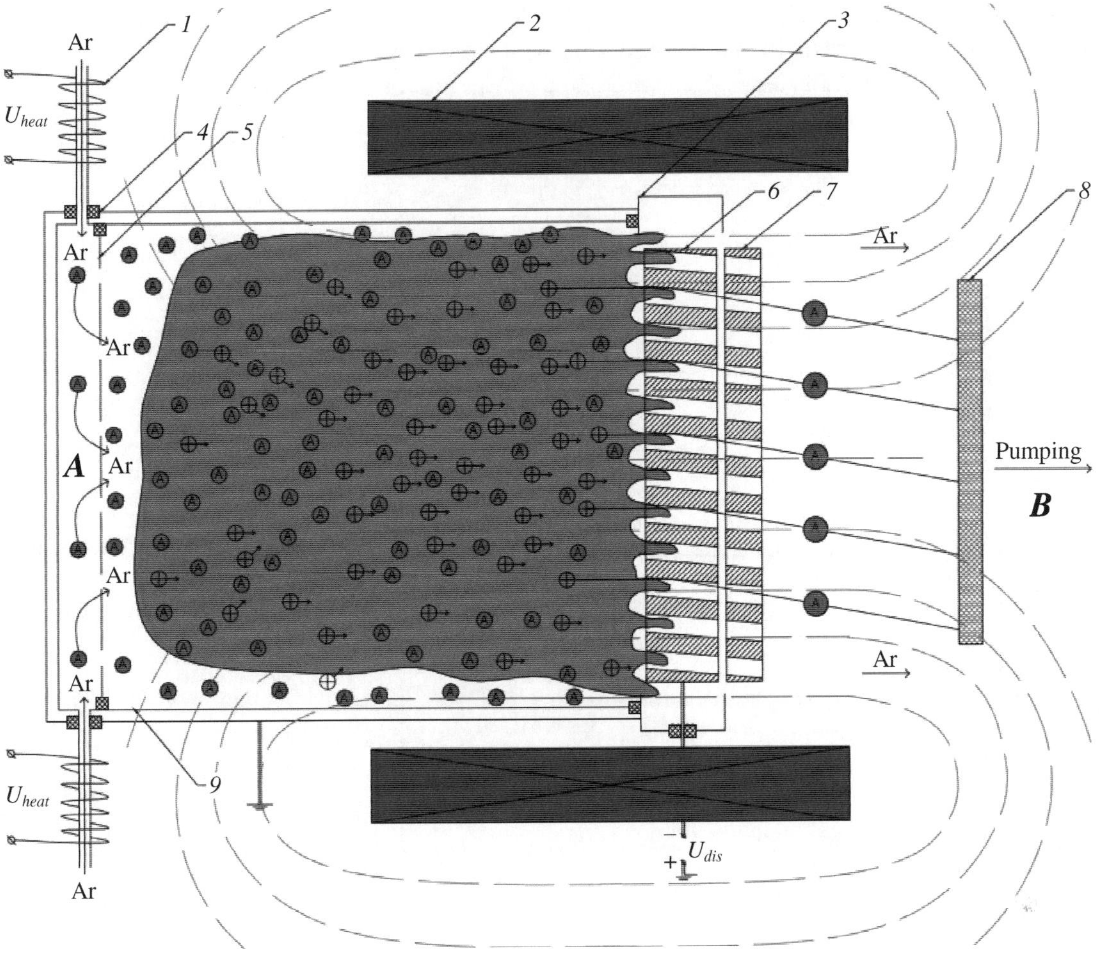

Figure 1. Design of a gas-discharge source of fast neutral particles.

When supplying the working gas into the source volume and applying the negative voltage to the main cathode the gas discharge is ignited, which occupies the entire volume of the source. In the gas discharge are generated the ions that are accelerated by the cathode voltage drop. Neutralization of the charged particles occurs mostly during the reflection of ions from the main cathode, but a small part of ions is neutralized due to the resonant charge exchange on the gas target. As a result of the neutralization process is formed an accelerated flow of neutral particles.

Development of a controlled gas-discharge source of fast neutral particles allowed proceeding to the experimental studies of its performance characteristics. To determine the minimum working pressure of the source has been investigated the dependence of the gas discharge ignition voltage from the pressure in the vacuum chamber $U_{ign} = f(pd)$ at a constant current of 80 mA. The obtained dependence is shown in figure 2.

Analyzing the obtained results can be concluded that the minimum possible pressure at which the source can operate is 0.1 Pa at a discharge voltage of 4.2 kV. Also from the dependence shown in figure 2 it can be seen that at a pressure of 2 Pa or more the discharge ignition voltage does not exceed 800 V and decreases linearly with increasing pressure in the working chamber. Based on the obtained results it was determined the minimum working pressure of the source, which is in the range of 1–2 Pa.

For analyzing the effectiveness of using in the design of a controlled gas-discharge source of fast neutral particles of a shielding disk was investigated the dependence of the ion current incoming at the workpiece from the pressure in the vacuum chamber. The study was carried out with the presence of a disk in the design and without it. The obtained dependences are shown in figures 3–4.

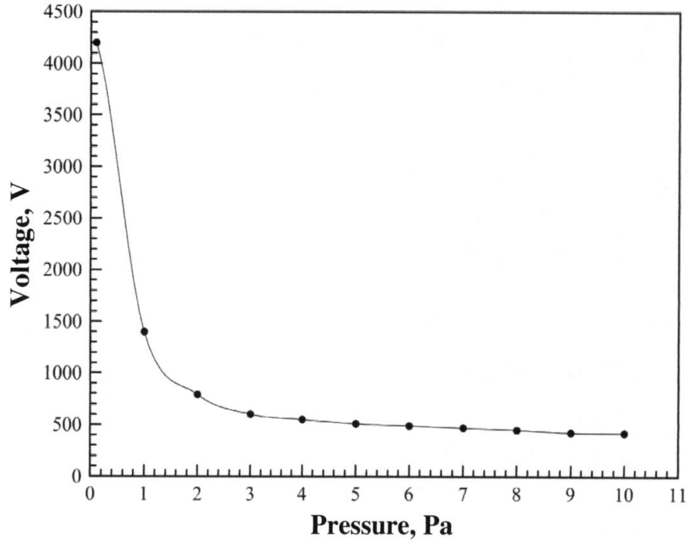

Figure 2. Dependence of the discharge ignition voltage from the pressure in the vacuum chamber.

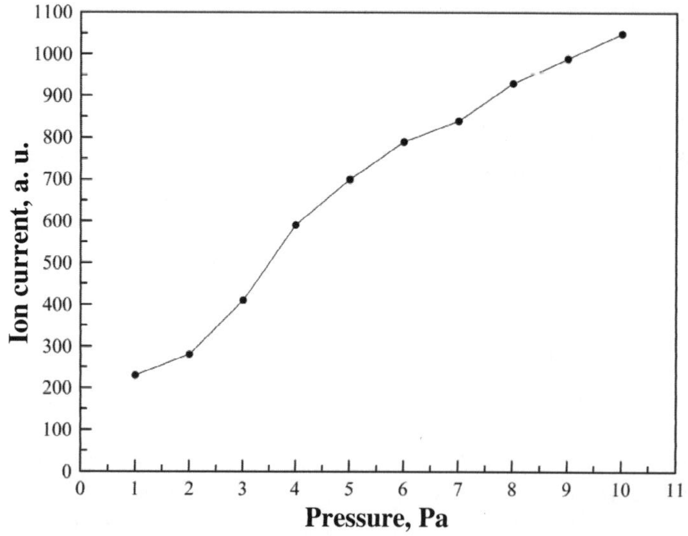

Figure 3. Dependence of the ion current from the pressure in the vacuum chamber without the use of the shielding disk.

As the treated surface in experiments was used a metal probe, having a form of a rectangular plate. Taking into account that the value of the ion current strongly depends on the shape and geometrical dimensions of the probe, and also on the distance between the source and the treated surface, the measurements of the ion current are given in arbitrary units.

Analysis of the obtained data showed that when using the shielding disk the magnitude of the ion current decreases, and also when lowering pressure in the vacuum chamber the value of the ion current decreases almost linearly. Therefore, the use of the shielding disk in the design of the source is a very effective solution for minimizing the amount of the ion current, and, consequently, its effect on the surface of a workpiece.

During the study of the source were obtained dependences of the differential pressure between the vacuum chamber and the volume of the source using two different prototypes of the gas distribution disks. The first prototype of the gas distribution disk have the holes drilled at a distance of 20 mm from the walls of the vacuum volume, for the second prototype this distance is 5 mm. Obtained dependences are presented in figure 5.

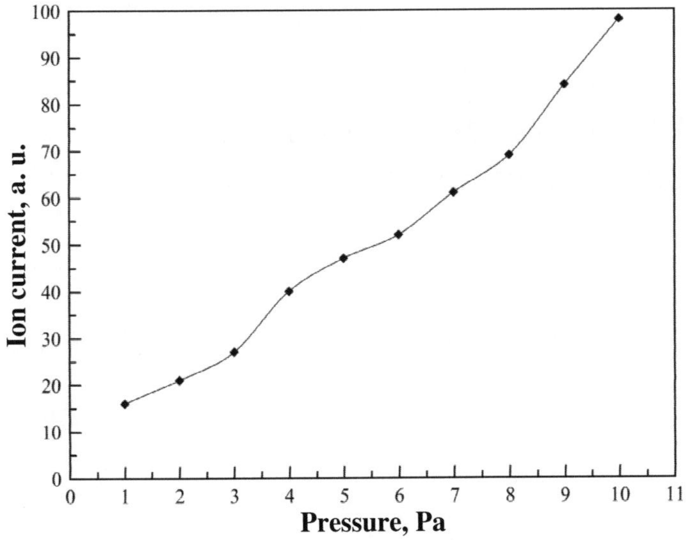

Figure 4. Dependence of the ion current from the pressure in the vacuum chamber with the use of the shielding disk.

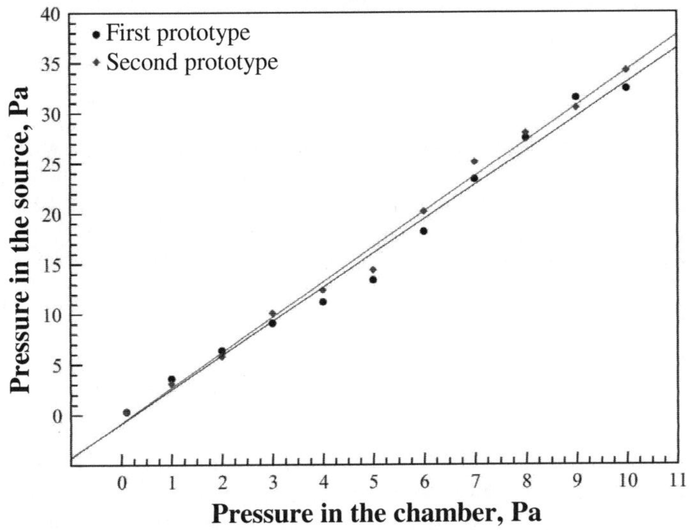

Figure 5. Dependence of the pressure in the source volume from the pressure in the vacuum chamber.

Analysis of the results showed that the use of the gas distribution disk allows creating a pressure differential between the source volume and the vacuum chamber, moreover the pressure in the source is 4 times higher than the pressure in the vacuum chamber. For the gas distribution disks having a different radial location of the holes the results are about the same.

The final phase of the study of a controlled gas-discharge source of fast neutral particles was a finish cleaning of a dielectric part. As the object of cleaning were used samples of borosilicate glass, with a deposited metal layer. After exposure to the flux of neutral particles from the glass surface was completely removed the metal contamination, which cannot be previously detached mechanically. Thus, it can be concluded that the use of a developed controlled gas-discharge source of fast neutral particles for cleaning of the dielectric and semiconductor surfaces is very effective.

References

[1] Barchenko V T, Bystrov Yu A and Kolgin E A 2001 *Ion-plasma technologies in electronic manufacturing* (Saint Petersburg: Energoatomizdat)
[2] Kostrin D K and Lisenkov A A 2016 *Materials Science Forum* **843** 278–83
[3] Shimokawa F, Tanaka H, Uenishi Y and Sawada R 1989 *J. Appl. Phys.* **6** 2613–8

IOP Conf. Series: Journal of Physics: Conf. Series **872** (2017) 012022 doi:10.1088/1742-6596/872/1/012022

Composition control of PZT thin films by varying technological parameters of RF magnetron sputter deposition

V P Pronin[1], D M Dolgintsev[1], I P Pronin[2], S V Senkevich[2], E Yu Kaptelov[2] and A Yu Sergienko[3]

[1] Faculty of Physics, Herzen State Pedagogical University, Kazanskaya (Plekhanova) ul. 6, St.-Petersburg, 191186, Russia
[2] Ioffe Institute, Russian Academy of Sciences, Politekhnicheskaya ul. 26, St. Petersburg, 194021, Russia
[3] Institute of Education Management of the Russian Academy of Education, ul. Chernyakhovskogo 2, St. Petersburg, 191119, Russia

E-mail: pronin.v.p@yandex.ru

Abstract. The article presents the effect of technological parameters of RF magnetron sputtering on the concentration of components of thin-film ferroelectric structures based on lead zirconate titanate PZT in the region of the morphotropic phase boundary. It is shown that by changing the distance from the target to the substrate and the pressure of the working gas mixture Ar + O_2, it is possible to vary the composition of the deposited thin layers.

1. Introduction

Solid solutions of zirconate-titanate lead are now one of the main materials of modern piezo-engineering [1]. Anomalously high values of the dielectric and piezoelectric characteristics of structures based on them are observed in the region of the morphotropic phase boundary (MPB), which separates the rhombohedral and tetragonal modifications of the ferroelectric phase (figure 1). Recently, these anomalies have been attributed to the existence of an intermediate monoclinic phase in the MPB region [2]. Particular attention is paid to the possibilities of forming thin-film structures based on PZT with extremely high piezoelectric characteristics. At the same time, the formation of such structures is associated with the need for fine-tuning the elemental composition of the PZT in the region of MPB. For practical purposes it is also important to obtain self-polarized thin films in which macroscopic polarization, comparable in magnitude to spontaneous polarization, is formed directly in the process of their formation [3–5]. In a number of works it is assumed that the appearance of a self-polarized state in thin PZT layers is associated with an excess content of lead in the form of an oxide [4–6]. The main goal of this work was to study the fine tuning of the composition of PZT films (changes in the ratio of Zr/Ti atoms and excess Pb atoms) in the region of MPB using the RF magnetron method of their formation with variation of technological parameters.

2. Experimental procedure

The studied samples of ferroelectric PZT films with a thickness of 500 to 800 nm were formed by a two-step procedure [7]. In the first stage, the films were deposited by the RF magnetron sputtering method in an argon-oxygen atmosphere on a silicon substrate coated with a layer of platinum 80 nm thick at ~150 °C. Ceramic targets (1 – x)$PbTiO_{3-x}PbZrO_3$ of compositions corresponding to the MPB region (x = 0.53 and 0.54) were used for sputtering. Variations in the composition of the deposited films were carried out by: a) changing the pressure of the working gas mixture in the range 2...8 Pa; b) changing the target-substrate distance in the range from 30...70 mm. In the second stage, the formed films were annealed in air in the range 580...600 °C for 1 h.

The structure of the films was determined using an optical microscope Nikon Eclipse LV150, an atomic force microscope Ntegra Prima (NT-MDT, Zelenograd) and a scanning electron microscope Zeiss EVO-40. The electron X-ray microanalyzer based on energy-dispersive analyzer INCA with the electrons energy about 12 keV was used for the composition analysis. Dielectric measurements were carried out using the E7-20 immittance and the modified Sawyer Tower scheme.

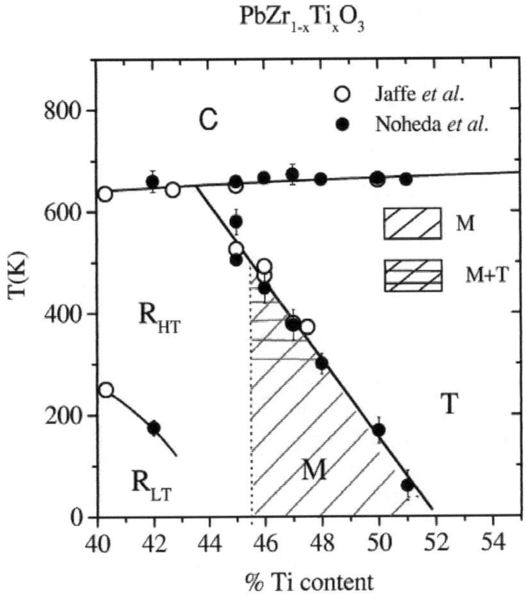

Figure 1. Phase diagram of PZT in the region of the morphotropic phase boundary: T – the region of the tetragonal phase; R – the region of the rhombohedral phase; M – the region of the monoclinic phase; C –region of the cubic phase (according to [1–2]).

3. Experimental results and discussion

Changes in the composition of deposited films with a variation in the pressure of the working gas and the distance between the target and the substrate are shown in figures 2 and 3. From figure 2(a) that an increase in the working gas pressure (x = 0.46) led to a significant change in the concentration of lead in the films (of the order of 30 %). The Pb/(Ti + Zr) dependence was a monotonically increasing curve, the initial value of which at 2 Pa corresponded to a composition with a strong lack of lead relative to the stoichiometric composition PZT – Pb/(Ti+Zr) = 0.93. Because of this, the crystallization of the perovskite phase occurred only in local regions of films islands, whose formation and growth was possible due to the migration of missing lead atoms to the interphase boundary of pyrochlore-perovskite during high-temperature annealing. At 8 Pa, the amount of excess lead was about 25 %, and the formed film was a composite structure consisting of PZT perovskite regions and PbO microinclusions. Thus, it turned out that the variation of the working gas pressure can change the relative content of lead atoms in a very wide range of concentrations.

The reason for such changes in lead content is related to the processes of scattering of sputtered target atoms in a gas plasma under conditions of a specific geometry of the mutual location of the circular zone of erosion of the ceramic target and substrate (figure 4). Analysis of literature data shows that in the investigated pressure range of the working gas medium, the so-called thermalization of lead atoms occurs when the directed motion of atomized atoms begins to be replaced by a diffusion motion characterized by an isotropic form of atom propagation [8–9].

IOP Conf. Series: Journal of Physics: Conf. Series **872** (2017) 012022 doi:10.1088/1742-6596/872/1/012022

In contrast to a strong change in the lead content with increasing working gas pressure, the ratio of Zr/(Ti + Zr) atoms varied in a relatively small range - no more than 2.5 % (figure 2(b)). The dependence of Zr/(Ti + Zr) on the pressure of the working gas was a curve with a maximum. However, such changes are sufficient to scan the composition of practically the entire expected MPB region at room temperature (figure 1). Of practical interest is the right-hand side of the curve, which corresponds to perovskite films with an excessive lead content and does not contain inclusions of the low-temperature pyrochlore phase. The presence of a maximum on the curve Zr/(Ti + Zr) indicates the presence of at least two different mechanisms that lead to a change in the composition of the films. The left-hand side of the curve (at low pressures of the working gas) can be connected with the final stage of the thermalization of zirconium atoms, and the right-hand side can be related to the difference in the redox potentials of zirconium and titanium atoms.

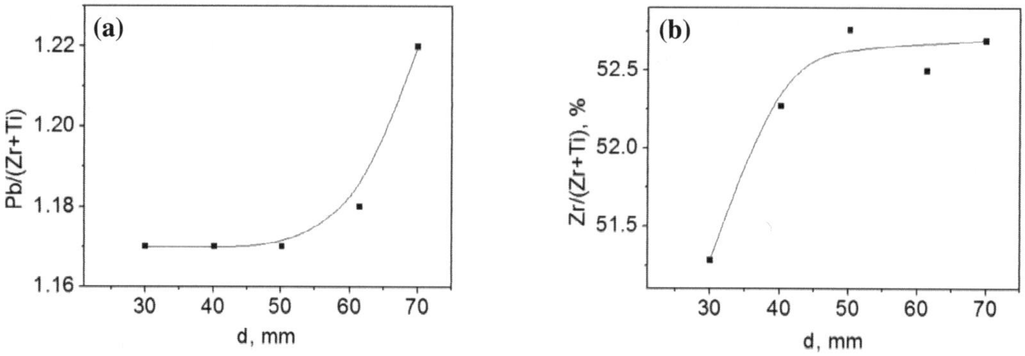

Figure 2. Graphs of the change in the ratio of the Pb/(Ti + Zr) **(a)** and Zr/(Ti + Zr) **(b)** atoms in thin PZT films prepared by sputtering a ceramic target $x = 0.46$ with the addition of 10 mol. % PbO at a pressure variation of the working gas. The film thickness is 800 nm.

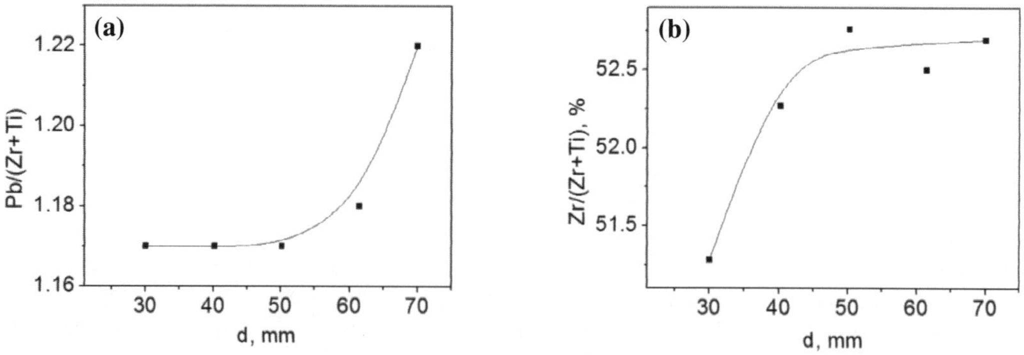

Figure 3. Graphs of the change in the ratio of the Pb/(Ti + Zr) **(a)** and Zr/(Ti + Zr) **(b)** atoms in thin PZT films prepared by sputtering a ceramic target $x = 0.47$, 500 nm thick from the distance (d) between the target and the substrate. Working gas pressure is 8 Pa.

On the basis of the approach on the thermalization of sputtered atoms of the ceramic target PZT, it was assumed that the composition of the deposited films can vary with a change in the distance (d) from the target to the substrate. In figure 3 presents the averaged results of the change in the composition of PZT layers deposited at a working gas pressure of 8 Pa, varying the distance from the target to the substrate in the range $d = 30$–70 mm. To reduce the lead content, a ceramic target of the stoichiometric composition PZT with $x = 0.47$ was used. In the process of high temperature annealing, single phase perovskite films with a thickness of $\approx$500 nm were obtained.

It is seen that with increasing d, the content of lead atoms Pb/(Ti + Zr) increased by approximately 5 %, from 1.17 to 1.22 (figure 3(a)). The behavior of the curve indicates that the atomic thermalization process is not complete for lead atoms, while a slight increase in the Zr/(Ti + Zr) atoms content by ≈1 % (figure 3(a)) is determined, apparently, by diffusion processes in the gas plasma of titanium atoms.

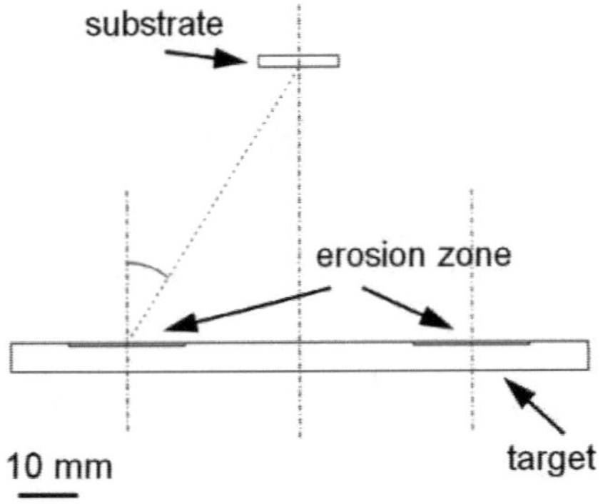

Figure 4. Diagram of the mutual arrangement of the substrate and the erosion zone of the target in an RF magnetron sputtering installation.

In films deposited from a target with a stoichiometric composition (x = 0.47), volt-capacitive characteristics and dielectric hysteresis loops were studied, and according to the asymmetry (the value of the internal electric field), one can judge qualitatively the degree of unipolarity of the formed films (figure 5). Preliminary results indicate that a unipolar state was observed in all the samples studied, and the value of the internal field was 17–25 kV/cm. However, in order to obtain an unambiguous regularity, additional research is needed.

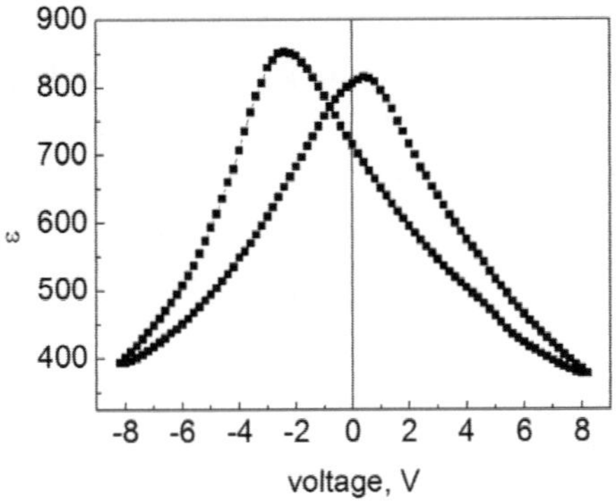

Figure 5. Voltage-capacitor (ε-V) curve for thin-film composition $x = 0.47$ (d = 30 mm).

The crystalline structure analysis of annealed PZT films by the scattered electrons diffraction method in the scanning electron microscope showed the presence of monoclinic phase in the films with the concentration increasing with the decrease of working gas pressure and the decrease of target-template distance.

4. Conclusions

The conducted studies showed that the change in the technological parameters of the RF magnetron sputtering (by varying the working gas or the distance between the sputtered target and the substrate) can, within a wide range (several tens of percent), change the lead content in thin PZT films whose composition corresponds to the morphotropic phase boundary, and also within several percent smoothly change the ratio of zirconium and titanium atoms. This makes it possible to search for optimal PZT compositions, and also to form thin films with a given composition gradient along their thickness.

Acknowledgments
This work was financial supported by the Ministry for Education and Science (Russian Federation) (Grant No 16.2811.2017/PCh) and grant RFBR № 16-02-00632.

References

[1] Jaffe B, Cook W R and Jaffe H 1971 *Piezoelectric Ceramics* (London: Academic Press)

[2] Noheda B, Cox D E, Shirane G, Guo R, Jones B and Cross L E 2001 *Phys. Rev. B* **63** 014103
[3] Sviridov E, Sem I, Alyoshin V, Biryukov S and Dudkevich V 1995 *Mater. Res. Soc. Symp. Proc.* **361** 141–6
[4] Kholkin A L, Brooks K G, Taylor D V, Hiboux S and Setter N 1998 *Integrated Ferroelectrics* **22** 525–33
[5] Afanasjev V P, Petrov A A, Pronin I P, Tarakanov E A, Pankrashkin A V, Kaptelov E Yu and Graul J 2001 *J. Phys.: Condensed Matter* **13** 8755–63
[6] Suchaneck G, Sandner T, Deineka A, Gerlach G and Jastrabik L 2004 *Ferroelectrics* **289** 309–16
[7] Pronin I P, Kaptelov E Ju, Senkevich S V, Klimov V A, Zajceva N V, Shaplygina T A, Pronin V P and Kukushkin S A 2010 *Physics of Solid State* **52** 124–8
[8] Volpyas V A and Kozyrev A B 2011 *J. Exp. Theor. Phys.* **113**(1) 172–9
[9] Volpyas V A, Tumarkin A V, Michaylov A K, Kozyrev A B and Platonov R A 2016 *Technical Physics Letters* **42**(14) 87–93

Increase of adhesion of conductive films on dielectric substrates by means of electric field

N S Pshchelko, E G Vodkaylo and B D Klimenkov

Department of General and Applied Physics, Saint-Petersburg Mining University, 199106, Saint Petersburg, Russia

E-mail: nikolsp@mail.ru

Abstract. Films of aluminium and copper produced by vacuum deposition obtained on the glass substrates were investigated. Improve of the efficiency of adhesion of conductive layers was made by the use of ponderomotive forces of an electric field (electroadhesive effects). It is found that thermal and electric treatment of films make possible to increase the strength of the adhesive bond of the film with the dielectric substrate.

1. Introduction

When films are produced by vacuum deposition, the physical contact of the film with the substrate is achieved relatively simply, but the adhesive strength of the compounds obtained in this case is not always satisfactory. An example is the production of film coatings in thin-film technology of the microwave range, where films of materials with thickness 10–15 μm are required, have poor adhesion to substrates, especially in a number of special applications where the adhesion strength of connections have to be particularly high. Improving the adhesion ability can be influenced by the electric field together with the heat treatment of the films [1].

In this paper, we consider ways to improve the efficiency of adhesion of conductive layers on ion based dielectric substrates on the use of ponderomotive forces of an electric field. The theoretical substantiation of the methods and methods for eliminating the electroadhesive effects is given in [2].

Electro-adhesion is the result of the electrostatic (ponderomotive) attraction of electrified bodies, in which charges of different signs are separated by a gap between contacting surfaces or a thin layer of one of the fastened objects depleted of charge carriers and therefore having an increased resistance adjacent to the contact interface. Most of the applied voltage therefore falls on this narrow layer, that is the reason for the appearance of large electrostatic fields and forces. To obtain greater attraction forces at the points of actual contact, the migration polarization observed in ionic dielectrics can be used. Migration polarization is effect of drift of the weakly fixed ions in the dielectric to the corresponding electrode under the action of the electrostatic field forces.

2. Research technique

Copper and aluminium were studied as coating materials. All these materials are widely used, they have different adhesion properties, which is important for a more complete study of the possibilities of the method of increasing adhesion. Films of copper and aluminium were deposited by thermal evaporation in a vacuum. The argon pressure at the time of deposition was of the order of $4 \cdot 10^{-3}$ mm Hg. The film thicknesses were 1.5–2 μm.

Experiments related to the study of the influence of the electric field on the adhesion of thin metal films were carried out using device for obtaining uncontrolled electroadhesive contacts (figure 1) [3]. To quantify the adhesion, the Krotyova and Deryagin adhesimetr was used. The detachment A of the film is found according to the expression [4]

$$A = \frac{mg}{b}(1 - \cos\alpha), \tag{1}$$

where m – load mass; g – gravity acceleration; b – width of the peeled film; α – angle at which separation starts.

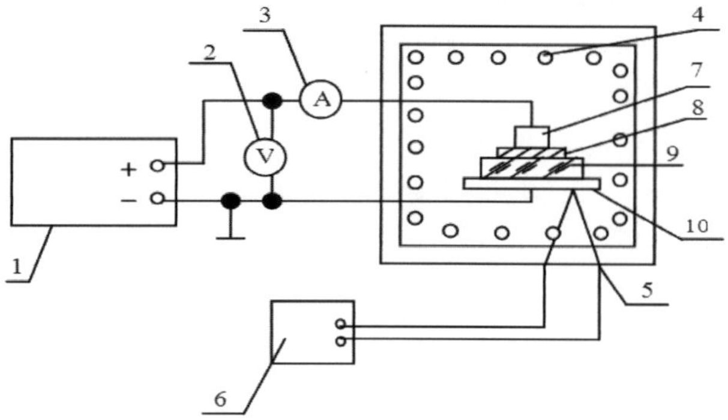

Figure 1. Scheme of installation for the obtaining of an electroadhesive connection: *1* – power supply; *2* – voltmeter; *3* – milliammeter; *4* – muffle furnace; *5* – thermocouple; *6* – potentiometer; *7*, *10* – electrodes; *8* – metal (semiconductor); *9* – insulator.

With high adhesion of the film to the substrate, only a partial peeling of the film from the substrate during measurements was observed. In this case, the relative share of the waste area was calculated based on the results of visual observations and the tear-off work calculated according to (1) was divided by this coefficient. Due to the fact that the value $k < 1$, the value obtained by the detachment was more than by (1). Thus, the detachment work was determined by the formula

$$A = \frac{mg}{b}(1 - \cos\alpha)\frac{S_0}{S_{sep}}, \tag{2}$$

where S_0 – total tear-off area, i.e. area of adhesive tape; S_{sep} – actually tearing off film area.

The possibility of improving the adhesion of films directly in the process of their condensation is of great interest. One of the measures used to increase the adhesion of films during their condensation is the heating of the substrate. However, this measure is in some cases insufficient. In some methods of applying film coatings, for example, ion-plasma sputtering "with displacement" presupposes the supply of a negative potential to the substrate relative to the plasma. Positive plasma ions bombard the substrate throughout the film deposition process, which allows higher adhesion values. The disadvantage of the method is that the negative potential prevents the formation of a transition oxide layer, which in some cases leads to insufficient adhesion [5]. By feeding a small positive potential at the first stages of film formation, it is possible to circumvent these difficulties. It can be seen that in this method, the potential difference is created between the plasma and the substrate, and not at the interface between the film and the substrate, and therefore is not achieved in all cases of the required adhesion value. Therefore, in the developed method, during the vacuum deposition of the film, a positive potential is supplied directly to the surface on which the film is deposited. To do this, a small contact spot must be made on this surface in advance, to which it is possible to connect a pressure contact with a positive potential with respect to the opposite surface of the substrate. The experiments were performed according to the scheme shown in figure 2.

3. Experimental results

The deposition of the films was carried out from a tungsten evaporator in a vacuum at a pressure of ~10^{-5} mm Hg. With a speed of ~0.015 μm / s. The thickness of the films was about 0.2 μm. The glass was used as the substrate material.

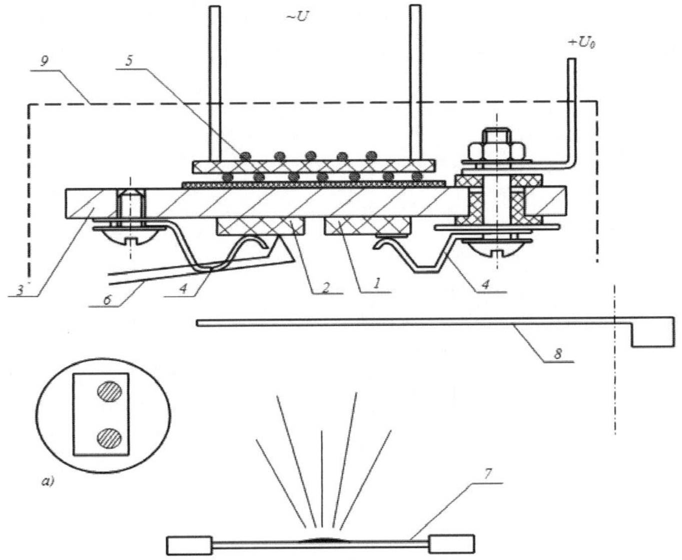

Figure 2. Tooling for the use of electro-adhesive forces in the process of vacuum film deposition: *1*, *2* – substrates; *3* – heat equalizing base; *4* – spring; *5* – heater; *6* – thermocouple; *7* – evaporator; *8* – damper; *9* – heat shield.

As an example, figure 3 shows some dependencies demonstrating the possibility of increasing adhesion by the proposed method. Adhesion of the films was measured at room temperature.

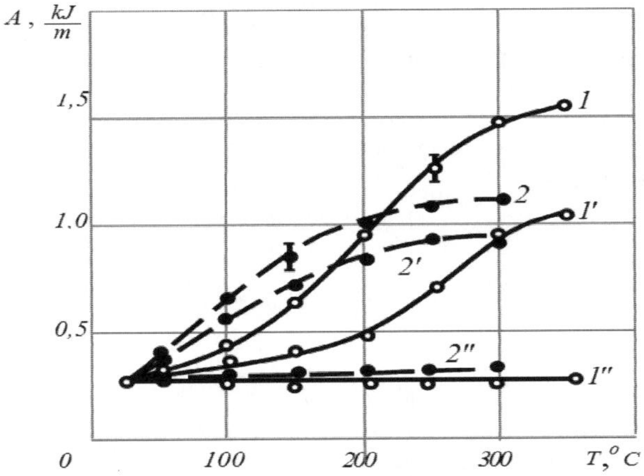

Figure 3. Dependence of the separation of copper (dashed) and aluminium (solid lines) films on the pane of glass from the temperature of the electroadhesion treatment. Holding time t = 10 min – *1*, *2* (U = 300 V) and *1*'', *2*'' (U_0 = 0 V); t = = 5 min – *1*', *2*' (U_0 = 300 V).

The increase in the stress applied to the film and substrate, as can be seen from figure 4, leads to a significant increase in the adhesion of the films to which the field was applied, with respect to those samples that experienced only heating and, in practice, did not change their adhesion.

The effect of the electric field on the formation of the film-substrate adhesion is determined by the same factors that affect the adhesion of the already applied films, the appearance of a large mechanical pressure at the film-substrate interface, due to the peculiarities of the migration polarization of the glass and the formation of an intermediate oxide layer between the film and a substrate was studied. However, the effect of the electric field on adhesion at the moment of film condensation turned out to be more significant than on the adhesion of already applied films. Confirmation of this fact was carried out as follows: the voltage was applied to the film after condensation under the same conditions (vacuum, temperature, voltage), before removing the substrate and breaking the vacuum. The increase in the adhesion of the film to the substrate in this case was much less than for the samples, the condensation of the film on which was produced in an electric field. The imposition of an electric field upon condensation of the film apparently leads to an additional activation of the surface of the substrate and, thus, to the appearance of a large number of chemical bonds between the adsorbed metal atoms and the substrate material. Electro-adhesive forces cause metal atoms to be attracted to the substrate more efficiently and occupy places with a minimum of potential energy, that is, deeper potential wells, and thereby increase the strength of the film-substrate joints.

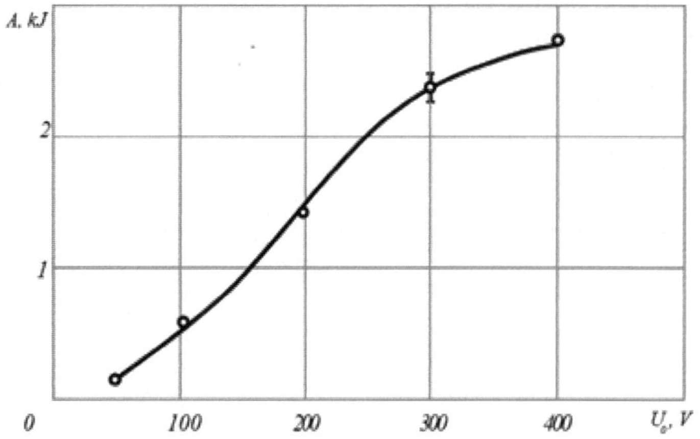

Figure 4. Dependence of the adhesion of an aluminium film on voltage under electroadhesive treatment. Substrate temperature $T = 120$ °C.

4. Conclusions

Thermal treatment of films in combination with the impact of an electric field makes it possible to increase or, if necessary, reduce the strength of the adhesive bond of the film with the dielectric substrate. Change the adhesion of films can be both – after application, and during the process of condensation of the film on the substrate.

References

[1] Galembeck F and Burgo T A L 2017 *Chemical Electrostatics* (Springer)
[2] Pshchelko N S 1986 *Conference (Rostov-on-Don)* p 62
[3] Pshchelko N S, Moshnikov V A and Sevryugina M P 2013 *Smart Nanocomposites* **4** 2
[4] Deryagin B V, Krotova N A and Smilga V P 1973 *Adhesion of Solid Bodies* (Moscow: Nauka)
[5] Pronin I A, Karmanov A A and Moshnikov V A 2015 *Physics of the Solid State* **57(12)** 2373–81

IOP Conf. Series: Journal of Physics: Conf. Series **872** (2017) 012024 doi:10.1088/1742-6596/872/1/012024

High-power sputtering employed for film deposition

V I Shapovalov

Saint-Petersburg Electrotechnical University "LETI", Saint-Petersburg, Russia

E-mail: vishapovalov@mail.ru

Abstract. The features of high-power magnetron sputtering employed for the films' deposition are reviewed. The main physical phenomena accompanying high-power sputtering including ion-electron emission, gas rarefaction, ionization of sputtered atoms, self-sputtering, ion sound waves and the impact of the target heating are described.

Magnetron sputtering (MS) is a well-known technology for the deposition of films of various material used in a wide range of applications. Two areas of performance improvement are being developed in the MS technology. The first one is based on an increase in the degree of ionization of the sputtered flux [1–7], the second one is based on an increase in the temperature of the target surface up to the melting temperature [8–13]. The methods that ensure a high degree of ionization of sputtered atoms are called ionized physical vapor deposition (IPVD) [1]. A very high density of charged particles, which increases with increasing power density on the target, is common to all IPVD methods. But the increase of power density is limited by the melting temperature of the target material.

Initially, the processes of the IPVD-type magnetron sputtering were based on the use of a conventional direct current magnetron with an additional HF generator that induces inductively coupled plasma in a vacuum chamber. In other cases, IPVD was achieved by applying a single-polarity pulse of high power, low frequency and low pulse ratio. This method was called high power pulsed magnetron sputtering (HPPMS) [14]. There are several options of HPPMS technology. A very short (tens of microseconds) pulses with very large amplitudes (more than 600 V) at a frequency of more than 50 Hz are applied to the target in high power impulse magnetron sputtering (HiPIMS). In order to distinguish this technique from other impulse magnetron processes, it is suggested in [15] that the peak power exceeds the time average by 2 orders in HiPIMS methods.

Sometimes a DC power supply is used to keep the magnetron sputtering at direct current (PMSdc – pulse magnetron sputtering direct current) with superimposition of powerful pulses with high ratio. This method is called HiPIMS with pre-ionization. Other approaches include pulse modulation in such a way that impulses (several hundred microseconds) are delivered to the target with a moderate power level (typical for DCMS) at the initial stage. Then a high power pulse is followed (from several hundred microseconds time interval to millisecond). This method is called modulated pulse power (MPP) [1].

Figure 1 shows the plane for the platinum target in the coordinates "peak power density on the target q_t – packing factor of *Dc* (reciprocal of pulse ratio)". The area $q_t < 0.1$ kW/cm^2 is typical for the so-called pulsing PMSdc, which includes bipolar asymmetric pulsed discharges operating in the medium-frequency band of 10–250 kHz. All discharges above the PMSdc limit are called HPPMS. In this range, the high peak power should be compensated by a lower pulse ratio. It gives the limiting power density line for the rectangular shape of the pulses shown in figure 1. The range for HiPIMS is defined by the limit $q_t > 1.0$ kW/cm^2. The pulses in MPP methods tend to have a low power density level, often in the PMSdc range, followed by a stronger pulse with an intermediate density ($0.1 < q_t <$ < 1.0 kW/cm^2) or even in the HiPIMS range.

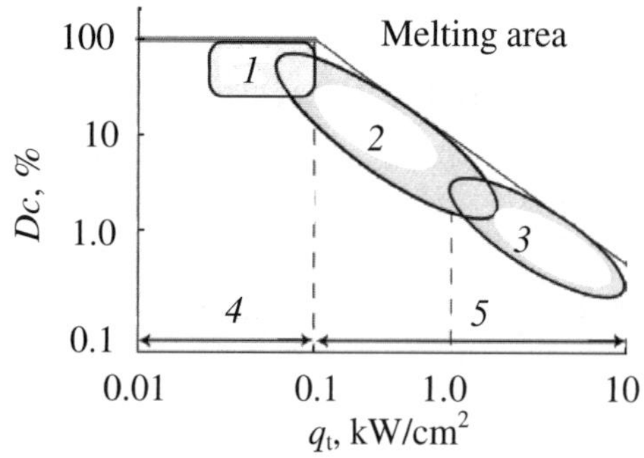

Figure 1. Types of pulse discharges: *1* – PMSdc; *2* – MPP; *3* – HiPIMS and range: *4* – PMSdc; *5* – HPPMS [1].

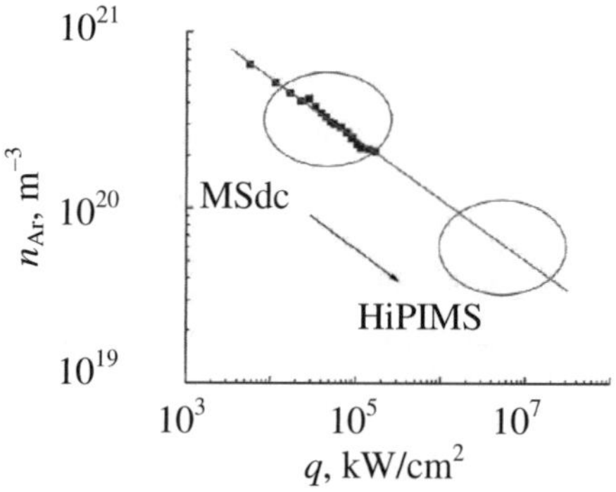

Figure 2. The argon density n_{Ar} at magnetron sputtering depending on the power density on the target q [15].

HiPIMS systems usually operate at a voltage of 500–2000 V, current density up to 3–4 A/cm^2, a peak power density of 0.5–10 kW/cm^2, a pulse time of 50–100 μs, a pulse-recurrence frequency of 50–5000 Hz, and a duty factor of 0.5–5% (or pulse ratio of 20–200). The average power remains unchanged although the peak power is 2–3 order-greater than that normally used in PMSdc. The range of parameters for HiPIMS is much wider than there are in the dc, rf and MPP systems. There is more flexibility in controlling the deposition of films due to the additional parameters such as the duration and frequency of pulses in addition.

The main motivation for using of HiPIMS is to achieve conditions that lead to high density of sputtered atoms and high ionization of the sputtered material. A partially ionized flux of sputtered material provides a reduction in the film growth rate as compared to other sputtering methods at a equal average power (table 1).

Table 1. Ratio of films deposition rates [16].

Target	Ti	Cr	Cu	Al	Ta	Zr	C
$\upsilon_{HPPMS}/\upsilon_{MSdc}$, %	15–75	29	37–80	35	22–40	15	77

Physical reasons providing the advantage of high-power sputtering are as follows:
- linear increase of the sputtering coefficient with increasing ion energy;
- change of the sputtering coefficient due to the types of particles changes participating in sputtering;
- increase of film density and self-sputtering on the substrate;
- changes of the keeping conditions of charged particles near the target at current intensity;
- superposition of sputtering and evaporation of the target.

Let's take a closer look at basic physical phenomena accompanying a high-power sputtering.

1. Emission. Ion-electron emission plays a decisive role in the formation of the magnetron discharge. When the γ emission coefficient γ is approximated, the expression [17] is used for many materials at $E_i < 600$ eV

$$\gamma = 0.032(0.78E_{0\to1} - 2\varphi), \quad (1)$$

where $E_{0\to1}$ – the first ionization potential, eV; φ – work function, eV.

To describe the current-voltage characteristics of the magnetron discharge and to establish its relation to the pressure of the working gas, an effective coefficient of ion-electron emission was introduced in [17] γ_{eff}:

$$U \sim 1/\gamma_{eff}, \quad (2)$$

which is proportional to the ionization probability $P(p)$:

$$\gamma_{eff} = P(p)\gamma, \tag{3}$$

the value of P is proportional to the pressure:

$$P(p) \sim p. \tag{4}$$

It follows from (2)–(4) that when the pressure increases, the voltage on the discharge must decrease. The cause-and-effect relationships reflected by expressions (2)–(4) are also valid for describing the current-voltage characteristics of the magnetrons with hot targets. In these cases it is necessary to use the results of studies where it was found that there's a heating and density decrease near the target during sputtering because of collisions between sputtered atoms and neutral particles of the working gas.

2. Rarefaction. Numerous studies have made it possible to establish that because of collisions between sputtered atoms and neutral particles of the working gas during sputtering, the heating and rarefaction arise [1]. The decrease of gas density was established as rarefaction, since a decrease in gas density is equivalent to a pressure decrease. Sometimes the term sputtering wind is used to describe this process. The wind of spraying has a direction from the target. This decrease in the density of the working gas leads to a decrease in the number of ions, which reduces the sputtering rate and, consequently, the rate of the film deposition. The data for MSdc in figure 2 show a typical example of a change in a gas density observed in front of a copper target [15]. Extrapolation of line to the high density area that is characteristic of HiPIMS shows possible very strong rarefaction which is confirmed by many authors.

3. Ionization of sputtered atoms. The electron density close to the target is of the order of 10^{16} m^{-3} in a conventional direct current magnetron. The fraction of ionized atoms in the sputtered flux has only a few percent under these conditions. High ion density without target overheating can be achieved with HPPMS or HiPMS with a low filling coefficient (about 1 %). In this case, the peak power on the surface of the cathode can reach several kilowatts per square centimeter and the peak electron density can be two orders higher. At the same time, the share of ionized atoms in the incident sputtered flux can increase up to 100 % [1].

4. Self-sputtering. The high degree of ionization of sputtered atoms affects the film deposition process. Some of them, due to fast Bohm diffusion in a magnetic field, will fall on the substrate. The other part can return to the target area. These accelerated in the dark cathode space ions will cause the sputtering of the target, which is called self-sputtering. In this case, the target can go into a stationary mode (it is often called self-sustaining) of self-sputtering, when the inert gas can be turned off. The condition of such a state is written in the form

$$\alpha\beta S > 1, \tag{5}$$

with α – the ionization probability of the sputtered atom; β – the probability of the ionized atom returning to the target; S – the self-sputtering coefficient. Since α <1 and β <1 it follows from equation (5) that self-sustained self-sputtering is attainable only in the case of S >> 1. The effect is experimentally observed for a few materials for this reason that have a high sputtering coefficient.

5. Ion sound waves. Ion sound waves can arise in the plasma at the form of longitudinal quasi-electrostatic waves associated with ion oscillations. These waves can be excited in isotropic nonequilibrium plasma, where the electron temperature is much higher than the ion temperature [18].

In the case when the electron and ion temperatures have a small difference, the phase rate of ion sound waves is close to the thermal ion rate. It leads to a strong collisionless decay due to the energy transfer to the plasma particles.

When the amplitude of the ion sound waves increases, nonlinear effects strengthen and lead to a steepening of the wave front. However, the dispersion prevents steepening and ultimately can lead to the emergence of a special class of nonlinear ion sound waves – the so-called ion-acoustic solitons, which are stable solitary disturbances of ion density propagating in space.

6. Evaporation of the target. During high-power sputtering the target surface can be heated to the temperature at which the flux of evaporated material increases up to significant value, and its density

is given by the Hertz–Knudsen equation [10]. It can lead to a significant heating of the substrate and an increase in the growth rate of the film.

Acknowledgments
The author thanks Russian Science Foundation for their support on the study (grant 15-19-00076).

References

[1] Gudmundsson J T, Brenning N, Lundin D et al. 2012 *J. Vac. Sci. Technol. A.* **30** 030801
[2] Drache S, Stranak V, Herrendorf A-P et al. 2013 *Vacuum* **90** 176–81
[3] Zhao X, Jin J, Cheng J-C et al. 2015 *Vacuum* **118** 38
[4] Vitelaru C, Lundin D, Brenning N et al. 2013 *Appl. Phys. Lett.* **103** 104105
[5] Huo C, Lundin D, Raadu M A et al. 2013 *Plasma Sources Sci. Technol.* **22** 045005
[6] Krysa J, Zlamal M, Kment S et al. 2014 *Chem. Eng. Trans.* **41** 379–84
[7] Holtzer N, Antonin O, Minea T et al. 2014 *Surf. Coat. Technol.* 19192
[8] Raman P, Shchelkanov I A, McLain J and Ruzic D N 2015 *J. Vac. Sci. Technol. A.* **33** 031304
[9] Gupta R, Pandey N, Behera L et al. 2016 *AIP Conf. Proc.* **1731** 080005
[10] Shapovalov V I, Karzin V V and Bondarenko A S 2017 *Phys. Lett. A.* **381** 472–5
[11] Raman P, Shchelkanov I A, McLain J et al. 2015 *J. Vac. Sci. Technol. A.* **33** 031304
[12] Niu G-J, Li X.-C, Xu Q et al. 2015 *Nucl. Instr. Meth. Phys. Res.* **349** 45–9
[13] Cormier P A, Balhamri A, Thomann A L et al. 2014 *Sur. Coat. Technol.* **254** 291–7
[14] Alami J, Bolz S and Sarakinos K J 2009 *Alloys and Compounds* **483** 530–4
[15] Anders A 2011 *Sur. Coat. Technol.* **205** S1–9
[16] Sarakinos K, Alami J and Konstantinidis S 2010 *Sur. Coat. Technol.* **204** 1661–84
[17] Depla D, Heirwegh S, Mahieu S et al. 2007 *J. Appl. Phys.* **101** 013301

Influence of ion bombardment of a substrate on the quality of vacuum-plasma Ti-TiN coatings

S R Shekhtman and M Sh Migranov

Ufa State Aviation Technical University, Ufa

E-mail: shex@inbox.ru

Abstract. The characteristics of the quality of multi-layer vacuum ion-plasma coatings Ti-TiN such as roughness and adhesion strength are investigated. It is shown that additional ion bombardment promotes the production of high-quality coatings. Ion bombardment has a significant effect on the state of the surface layer of the metal. In the process of ion bombardment, conditions are created for the formation of active adsorption centers and the formation of a fine-grained structure, nanoscale grains and layers.

1. Introduction

Characteristics of the quality of a surface with a vacuum ion-plasma coating (VIPC) under ion bombardment conditions largely depend on the initial surface, as well as on preliminary preparation before coating deposition. Preliminary preparation allows to clean the surface of all kinds of contaminants, and includes modification, thermal activation of the surface and the formation of active nucleation centers. That, first of all, promotes high adhesion of coatings. Existing traditional methods of preliminary preparation of the substrate surface (mechanical, electrochemical and chemical polishing, ultrasonic cleaning, combined methods) do not always guarantee a high degree of purification [1, 2].

The purpose of this work was to investigate the effect of additional ion bombardment (AIB) on the stage of the surface pre-cleaning and deposition of coatings on the quality characteristics of the synthesized VIPC of Ti-TiN system.

2. Results and discussion

Heat-resistant steel of the martensitic class of Russian brand 13H11N2V2MF was chosen as the object of research. Coatings with additional ion bombardment were applied to flat samples with a diameter of 20 mm, a thickness of 3 mm. Samples were polished.

Synthesis of multilayer coatings was implemented on the upgraded installation of NNV-6.6-I1 using the following technology, including [3, 4]:

– preparation of the surface for spraying;
– preliminary cleaning and heating of the surface using the plasma generator “PINK”;
– formation of a multilayer coating of the Ti-TiN system;
– cooling of the vacuum chamber for a predetermined time.

To realize additional ion bombardment, a plasma generator “PINK” was used. It realizes a non-self-sustaining high-current diffusion discharge (NSDD). The generator was used both at the stage of preliminary treatment and activation of the surface, and for the synthesis of protective coatings in the plasma assisting mode. Use of the "PINK" allows to significantly increase the adhesive strength and to form a better coating, as well as to reduce the temperature of the formation of layers. On the stage of coating deposition, application of the plasma generator provides significant additional acceleration to the ions of the plasma-forming gas and also promotes the acceleration of diffusion processes in the near-surface layers of the substrate. Plasma assisting involves the bombardment of ionic coatings during their formation and growth by high-energy ions, providing a significant improvement in the adhesion strength, microstructure and stoichiometry of coatings, as well as the formation of structures with pronounced anisotropy of properties, reducing the deposition temperature without deteriorating the level of physicomechanical properties of the surface layer [5–7].

The formation of multilayer coatings is realized by successive deposition of Ti and TiN from an arc discharge plasma. The plasma generator "PINK" was used for simultaneous plasma assisting in the deposition of coatings. The uniformity of the distribution of the thickness of the formed coatings is achieved by planetary rotation of the work part simultaneously around the axis of the table and relative to its own axis [7, 8].

Conducting an AIB with a plasma generator "PINK" allows:

– to ensure high energy efficiency of the process of generation of low-temperature bulk plasma;

– to reduce the fraction of the microdrop fraction in the plasma flow of vacuum electric arc evaporators;

– to produce plasma cleaning, etching and activation of the surface without spraying by vapors of the cathode material;

– to provide complex processing of products in a single vacuum cycle, including processes of final cleaning, activation, electron-ion-plasma nitriding of the surface and spraying of functional coatings with plasma assisting;

– to ensure the formation of micro- and nanostructured coatings with high hardness, increased wear resistance, improved corrosion resistance.

Ionic purification and activation of adsorption centers on the processed surface of NSDD is carried out in an inert gas (argon) medium under following conditions: the pressure is 10^{-1} Pa, displacement on a substrate is negative (900–1100 V), discharge current is 30–40A, processing time is 1...2 min. The end of the processing was fixed by stopping the occurrence of micro-arcs. This made it possible to avoid the introduction of plasma ions during the process of modifying and activating the surface, including the pre-cleaning.

The thickness of the deposited multilayer coatings, according to microstructural analysis, was 5...7 μm. These data are consistent with the results of calculations of the weighting method. The use of a multilayering of coating ensures:

– smooth reduction of residual stresses in thickness;

– decrease in porosity and defectiveness of the coating;

– increase in the integral thickness of the coating without reducing the performance characteristics;

– providing a smooth transition from a soft base to a hard functional layer.

Modification of the surface and its activation by means of a plasma realized by the plasma generator allows to increase the rate of spraying the surface, to reduce the processing time, to reduce the temperature of the coating formation, and to achieve uniform heating of the surface, which in turn ensures high adhesion strength of the VIPC.

To analyze the properties of the formed surface layer, microhardness studies were carried out. The tooling "PMT-3" was used. The measurements were carried out by successively pressing a diamond pyramid 136° with a square base with loads of 0.2, 0.5, 1.0 and 0.20 N. The stress lasted for 5–10 s. The diagonals of the prints were measured by a 30-hapiobjective OE-6 with aperture A = 0.65 and a 15-fold eyepiece micrometer AM9-3. It ensured a total magnification of 485 times. The microhardness value was identified from the tables by the lengths of the measured diagonals of the prints and the value of the applied load.

The depth of penetration of the indenter was calculated as the ratio of half the diagonal of the print to half the tangent of 136°. To estimate the true value of the measured value, the arithmetic mean was used.

For a sufficiently reliability of average result, the number of measurements was increased to 7. At different points of the sample, when estimating the arithmetic mean of the microhardness, the tests showed the same results (table 1).

Table 1. Results of the study of the microhardness of coatings.

Material	Coating type	Microhardness, GPa
13H11N2VMF	Initial condition	16.3
	Multilayer coating of Ti-TiN (3 layers)	17.8
	Nanostructured multilayer coating	25.7

The nanostructured multilayer coating was formed from nanometer layers (up to 100–150 nm) by the number of layers up to several tens. Due to the peculiarities of the structure, the ultra-thin layer thickness and their large number, as well as the possibility of dense interfacing layers of different materials, these coatings combine the qualities of layered systems and the specific properties of nanoobjects.

The main influence of the effect of AIB in the process of modification and ionic cleaning on the properties of coatings is formed at the stage of coating initiation due to leveling of stresses upon exposure to the substrate and transformation of the crystal structure. The ion flux forms the point defects on the surface, these defects are active adsorption centers.

In comparison with the original polished surface of the substrate, its roughness after the synthesis of the coating increased. This is due to the peculiarities of the formation of layers under the conditions of additional ion bombardment, since active adsorption centers are formed and the ions of the coating material are deposited on the protrusions of the microrelief of the surface of the material being processed. Adhesive strength is determined by:

– composition, thickness, structure, residual stresses, that is, coating parameters;

– composition, microgeometry of the surface, type of preliminary machining, – i.e. parameters of the substrate and the type of its preparation;

– type of cleaning, preheating, – i.e. preparation of the surface for spraying;

– substrate temperature, accelerating voltage, gas composition and pressure, condensation rate and arc current, – i.e. technological modes of deposition of coatings.

Studies of adhesion strength were carried out by various methods. So, the study by bending samples using the VIAM technique showed excellent adhesion of the coating with respect to the main material. A study of a diamond pyramid imprint with an increase using an optical microscope showed that there was no cracking of the coating material near the exposure zone, which also indirectly indicates a high adhesion of the coating.

Investigations of the adhesive strength of deposited coatings, which were carried out by the VIAM method by bending samples at 90°, showed that the coatings have a high adhesion to the main material. In addition, adhesion was investigated by indenting the diamond pyramid indentor and studying its imprints. Indirect evidence of high adhesion of the coating to the main material was the absence of cracking of the coating material near the impact zone.

Figure 1 shows a photograph in the optical microscope of the beginning of the destruction of the Ti-TiN coating with an increase in the load on the indenter during the scratching process.

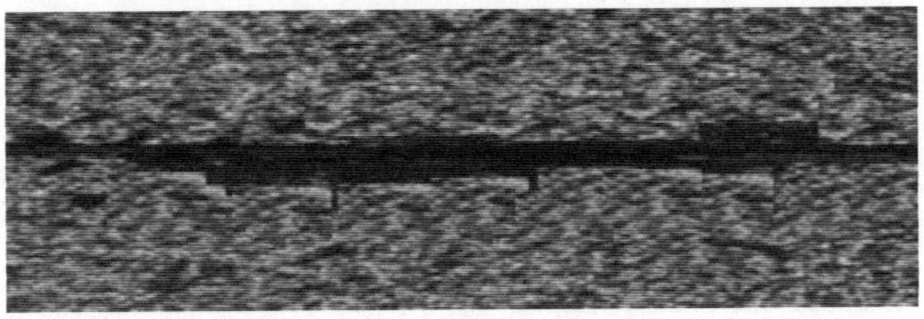

Figure 1. Evaluation of the adhesion strength of multi-layer Ti-TiN coatings by scratching.

The study of the adhesion strength of coatings on a scratch device by scratching method showed satisfactory adhesion. The observed minor cleavages along the edges of the scratch correspond to the requirements. The load on the tip was 3.0 N. In addition, the effect of surface pretreatment and the quality of preparation of the substrate surface was investigated by ionic cleaning directly in the vacuum chamber. In this case, the adhesive strength of the coatings on substrates was evaluated. It has been established that with a cleaning duration in a vacuum chamber of less than 30...45 s, the adhesion strength of Ti-TiN coatings is about 20 to 30 MPa.

Coatings formed at the minimum values of the duration of the ion-cleaning process are characterized by poor quality up to partial peeling off the substrate. This is explained by the fact that the removal of oxide films from the surface does not occur, and the temperature is insufficient for the effective course of plasma-chemical reactions. At optimal values of the cleaning duration, the coating has good characteristics of the adhesive interaction of the substrate and the coating. With an increase in the duration of cleaning, surface overheating and a decrease in adhesion strength are possible. Further overheating of the surface can lead to the formation of a columnar structure of the coating.

AIB has a significant effect on the physico-chemical state of the surface layer of the substrate. In the process of ionic exposure, surface contamination is removed. The substrate remains atomically pure until the formation of the main coating layer.

The process of ion cleaning and surface activation under the conditions of additional ion bombardment, realized by the plasma generator "PINK", ensures the production of high-quality vacuum-plasma coatings.

References

[1] Uglov V V, Prikhodko Zh L and Khodasevich V V 2003 *Physics and Chemistry of Material Processing* 48–52
[2] Shekhtman S R and Suhova N A 2016 *Materials Science Forum* **870** 113–7
[3] Shekhtman S R and Suhova N A 2016 *Journal of Physics: Conference Series* **729**
[4] Shekhtman S R, Sukhova N A and Yansaitova M I 2017 *Quality. Innovation. Education* **1** 40–5
[5] Shekhtman S R and Sukhova N A 2016 *Bulletin of UGATU* **3** 44–8
[6] Ismagilova L A and Suhova N A 2016 *Procedia Engineering* **150** 849–53
[7] Suhova N A 2002 *Bulletin of the Ufa State Aviation Technical University* 211–16
[8] Suhova N A 2014 *Science Production Annual scientific and technical collection (Ufa)* pp 88–97

On the influence of Ti-Al intermetallic coating architecture on mechanical properties and wear resistance of end mills

E L Vardanyan, V V Budilov, K N Ramazanov and Z R Ataullin

Ufa State Aviation Technical University, K. Marx Street 12, Ufa, Russia, 450000

E-mail: vardanyaned@gmail.com

Abstract. Thin-film wear-resistant coatings are widely used to increase life and efficiency of metal cutting tools. This paper shows the results of a study on the influence of architecture (number, sequence and thickness of layers) of wear-resistant coatings on physical, mechanical and operational properties of end mills. Coatings consisting of alternating Ti-Al/Ti-Al-N layers of equal thickness demonstrated the best physical and mechanical properties. Durability of coated tools when processing materials from chromium-vanadium steel increased twice as compared to uncoated tools.

1. Introduction

One of the main ways to improve the efficiency of metal cutting tools is to increase their life through the use of thin-film wear-resistant coatings consisting, for example, of various refractory compounds: TiN, TiCN, TiAlN, ZrN, ZrCN, ZrHfN, CrN [1]. Increased life is commonly assessed by increased hardness of the surface layer. However, when parts and tools interact, tool wear also depends on elasticity and resistance to deformation of their surface layers. Therefore, materials with optimum hardness and increased tribological properties (low friction factor, minimum run-in duration, minimum heat release in friction, minimum wear of exposed surfaces) should be referred to wear-resistant coatings. Physical, mechanical and operational properties of wear-resistant coatings greatly depend on their chemical and phase composition as well as on architecture (sequences of alternating layers, layer thickness ratio, etc.). Besides, the use of modern wear-resistant coatings should ensure maximum adhesion to the base material.

Surface properties of tool materials can be directionally modified with high efficiency by application of functional coatings to working surfaces of the cutting tools from vacuum-arc discharge plasma. Recently, Ti-Al intermetallic coatings have become of great interest [2, 3]. The issue yet is to identify optimal parameters of wear-resistant coatings (phase composition, architecture) and the technology of their application to improve operational properties of cutting tools [1].

Therefore, this paper aims to study the influence of architecture (number, sequence and thickness of layers) of wear-resistant coatings on physical, mechanical and operational properties of end mills.

2. Experimental methods

Coatings were applied on a modernized facility NNV-6.6-I1 [3]. The developed technology for the application of intermetallic coatings is described in detail in [3]. Four types of coatings were chosen to study the influence of coating architecture on physical, mechanical and operational properties of end mills. Schematic coating architectures are shown in figure 1.

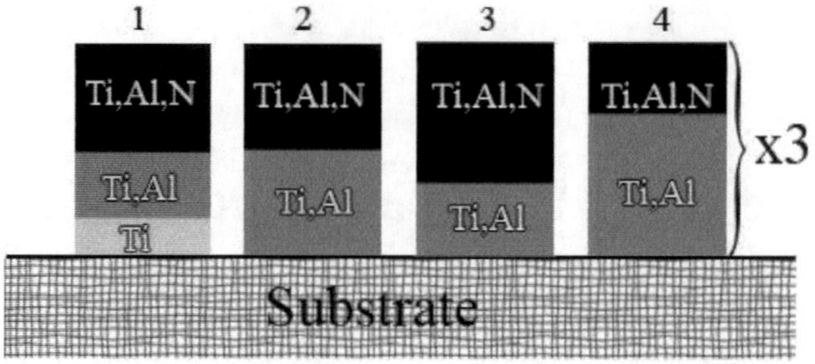

Figure 1. Schematic architectures of Ti-Al intermetallic coatings.

The thickness of coatings was determined from well parameters. Wells were made with the CSM Calotest. A scratch-tester (CSM Instruments) was used to measure adhesive strength and scratch resistance, as well as to clarify the mechanism of coating destruction. Scratches were applied on the coating surface under a continuously increasing load with a diamond spherical indenter of the Rockwell C type with a curvature radius of 200 μm. The microstructure and chemical composition of the sample surfaces were examined by scanning electron microscopy (SEM).

Parts from chromium-vanadium steel were treated on a milling machine in the tool shop of the Ufa Aggregate Production Association. Tests were performed on end mills made from P6M5 high speed tool steel with and without Ti-Al intermetallic coatings applied with different technologies. The mills were treated on a machine vertical console and milling 6T12 at a spindle rotation speed of 100 rpm and feed rate of 20 mm/m.

3. Results and discussion

The thickness of the coatings on the samples positioned perpendicular to the flow was 2.2–2.5 μm, and on the samples positioned parallel to the flow was 3.5–3.8 μm. The surface microstructure of beveled cuts on samples with different coatings is shown in figure 2.

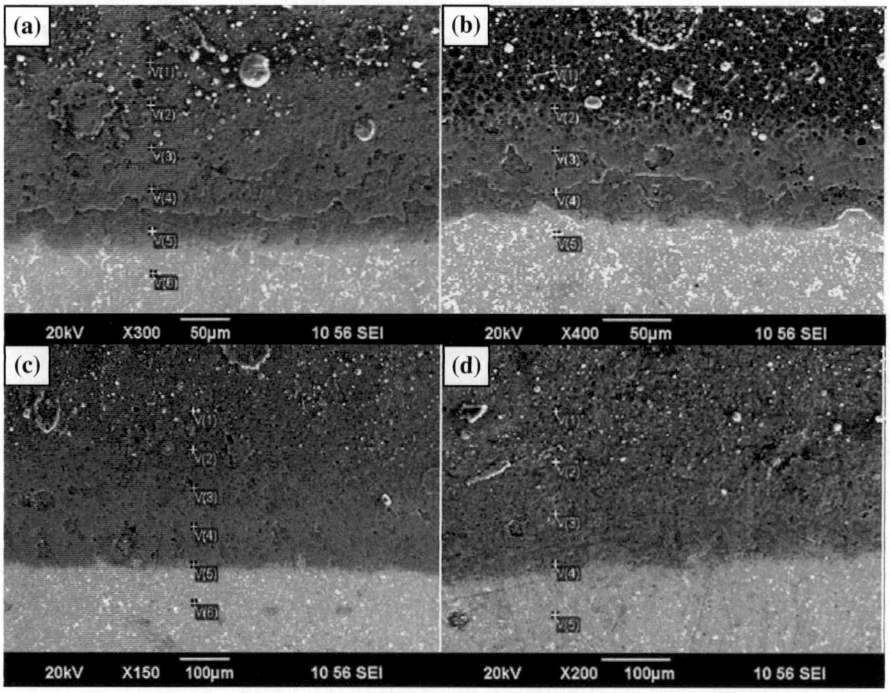

Figure 2. SEM images of beveled cuts on P6M5 steel samples with different coatings: **(a)** – 1; **(b)** – 2; **(c)** – 3; **(d)** – 4. V(1–6) – measuring points for the chemical composition of the coatings.

Boundaries between the solid nitride layers (Ti-Al-N) and the softer intermetallic layers (Ti-Al) are clearly visible in the microstructure of the coated sample in figure 2(a). Similar situation is observed in the microstructure of the sample in figure 2(b). In the sample in figure 2(c) nitride layers are twice thicker than the softer intermetallic layers and the microstructure demonstrates a continuous surface with sparse pores of small size, almost no obvious boundary between the layers is observed. The microstructure of the 4-th group of samples differs from the previous one to a small extent, since this technology differs only in the ratio of the thicknesses of the layers.

Chemical composition of the sample surfaces measured at different depth (table 1) testifies to the difference in the architecture of coatings obtained using different technologies.

Table 1. Chemical composition of coatings.

Measuring	1				2				3				4			
point	N	Al	Ti	Fe	N	Al	Ti	Fe	N	Al	Ti	Fe	N	Al	Ti	Fe
1	11	18	68	0	14	21	63	0	17	25	56	0	10	25	62	0
2	9	15	73	0	17	20	60	1	10	34	55	0	8	25	67	1
3	11	11	76	1	17	21	60	1	12	27	59	1	10	20	70	2
4	7	19	72	2	16	14	65	5	19	23	56	2	15	15	55	13
5	15	16	47	15	0	0	0	85	16	15	32	32	0	0	0	85
6	0	0	0	89	–	–	–	–	0	0	0	85	–	–	–	–

The results of scratch tests are given in table 2. They show that the strength of adhesion of the coatings to the base material is high for all samples. However, physical and mechanical properties of the coatings differ to a great extent.

Table 2. Maximum indentation depth (h_{max}), critical load (L_c) and calculated coefficient of recovery (W_e).

Technology		Maximum indentation depth h_{max}, μm	Critical load L_c, H	Coefficient of elastic recovery W_e, %
1	Perp.	8	20	92
	Par.	7	15	76
2	Perp.	5	23	95
	Par.	4	21	92
3	Perp.	8.5	18	50
	Par.	7	20	60
4	Perp.	8.5	18	72
	Par.	8.5	15	50

Increased mechanical properties are observed in the samples processed by the second technology: the maximum elastic recovery W_e up to 95 % at the minimum indentation depth h_{max} = 5 μm and the maximum critical load L_c = 23H. Six milling cutters coated with Al-Ti intermetallics with different alternating layers were tested in the tool shop of the Ufa Aggregate Production Association. Images of the instruments during operation and after testing are shown in figure 3. Test results showed that the greatest increase in wear resistance of end mills was observed in the tool prepared by the second technology (figure 3(c)). The tool's durability was 90 min, which is twice the life of the uncoated tool and 1.5 times longer in comparison with technologies 1 and 3.

4. Conclusion

The experiments have shown that the architecture of multilayer coatings affects physical and mechanical properties. The highest increase in physical and mechanical properties was observed in the coating consisting of alternating Ti-Al/Ti-Al-N layers of equal thickness. Operation tests were the greatest for the mills with the coatings of the second architecture. Durability of the tools when processing materials from chromium-vanadium steel increased twice as compared to uncoated tools and by 1.2–1.5 times in comparison with coatings with a different architecture.

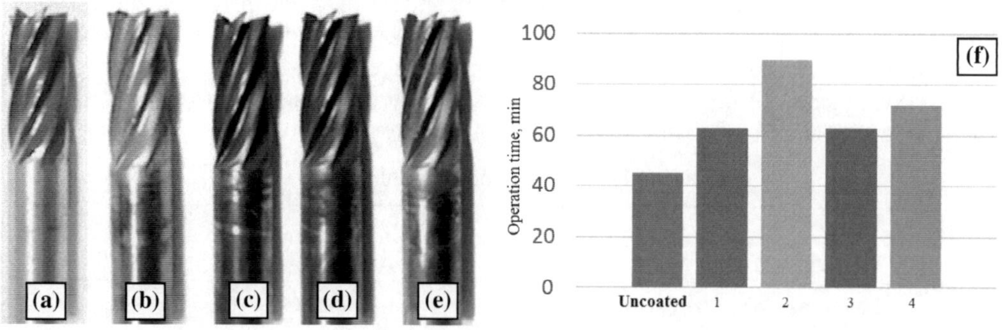

Figure 3. Instruments after operation tests: **(a)** – uncoated; **(b)**, **(c)**, **(d)**, **(e)** – with coatings of different architecture (1–4); **(f)** – test results.

References

[1] Topolyansky P A, Ermakov S A et al. 2013 *Metal working* **76** 28–54
[2] Musil J and Stupka P 2015 *J. Nanomanufacturing* **11** 78–93
[3] Vardanyan E L and Budilov V V 2016 *Surf. Invest: X-ray, Synchrotron and Neutron Techniques* **10** 728–31

Vacuum coating system for deposition of superconducting $W_xSi_{(1-x)}$ ultrathin films used in single photon detectors

D D Vasilev, E I Malevannaya and K M Moiseev

Bauman Moscow State Technical University, 2-nd Baumanskaya str. 5/1, 105005, Moscow, Russia

E-mail: d.d.vasiliev@yandex.ru, el.malevan@gmail.com, mkm430@yandex.ru

Abstract. VUP-11M coating system for thin films deposition in vacuum is described in this article. Technological features which appear at laboratory vacuum system modernization for implementation of dual-components $W_xSi_{(1-x)}$ films forming process by means of magnetron co-sputtering from two sources are shown. Totals of the system modernization and research results of obtained $W_xSi_{(1-x)}$ films are reported.

1. Introduction

In recent years development and manufacturing of single photon detectors is one of the most actual direction in microelectronics. Single photon detectors are indispensable in such areas as infrared wave length research, quantum cryptography, at weak laser radiation detecting, at LIC (large integrated circuit) testing, in radio astronomy observing, fluorescence radiation measurements with high time resolution and others [1]. Along with existing types of single photon detectors (photomultipliers, avalanche photodiodes, etc.) which differ by photon detecting principle, during last decade the new type of detectors has appeared. It is superconducting single photon detectors (SSPD) [1]. This type of detectors is oriented on the spectral range from 0.6 to 2.3 μm. SSPD is characterized by low jitter (less 40 ps), low dark count rate (less 10 GHz) and has quantum efficiency more 85 % [2]. In contrast of avalanche photodiodes and photomultipliers SSPDs have higher performance wherein its working temperature isn't ultra-low as transition edge sensors working temperature and is around 4.2 K.

At present time SSPDs are produced on the base of ultrathin NbN film which thickness is about 4 nm [1]. Due to such small thickness radiation absorption of film itself is limited and that decrease detecting quantum efficiency. However coating thickness increasing, i.e. conductor cross-section increasing, also reduce this parameter because hot spot which arises after photon absorption can't guaranteed to cause the junction to normal state on the whole cross-section of conductor. One of material which could replace NbN in SSPD is $W_xSi_{(1-x)}$ [3]. Using this material it is possible to achieve quantum efficiency close to 100 %.

2. Features of $W_xSi_{(1-x)}$ coatings forming

As it was founded earlier content percentage of tungsten in these films has been determined as 71 ± 10 % [4]. At such quantity of tungsten $W_xSi_{(1-x)}$ coating which thickness is 500 nm goes over to superconducting state at the temperature ~5 K [4]. Precise components' ratio for achievement of higher critical temperature can be found only by means of empirical path. Wherein required coating thickness depends on SSPDs operation features and shouldn't exceed 10 nm.

To achieve the high output parameters at the technology development the following requirements for the vacuum coating system appear. This requirements are the result of the features of $W_xSi_{(1-x)}$ thin films forming and they are:

– necessity to vary components ratio to find the best ratio value;

– possibility to change of process parameters (power source capacity and distance to technological source) for obtaining of controlled and reproducible coatings deposition rates;

– ultimate pressure not more than $3 \cdot 10^{-5}$ mbar to ensure the suitable purity of residual environment in vacuum chamber;

– accuracy of the beginning and ending of the deposition process because films' thickness is several nanometers.

3. Purpose and tasks

The purpose of this work is realization of ultrathin superconducting $W_xSi_{(1-x)}$ coating forming process on VUP-11M vacuum coating system [5] with reproducible results. VUP-11M vacuum coating system was developed in the "Elion processes and nanotechnologies" laboratory of "Electronic technologies in machine building" department of the Bauman Moscow State Technical University. The basic version of this system include 4" planar cylindrical magnetron sputtering source, arc evaporator and grid ion source for surface ion treatment. The tasks within the framework of this theme are:

1) to realize $W_xSi_{(1-x)}$ thin films deposition on VUP-11M system by magnetron sputtering;
2) to modernize the vacuum chamber with heat and cooling system;
3) to design and produce special shutters for the reproducibility improving of the coating forming process.

4. Modernization of the technological system

Deposition of the two-component thin films by magnetron sputtering can be realized with using of two ways. The first one is sputtering of the two-component alloy target or a mosaic target (figure 1(a)) using just one sputtering source. The second one is to sputter homogeneous targets using two separate sputtering sources located at the same plane or at an angle to each other (figure 1(b)).

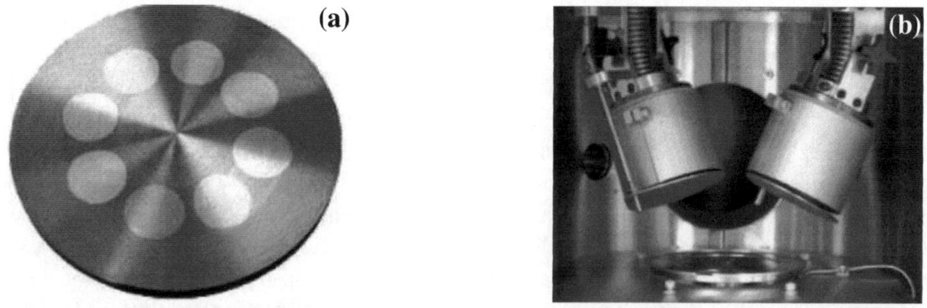

Figure 1. Two-component coatings deposition: **(a)** – using alloy or mosaic target and single magnetron source; **(b)** – using two magnetron source and homogeneous targets.

Using of the alloy or composite targets doesn't allow controlling components ratio in the coating at working out of its forming technology. Besides in process of magnetron operation for a long time the components composition in the erosion zone and on the substrate surface become different [6]. Using of two magnetron sources located at the same plane doesn't allow reaching the high coating uniformity both in thickness and components content in the film. Therefore it is expediently to realize the coating forming by means of simultaneous magnetron sputtering from two sources located at an angle to each other. This will allow to control the components ratio with maximize precision.

5. Installation of the additional magnetron sputtering source at the coating system

Initially vacuum coating system VUP-11M was equipped with one 100 mm size magnetron sputtering source installed at the left flange of the chamber. To realize deposition of WSi additional 50 mm size sputtering source need to be installed. The only possible place was at the top flange of the vacuum chamber (figure 2). For this purpose mechanical treatment of the flange has been carried out and special connector has been produced. This connector was designed so that it doesn't require any additional adaptors for magnetron fixing. Sealing of the magnetron was provided between connector and magnetron's tube so that allow to vary the distance between substrate and magnetron in the wide range from 75 to 135 mm and therefore to vary the deposition rate.

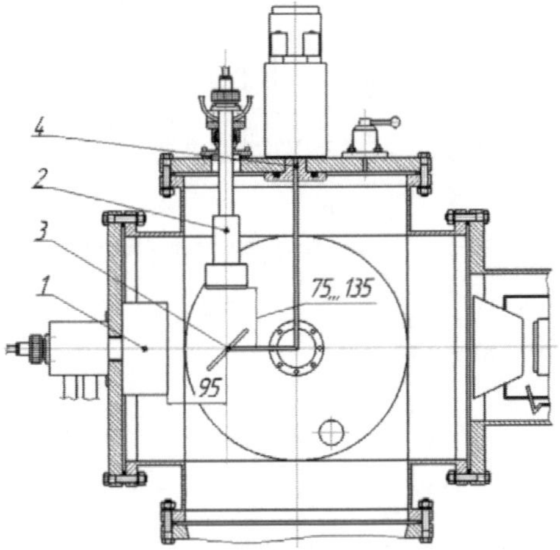

Figure 2. VUP-11M vacuum chamber after modernization: *1* – existing magnetron (Ø100 mm); *2* – installed additional magnetron (Ø50 mm); *3* – substrate holder; *4* – rotation input.

As a result of this modernization it became possible to obtain $W_xSi_{(1-x)}$ coatings with total deposition rate around ~0.47 nm/s [7]. Such deposition rate allows getting reproducible thin films which thickness of several nanometers. Magnetron sources are settled at the angle 90° between them. Silicon target is set up on Ø100 mm magnetron and the tungsten target is set up on Ø50 mm magnetron. Substrate is located on the substrate holder at the angle 45° to both of magnetrons targets (figure 2).

6. Modernization of heating and cooling system of the vacuum chamber

Heating and cooling system of the vacuum chamber has been modernized for improvement of residual vacuum. In addition to banded heater the flowing water heater has been installed. This heater provides water heating in copper tubes which are soldered to the vacuum chamber on all sides. Water circulation is carried out by water pump in a closed loop. Flowing heater allows to heat the vacuum chamber up to 60°C in a short time due to of tight contact between copper tubes and chamber walls. The banded heater provides temperature of the chamber walls at the level about 70–80°C. Cooling is realized at the same copper tubes by cool tap water supply. Chamber heating and cooling processes allow to obtain less pressure of residual gasses. This is confirmed by mass-spectrometry research (figure 3) [8].

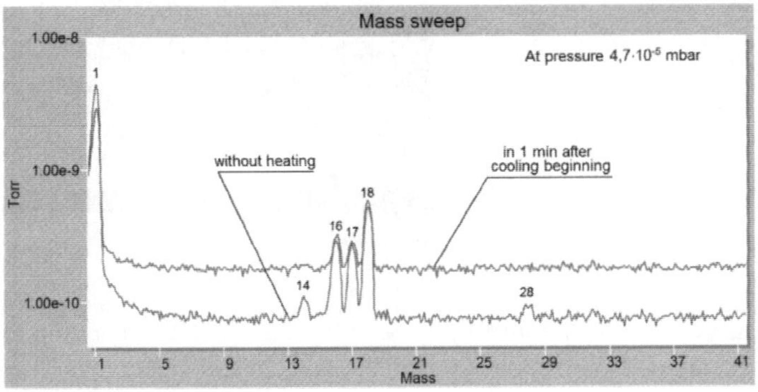

Figure 3. Meters data of mass-spectrometer.

7. Design of technological fixture

To ensure the reproducibility of $W_xSi_{(1-x)}$ thin films deposition suitable fixture has been designed. This fixture prevents deposition at the back side of substrate while operation of targets in idle mode and provides accuracy beginning and ending of the deposition process. The fixture is a system of shutters between both magnetrons and substrate holder. There are two ways for this shutters realizing. The first construction is stationary substrate holder with movable shutters between magnetrons and substrate and between two magnetrons (figure 4(a)); the second one is movable substrate holder with movable shutter between two magnetrons and stationary shutter between magnetrons and substrate holder (figure 4(b)).

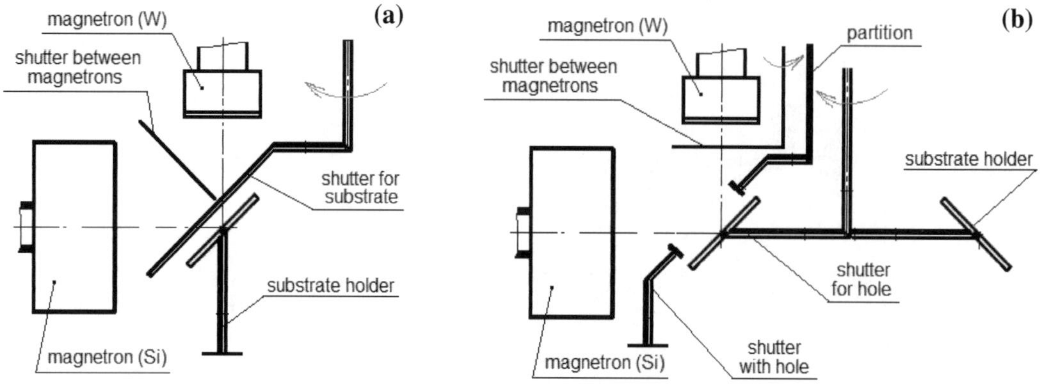

Figure 4. Variants of shutter fulfilment: **(a)** – with stationary substrate holder and mobile shutter; **(b)** – with mobile substrate holder and stationary shutter.

The first variant is easy in design and production but doesn't allow turning the substrate holder to others radially located technological sources, e.g. to ion source for ion pretreatment in the single vacuum cycle. Therefore the second variant of fixture construction has been realized (figure 5).

While pumping the substrate is located at maximal distance from bypass pump line and during targets training the substrate is out of plasma area. Wherein coating deposition process begins by substrate holder turning and placement of sample in sputtering zone. Process stopping occurs by the same way: by turning of substrate holder out from the sputtering zone. Because of this there is no necessity to track simultaneous magnetrons switching-off. Forming of last coating layers with another components ratio is also excluded.

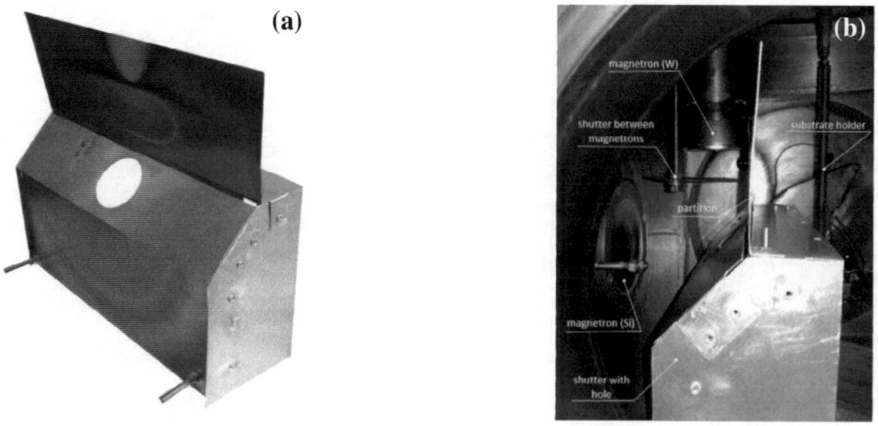

Figure 5. Shutter: **(a)** – appearance; **(b)** – inside VUP-11M system.

Due to dividing the space inside vacuum chamber into two zones there is no film deposition at the back side of the substrate during the targets training because of the hole in the shutter is closed by special plate. The cross-sputtering during targets training is excluded due to shutter between magnetron sources and the cross-sputtering during deposition is insignificantly enough due to very small time of it (~20 s).

8. Modernization results

1. Installation of the second magnetron source on the top flange of the chamber allows obtaining superconducting coatings by magnetron sputtering from two sources simultaneously at the VUP-11M coating system. There is opportunity to vary components ratio while the process. Wherein deposition rate is ~0.47 nm/s that allows to get thin films which thickness is in the range from 5 to 9 nm.

2. Improvement of the heating and cooling system provides decrease of ultimate pressure and obtaining more cleaner residual vacuum for coatings process. It has been confirmed by mass-spectrometry research.

3. Designed and produced system of shutters provides reproducible output coatings parameters what is confirmed by measurements of output parameters of $W_xSi_{(1-x)}$ thin films [9] and SSPD based on this films [10].

9. Conclusion

Modernization of VUP-11M coating system has allowed to realize reproducible forming technology of ultrathin superconducting WSi coatings with critical temperature about 3.7 K. Produced on the base of this films detectors shows quantum efficiency on the level of 65 %. Moreover opportunity to vary components ratio and deposition rates allows carrying out further researches.

References

[1] Minaeva O V et al. 2013 *High-speed operating single photon detector on the base of superconducting NbN thin film* (Moscow: Prometey)
[2] *Superconducting Single Photon Detecting System (SSPD)* (URL: http://www.scontel.ru/sspd)
[3] Zotova A N and Vodolazov D Yu *Single photon detectors of the electromagnetic radiation: operating principles and capable applications* (URL: http://pandia.ru/text/78/361/1640.php)
[4] Kondo S 1992 *J. Mater. Res.* **4** 853–60
[5] Moiseev K M 2014 *Nanoengineering* **8** 18–21
[6] Kalinin V N and Popova M G 2015 *Youth scientific and technical herald* **1**
[7] Vasiliev D D, Malevannaya E I and Moiseev K M 2015 *«Vacuum technic, materials and technology». Articles digest of the X International STC (Moscow)* pp 146–50
[8] Vasiliev D D, Malevannaya E I and Moiseev K M 2015 *Articles digest of the XX International scientific and technical conference «High technologies in Russian industry» (Moscow)* pp 208–12
[9] Vasiliev D D, Malevannaya E I, Moiseev K M, Seleznev V A, Divochiy A V, Vakhtomin Yu B, Morozov P V, Zolotov P I and Smirnov K V 2016 *«Vacuum science and technics». Articles digest of the XXIII STC with participation of foreign specialists (Moscow)* pp 157–62
[10] Seleznev V A, Divochiy A V, Vakhtomin Yu B, Morozov P V, Zolotov P I, Vasil'ev D D, Moiseev K M, Malevannaya E I and Smirnov K V 2016 *Journal of Physics: Conference Series* **737** 012032

IOP Conf. Series: Journal of Physics: Conf. Series **872** (2017) 012028 doi:10.1088/1742-6596/872/1/012028

Calculation of the distribution of film thickness on a rotating disk substrate in a sputtering system with long plate targets

O D Volpian[1], I M Drozd[2], A I Kuzmichev[2] and Yu A Obod[3]

[1] "Fotron-Auto" Scientific Manufacture Enterprise, 117105, Moscow, Russia
[2] Igor Sikorsky Kyiv Polytechnic Institute, 03056, Kyiv, Ukraine
[3] Moscow Technological University (MIREA), 119454, Moscow, Russia

E-mail: kuzmichev-kpi@ukr.net

Abstract. The results of calculation of the film thickness distribution of sputtered material deposited on a horizontal rotating disk substrate in a multi-cathode sputtering system with horizontal and inclined long plate targets are presented. The possibility of obtaining the high uniformity in the thickness distribution is shown, which makes it possible to recommend the similar system for use in precision optical technology.

1. Introduction

Ion sputtering systems are widely used for deposition of thin-film coatings, including coatings for optics. Nowadays, optical films are often deposited by the bipolar pulsed medium-frequency (10–400 kHz) sputtering of several cathode targets in the reactive gas medium that eliminates arcing on the targets and the "vanishing anode" effect [1, 2]. The use of several targets helps also to obtain practically uniform distribution of film thickness on stationary (non-moving) substrates [2], but even a small thickness non-uniformity (~ several percents) is unacceptable for precision optics. To solve the problem, one manages motion of the substrates in the sputtered substance flow, and it is desirable to move them along the simplest trajectory. In this paper, a sputtering system is analyzed with rotation of a disk substrate above long plate targets. Such a configuration is convenient for combining several systems within one installation for the purpose of depositing many layers from different materials. Sputtered surfaces in the system are formed by flat targets: horizontal ones, parallel to the surface of the substrate, and ones, inclined towards the substrate. The number of sputtered targets is not limited. They are arranged symmetrically with respect to the horizontal axis.

2. Calculation of mass transfer from targets to the deposition surface

Calculation of the mass transfer of the sputtered material is based on a number of generally accepted assumptions: the cosine distribution of sputtered atoms over the emission angles (φ), their collisionless motion in the gas, 100 % probability of condensation of particles depositing on the substrate, neglect of their diffusion over the substrate surface, the density of the deposited material is equal to the sputtered material density. We also assume that the targets are made of the same material, their length is unlimited, the sputtering rate of all parts of the target surface is the same. The substrate rotates about its vertical axis; the axis of rotation is perpendicular to the horizontal axis of symmetry of the sputtering system, both axes being in the same plane. At last, we consider the case with substrate rotation frequency much less than pulse rate of substrate power supply (that is the modulation of target voltage is not taken into account). The initial expressions for the calculation of the transfer of the sputtered material from the target surface element to the element of the deposition surface under these conditions can be taken from [3]. Then the thickness of the deposited film is determined by summing at a given point in the deposition surface the partial fluxes of particles coming from all points of the sputtered surfaces.

Figure 1(a) is a diagram for calculating mass transfer from the surface of horizontal targets. Here, x is the horizontal coordinate; dS_{sp} and dS_d are the elements of the sputtered surface (target) and deposition (substrate), respectively; *D-O-A-B* is the deposition surface (substrate plane); the vertical line passing through point *O* is the axis of substrate rotation; φ and Θ are the angles of emission and incidence of the sputtered particles, respectively; h is the distance between the target and the deposition surface; k is the distance between the sputtered surface element and the vertical axis; the long side of the targets is perpendicular to the plane of the figure.

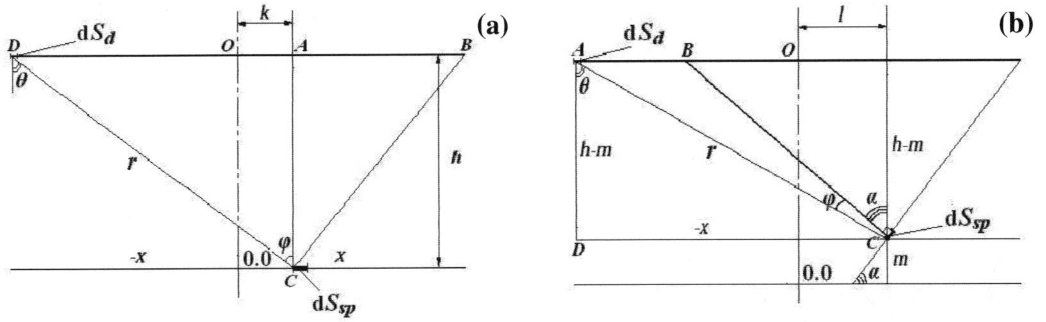

Figure 1. Diagrams for calculating mass transfer from the surface element of the horizontal target **(a)** and the inclined target **(b)** to the deposition surface.

Taking into account that $\varphi = \Theta$, we obtain an expression for the thickness of the material layer transferred from the target surface element (point *C*) to the element of the deposition surface (point *B* or *D*), where M_1 and ρ are the mass and density of the substance sputtered from the unit target length in the direction perpendicular to the plane of the figure. Then, to obtain the thickness due to the sputtering of the entire horizontal target, the resulting expression is integrated with respect to the variable $x(k)$ within limits of the transverse dimension of the target:

$$d(x)=\int_{k_1}^{k_2}\frac{M_1 dk}{\rho\pi h^2\left[1+\left(\frac{k-x}{h}\right)^2\right]^2}.$$

Figure 1(b) is a diagram for calculating mass transfer from the surface of inclined targets. Here α is the angle of target inclination; l is the distance between the sputtered surface element and the vertical axis; m is the height of the sputtered surface element above the horizontal target; the remaining symbols and parameters are similar to ones in figure 1(a). Expressing the distance r through the parameters h, l and m, as well as the angles α, φ, Θ, we obtain an expression

$$M_2'(x)=\frac{M_2}{\pi\left[(h-m)^2+(l-x)^2\right]}\cos\varphi\cos(\Theta)$$

for the mass of material deposited per unit area at different points of the deposition surface, when the mass M_2 is sputtered from the element of the inclined target with the coordinates l and m. To obtain the thickness of the deposited material by sputtering the surface of the entire inclined target, the latter expression is integrated with allowance for the relationship between the angles α, φ, θ and the coordinates of the target edges. To obtain the full thickness of the deposited material d_Σ, due to sputtering of all targets, the thickness values obtained by sputtering individual targets are summed taking into account the intensity of their sputtering (ion current density on them). In the future, we will analyze the profile of the relative thickness of the film D normalized to the maximum value of d_0, i.e. $D = d_\Sigma/d_0$.

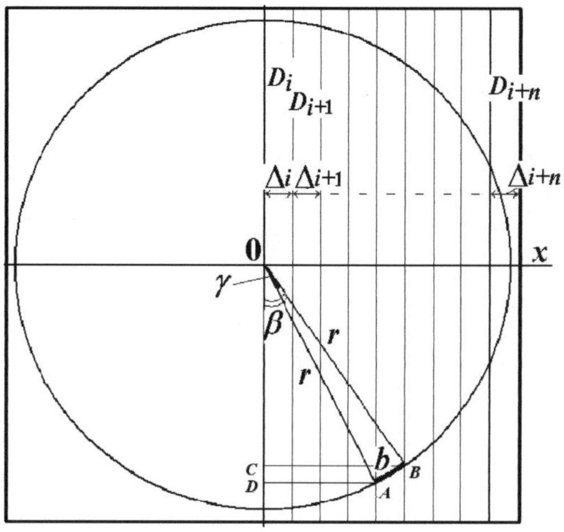

Figure 2. Scheme of the calculation of mass transfer to a rotating substrate.

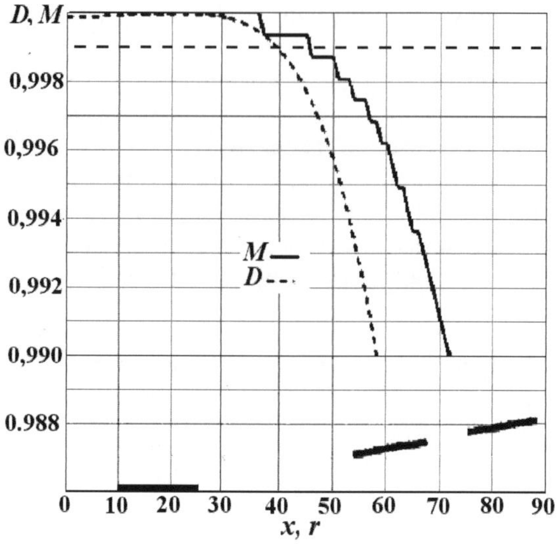

Figure 3. Film thickness distribution on the rotating (M) and non-rotating (D) substrates; h = 100.

3. Calculation of the thickness of the film on the rotating substrate

The above approach is suitable for calculating the film thickness on a non-rotating substrate. It is also used to calculate the thickness of the film on a rotating substrate, complemented by integration over the rotation angle. To this end, we divide the deposition plane along which the substrate moves on n narrow bands along the system axis of width Δ ($\Delta = \Delta_i = \Delta_{i+1}$), as in figure 2. Each band is considered as a source of deposited material with intensity proportional to the value D found above for the non-rotating substrate. Hence, the partial volume of the deposited material when a portion of the substrate goes over one of the bands, for example, over the band with number (i + 4), is proportional to $M_p(r)_{i+4} = D_{i+4}b$, where $b = r\gamma$ is the arc length limited by the band within the angle γ. Here we take into account the path length b of the substrate portion above the band (deposition zone) of width Δ, since the time of finding the portion of the substrate above this band depends on b and, in the final analysis, on the radius. Hence we obtain the expression $M_p(r)_{i+4} = D_{i+4}r\left(\arcsin\frac{5\Delta}{r} - \arcsin\frac{4\Delta}{r}\right)$. To determine the relative rate of depositing the partial volume of the material, we divide the latter expression by the linear velocity of the substrate points going over this band, where $v = r\omega$ is the angular velocity, and we get: $M_v(r)_{i+4} = \frac{D_{i+4}}{\omega}\left(\arcsin\frac{5\Delta}{r} - \arcsin\frac{4\Delta}{r}\right)$. Summarising the deposition of the partial layers upon passage over all the bands and taking into account the same angular velocity of the substrate rotation for all radii, we obtain the final expression for the relative thickness of the deposited film on the rotating horizontal flat substrate:

$$M_\Sigma(r) = \sum_{i=1}^{i=n} D_i\left(\arcsin\frac{i\Delta}{r} - \arcsin\frac{(i+1)\Delta}{r}\right).$$

Here for each radius there will be an own value of n. In what follows we shall use the values of M obtained by normalizing M_Σ to the maximum values.

4. Results of the calculation and discussion

Figure 3 shows the results of one of the calculations for the non-rotating and rotating substrates. The thick lines indicate the position and size of the sputtered surfaces of the targets. The horizontal dashed line shows the height of the substrate location above the targets; the geometric scale along the vertical axis corresponds to the horizontal scale. The calculations show that the uniformity of the thickness distribution improves with increasing width of the horizontal targets, an increase in the height of the location of the substrates above the targets, the correct selection of the ratio of the intensity of sputtering of individual targets. It is quite possible to obtain an inhomogeneity of thickness substantially less than 1 %.

5. Conclusions

Thus, the obtained results indicate the possibility of achieving a high uniform thickness of the coating on the substrates in the examined sputtering system with horizontal and inclined targets. This allows us to recommend this sputtering system to the application in the technology of precision optical films, in particular, in the manufacture of filters, mirrors and other elements from optical metamaterials [1]. The proposed sputtering system configuration allows processing a large number of substrates arranged in the row along the targets. The combination from several systems with targets from different materials allows producing multilayer coatings and coatings with a refractive index gradient [1]. The methodology of the presented calculation is used in the development of the design of new facilities for the coating of optical products.

Acknowledgments

The work was financially supported by a grant from the Ministry of Education and Science of the Russian Federation (Scientific project "Optical transistors based on metamaterials". Agreement on granting a subsidy from the Ministry of Education and Science of the Russian Federation dated September 29, 2016, No. 14.577.21.0219, Unique identifier of PNIER RFMEFI57716X0219).

References

[1] Volpian O D, Kuzmichev A I, Ermakov G F, Krikunov A I, Obod Yu A, Silin N V and Shkatula S V 2015 *J. Phys. Conf. Ser.* **652** 012009

[2] Frach P, Goedicke K, Winkler T, Gottfried, Walde H and Hentsch W 1995 *Surf. Coat. Technol.* **74-75** 85–91

[3] Glang R 1970 *Handbook of Thin Film Technology* vol 1, ed. L I Maissel and R Glang (New York: McGraw Hill) pp 77–82

Structure and mechanical properties of nanocomposite Ni-ZrO_2 films

M S Filatov[1], O V Stognei[1] and M S Antonova[2]

[1] Voronezh State Technical University, Voronezh, Russia
[2] Voronezh State University, Voronezh, Russia

E-mail: 36nord36@mail.ru

Abstract. The composite Ni-ZrO_2 films were produced by simultaneous reactive sputtering of two metal targets (Ni and ZrY). The phase ratio in the composites varies due to the change of the sputtering magnetron power. Depending on the partial pressure of oxygen one can obtain Ni-ZrO_2 composites (P_{Ar} = 1.05 Pa, P_{O2} = 0.01 Pa) or NiO-ZrO_2 (P_{Ar} = 1.05 Pa, P_{O2} = 0.04 Pa). The mechanical properties of the composites were studied using a nanoindentation system: the concentration dependences of hardness, Young's modulus and fraction of the elastic component in the total deformation of the film were established. It is established that all mechanical parameters have a higher value in the NiO-ZrO_2 composites compare with the Ni-ZrO_2 system.

1. Introduction

Zirconium dioxide (ZrO_2) is one of the most widely used ceramic materials for thermoprotective barrier and hardening coatings formation. Improvement of the zirconia properties is possible if structure of the material is transferred to a nanostructured state. It is assumed that a decrease of the grain size and an increase of the density of intergrain boundaries, respectively, allows to reduce the thermal conductivity of the dioxide coating, thereby improve its performance characteristics [1]. Another option to improve the mechanical properties and adhesion strength of zirconia-based coatings can be the creation of nanostructured composites metal-zirconia. The possibility of obtaining of the nanostructured composite material based on nickel and zirconium dioxide by magnetron sputtering of metal targets was investigated in the present work.

2. Results of the study

Samples of nanostructured composite Ni-ZrO_2 films with different ratio of metal and dielectric phases were obtained by high-frequency magnetron reactive sputtering. The formation of the composites was carried out by two magnetrons sputtering in the reactive atmosphere (Ar + O_2) two metal targets (Ni and $Zr_{92}Y_8$) simultaneously. The change of the metal concentration in the composites was made by change on the power of the magnetron sputtering the nickel target, while the power of the magnetron sputtering the alloy target ($Zr_{92}Y_8$) remained constant. The structure of the obtained films was studied using a D2 PHASER X-ray diffractometer from BRUKER. The chemical composition was determined by electron-probe X-ray spectral microanalysis on a X-ray scanning JXA-840 microanalyzer. Two series of deposition process with different oxygen partial pressure were carried out. All the obtained composites were 1–1.5 μm thick and were deposited on a glass substrates.

Figure 1 shows the X-ray diffraction patterns of NiO-ZrO_2 composites obtained with high oxygen pressure of 0.04 Pa. The argon operating pressure was 1.05 Pa.

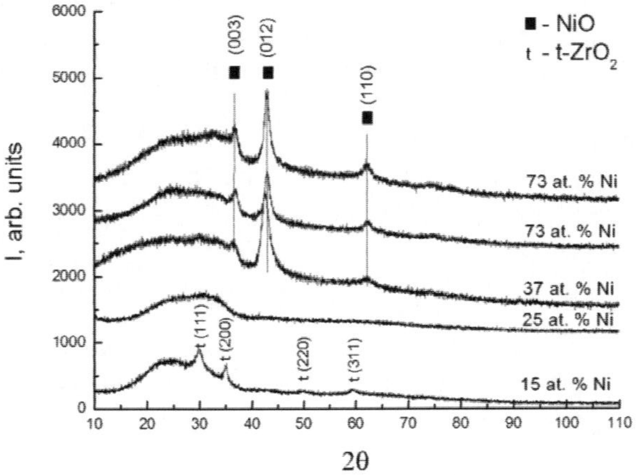

Figure 1. X-ray diffraction of NiO-ZrO_2 films obtained in the regime with high oxygen pressure (0.04 Pa). Ni concentration is shown in at. %.

The only peaks from the tetragonal zirconia as well as the halo from the glass substrate (in the angular interval 20–30 deg) are present on the X-ray diffraction pattern from the composite with low nickel concentration (15 at. %). With increasing of the nickel concentration (25 at. % Ni) the diffraction peaks disappear and the film becomes X-ray amorphous. Apparently, this is a consequence of a significant reduction of the zirconia grains size caused by the presence of nickel, which begins to form a separate phase and prevents the growth of the dioxide crystals. A further increase of the nickel concentration (the range of 37–73 at. % Ni) leads to the appearance of peaks corresponding to nickel oxide. It is characteristic that diffraction peaks from zirconium dioxide in samples with high nickel content are not observed but the halo in the angular interval from 30 to 35 deg is preserved. This indicates the presence of a fine-grained zirconia phase since the halo is in the angles range where the main diffraction peaks of the dioxide are located. The presence of the nickel oxide phase instead of expected pure nickel indicates that the concentration of the reactive gas during the composites deposition is an excessive. In fact the composite NiO-ZrO_2 samples were obtained as a result of the first series of deposition.

The grains size of crystalline phases in the obtained samples were determined using a specialized licensed software package “DifracEva” and a Scherrer method. The average size of nickel oxide crystals was 8.5 nm which indicates the formation of nanostructured morphology in deposited samples.

Mechanical properties of the obtained composites were studied by using of the NanoHardnessTester nanoindentation system from CSM Instruments. The operation principle of the used technique is based on indenting of a geometrically certified indentor under the action of a specified profile of the normal force $P(t)$ and simultaneously recording the depth of its immersion into the material $h(t)$ [2]. This allows to determine several parameters characterizing the mechanical properties of the material. Figure 2 shows the dependences of hardness, Young's modulus (E) and the elastic deformation component of NiO-ZrO_2 composites on the metal concentration. With increasing of the nickel concentration (actually nickel oxide) the measured parameters do not change monotonically. For the hardness and elastic component of deformation a maximum is observed in the region of 25 at. % Ni (figure 2). The maximum values of the measured characteristics correspond to a composite with a “transition” (X-ray amorphous) structure. Apparently, this is a consequence of the finely dispersed structure of this composite. Since the initial partial pressure of oxygen used in the first series of the composites deposition was excessive (the metal phase was oxidized) the second series of depositions was carried out at a lower oxygen pressure (0.01 Pa). The results of X-ray diffraction analysis of the obtained films are presented in figure 3.

Despite the significantly lower partial pressure of oxygen (0.01 Pa) the evolution of the composites structure with an increase of the metal concentration follows the same pattern as in the composites of the first series of sputtering. The only peaks belonging to the high-temperature tetragonal modification of zirconia are presented on the X-ray diffractogram obtained from the composite with low nickel concentration (15 at. %). Increase of the nickel concentration to 25 at. % leads to the disappearance of peaks but halo in the region of 25–35 deg was appeared. Moreover, this halo is also observed in diffractograms from composites with a higher nickel content. This may indicates the presence of the finely dispersed zirconia phase in all obtained samples. The diffracion peaks from crystalline nickel are observed on the diffractograms obtained from composites with a higher nickel content (25–74 at. %). Thus, a decrease of the oxygen partial pressure during the sputtering of two metal targets leads to formation of composite Ni-ZrO_2 structure. The average size of nickel granules in these composites was 6 nm according to Scherrer method.

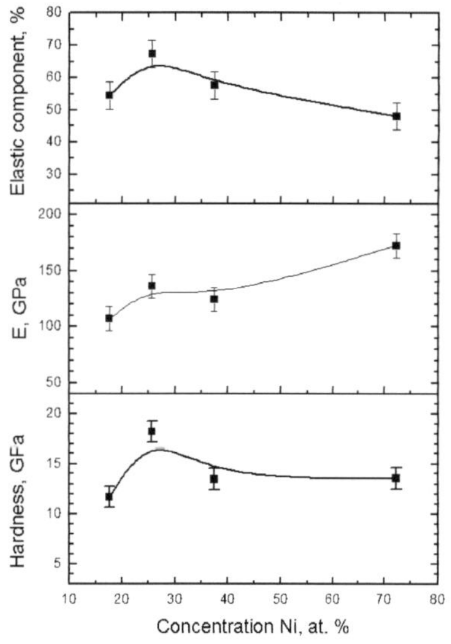

Figure 2. Concentration dependences of mechanical parameters of NiO-ZrO_2 films obtained in the regime with high oxygen pressure (0.04 Pa).

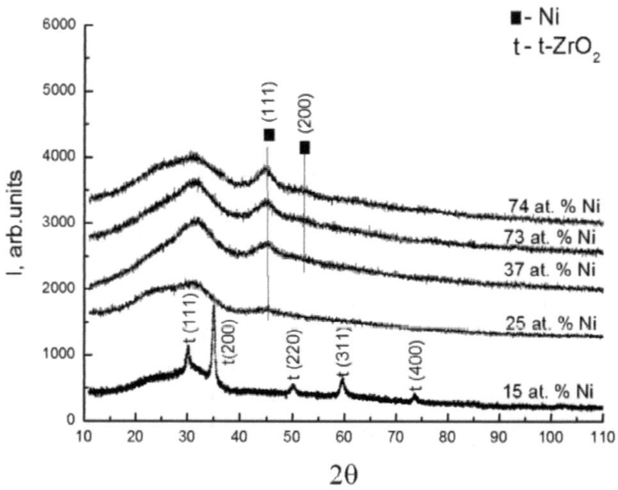

Figure 3. X-ray diffraction patterns of Ni-ZrO_2 films obtained in a regime with a low oxygen pressure (0.01 Pa). Ni concentration is shown in at. %.

A comparison of the mechanical properties of two studied type of composites indicates that the absolute values of Young's modulus and the hardness of oxide-oxide composites (NiO-ZrO_2) are higher than those of the metal-oxide composites (Ni-ZrO_2). This difference is particularly evident when the hardness values of the composites are being compared.

The hardness does not exceed 11.4 GPa in the Ni-ZrO_2 composites while the minimum value in the NiO-ZrO_2 composites is 11.7 and the maximum value reaches 18.2 GPa. Despite the fact that the properties of pure phases (Ni and ZrO_2) differ substantially from each other the concentration dependences of the mechanical parameters in Ni-ZrO_2 composite are weakly expressed (figure 4). The measured values of the parameters of $Ni_{15}(ZrO_2)_{85}$ and $Ni_{74}(ZrO_2)_{26}$ composites are slightly differ from each other.

3. Conclusions

Simultaneous reactive sputtering of nickel and $Zr_{92}Y_8$ alloy by high-frequency magnetrons in mixed atmosphere (Ar + O_2) allows to obtain composite films. Depending on the partial oxygen pressure a composite consisting of two oxide phases (NiO-ZrO_2) or a metal-insulator composite (Ni-ZrO_2) can be formed. The change of the power of one magnetron which sputter the nickel target allows to vary a phase ratio in very wide region in the composites forming during deposition process. It has been found that the hardness, the Young's modulus and the fraction of the elastic component in the total deformation have higher values in the NiO-ZrO_2 composites compare to the Ni-ZrO_2.

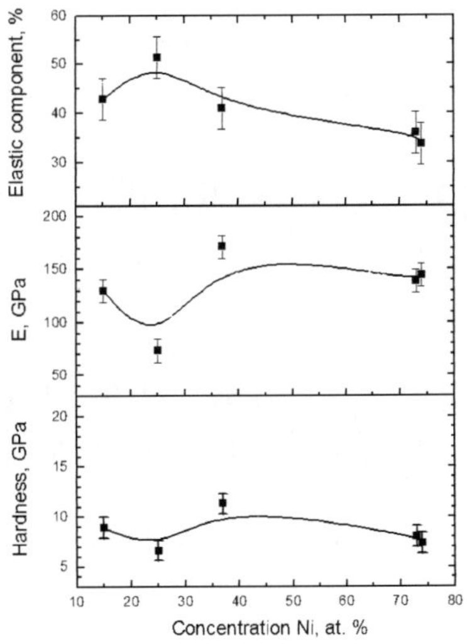

Figure 4. Concentration dependencies of mechanical parameters of Ni-ZrO_2 composites obtained in the regime with reduced oxygen pressure (0.01 Pa).

Acknowledgments
The work was supported by the Russian Foundation for Basic Research (grant No. 16-42-360778 p_a).

References

[1] Kabalov E N and Muboyadjian S A 2012 *Aviation materials and technologies* 3–20
[2] Golovin Y I 2008 *Solid State Physics* **12** 2113–42

Dependencies of photoelectric properties of SiC/Si structures grown by the method of atoms substitution on synthesis time

A S Grashchenko[1], **S A Kukushkin**[1], **A V Osipov**[1] **and N A Feoktistov**[1,2]

[1] Institute of Problems of Mechanical Engineering, Russian Academy of Sciences, 199178, Saint Petersburg, Russia
[2] Ioffe Physicotechnical Institute, Russian Academy of Sciences, 194021, Saint Petersburg, Russia

E-mail: asgrashchenko@bk.ru

Abstract. This paper is dedicated to an exploration of the photoelectric properties of Si–SiC structures grown by the substitution method on silicon substrates of (001) orientation. For the samples with the synthesis times of 40, 60, 90, 120 and 900 s, magnitudes of the saturation currents are determined and the coefficients of efficiency are calculated. The obtained dependencies of the photoelectric characteristics on the synthesis time are explained using the theory of formation of dilatation dipoles during the synthesis by the method of atoms substitution.

1. Introduction

Interest in the study of the photoelectric properties of SiC films on Si at the present time is steadily increasing [1], although these studies are at the initial stage. This is due to the fact that it is possible to obtain photosensitive structures operating at elevated temperatures and in conditions of high radiation. However the production of such photosensitive structures is currently hindered by a lack of high-quality epitaxial layers of SiC on Si. This is due to a large difference between the lattice parameters of Si and SiC, primarily. In the works [2–6] authors have developed a new method for the synthesis of nanoscale SiC films on Si. This method is based on the discovery of a new mechanism of relaxation of elastic stress during growth of epitaxial films by pre-embedding the ensemble of nanoobjects – dilatation dipoles in the lattice of the substrate. Dilatation dipole is a stable complex consisting of the attractive centers of dilatation of the carbon atoms in the interstitial positions of silicon atoms and silicon vacancies. In [7] photoelectric properties of heterostructures − substrate Si nanoscale film of silicon carbide (Si/SiC), grown by the method of substitution atoms were investigated. In this article we studied the structure of Si/SiC synthesized on *n*-Si (001) of KEF-4.5 type and structures grown on *p*-Si (111) substrates of KDB-10 type. The authors have shown that the most effective structure of heterojunctions of Si/SiC is formed on the substrates of *n*-type conductivity type KEF-4.5 (001) orientation.

The aim of this work is to study the dependency of photoelectric characteristics of SiC films on Si grown by the method of substitution atoms on the time of synthesis.

2. Materials and methods

In this work the photoelectric properties of the heterostructures Si/SiC are investigated. These heterostructures were grown by the method of substitution of silicon atoms on the carbon atoms inside the silicon substrate [2–6]. The films were synthesized on *n*-Si (001) substrates of KEF-4.5 type in atmosphere of carbon monoxide (CO) and silicon tetrahydride (SiH_4) at a temperature 1250 °C and at a pressure 0.5–0.7 Torr. Measurements of the photoelectric characteristics was carried out on five samples of Si/SiC grown at different synthesis times of, namely at 40, 60, 90, 120 and 900 s. To measure the photo-properties, we deposited round semitransparent nickel (Ni) contacts onto surface of the SiC films with a radius of 1 mm by magnetron sputtering through masks under heating to 200 °C. The transparency of the contacts, monitored relative to a glass reference sample, was found to be 30 %. The contact to Si was formed from conducting silver paste, which also served as a contact to the substrate holder. Thus, the closure of the contacts formed a closed electrical circuit (figure 1).

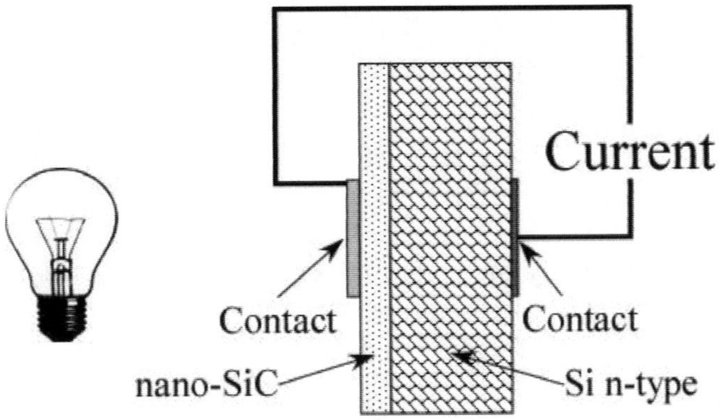

Figure 1. The circuit of the electronic circuit through the diode structure Si/SiC.

To study the photocharacteristics of heterostructures Si/SiC special electro-optic installation was constructed. The installation includes an OSRAM XBO 150W/4 xenon lamp emitting light in the spectral range 185–3500 nm, a quartz lens with a diameter of 50 mm and a focal distance of 100 mm, an iris diaphragm, V7-21A voltmeter and voltage source. With the help of this installation a light spot with a diameter of 2 mm was focused on the semitransparent contact formed on the surface of the SiC film. The intensity of the light was set by the iris diaphragm and controlled by a photodiode placed near the sample. The current–voltage (I–V) and load I–V characteristics of the samples were measured by a voltmeter and voltage source. Load I–V were determined at power of the light 100 mW/cm^2. All the electrical and optical characteristics were measured at room temperature T = 293 K. More details on this electro-optic installation and the studies on it are described in [7].

3. Results and discussion

The graph (figure 2(a)) shows I–V characteristics of heterojunctions SiC/Si grown by the method of atoms substitution with the times of synthesis 60, 90 and 900 s. It follows from the figure 2(a) that I–V characteristics of the investigated structures are of the same nature as of the diode. Note that I–V characteristics of the other samples had a similar appearance. The biggest difference of the forward and reverse currents was observed for the sample that was synthesized in 60 s. The initial regions of the forward I–V characteristics at low voltages are described by the dependence (1) with factor of not ideality $n = 4$:

$$j = j_s \exp\left(qV/nkT\right), \tag{1}$$

here, j is the current; j_s is the saturation current; q is the elementary charge; V is the voltage; k is the Boltzmann constant and T is the temperature. Figure 2(b) shows the dependence of the voltage-saturation current passing through an area of 1 cm^2 on the time of synthesis of SiC films on substrates Si. From this dependence it follows that the maximum of the voltage-saturation current was observed for structures that have been synthesized in 60–90 s. Hence, most efficient in terms of photovoltaic conversion structure is formed approximately 1 min.

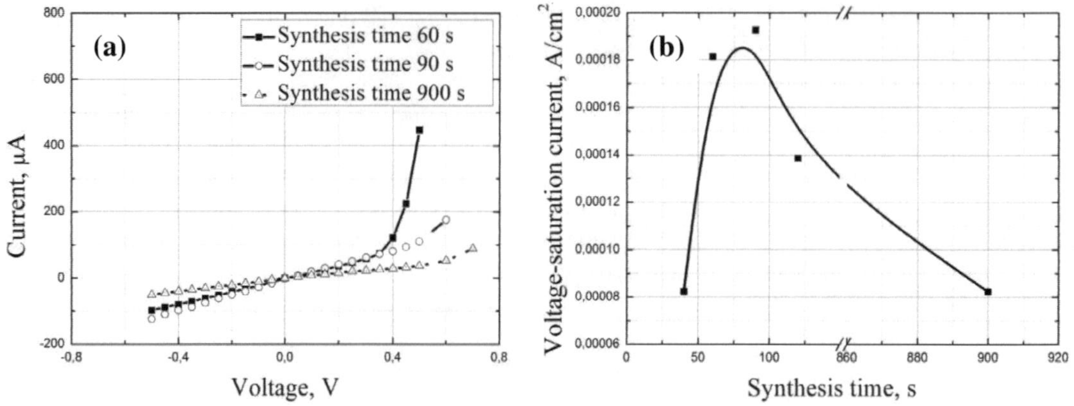

Figure 2. I–V characteristics of the heterostructures Si/SiC **(a)**; dependence of the saturation current on the time of synthesis of heterostructures Si/SiC **(b)**.

The graph (figure 3(a)) shows the load I–V characteristics of heterojunctions SiC/Si grown by the method of atoms substitution with the times of synthesis 60, 90 and 900 s. All the load dependencies of the structures of Si/SiC demonstrate the characteristic diode bend. According to the load curves values of the efficiency η, short-circuit current I_{SC}, open-circuit voltage U_{OC}, fill factor FF and maximum power output P_{max} of test specimen were calculated. The efficiency is determined by the formula (2)

$$\eta = FF \cdot I_{sc} U_{oc} / P, \tag{2}$$

here P – the radiation power incident on the structure Si/SiC. The dependence of the efficiency on the time of synthesis is depicted in figure 3(b). It should be emphasized that all the above values except for the fill factor are correlated with each other.

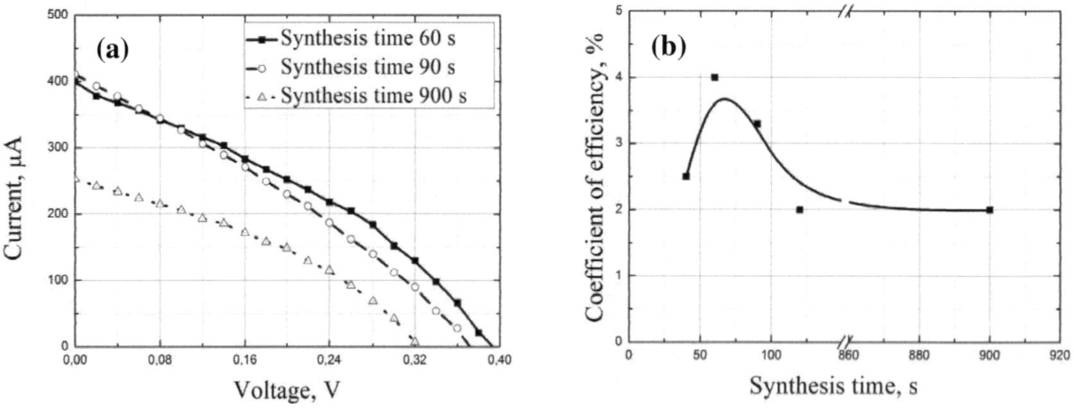

Figure 3. Load I–V characteristics of the heterostructures Si/SiC **(a)**; dependence of the coefficient of efficiency on the time of synthesis of heterostructures Si/SiC **(b)**.

Such dependencies of the photovoltaic characteristics on the time of synthesis are associated with features of the method of synthesis of structures of Si/SiC. According to the concept of the atomic substitution [4–6] at the initial stage of the growth within the matrix of a silicon substrate a part of the silicon atoms are replaced by carbon atoms. As a result, nano-objects, which the authors called dilatation dipoles are being formed. Dilatation dipoles represent the stable complexes of the two centers of dilatation, namely the carbon atom in interstitial position and the vacancy in silicon. It was discovered that the higher the concentration of dilatation dipoles, the higher conversion efficiency of sunlight with heterojunctions of Si/SiC. We established one-to-one correspondence between the time of synthesis SiC films, the maximum density of the dipoles and the efficiency of conversion of solar light into electrical energy.

4. Conclusions

We investigated photoelectric properties of heterostructures Si(001)/SiC grown by a new method of atoms substitution. The effect of strong dependence of saturation current and maximum efficiency of conversion of solar light heterojunction Si(001)/SiC on the time of synthesis of films and the concentration of dilatation dipoles is discovered. Time interval, which is the most effective for photopleasure structure of heterojunctions of Si/SiC is specified.

Acknowledgements

This work was supported by the RFBR (Grants №15-03-06155 and №16–29-03149-ofi). The experimental results were obtained using the unique scientific facility “Physics, chemistry and mechanics of crystals and thin films” (IPME RAS, St. Petersburg).

References

[1] Watanabe R, Tsukamoto T, Kamisako K and Suda Y 2017 *Journal of Crystal Growth* **463** 67
[2] Kukushkin S A, Osipov A V and Feoktistov N A 2008 *Patent of invention № 2363067*
[3] Kukushkin S A and Osipov A V 2008 *Physics of the Solid State* **7** 1238
[4] Kukushkin S A and Osipov A V 2013 *Journal of Applied Physics* **2** 024909
[5] Kukushkin S A and Osipov A V 2014 *Journal of Physics D: Applied Physics* **31** 313001
[6] Kukushkin S A, Osipov A V and Feoktistov N A 2014 *Physics of the Solid State* **8** 1507–35
[7] Grashchenko A S, Feoktistov N A, Osipov A V, Kalinina E V and Kukushkin S A 2017 *Semiconductors* **5** 621–7

IOP Conf. Series: Journal of Physics: Conf. Series **872** (2017) 012031 doi:10.1088/1742-6596/872/1/012031

The influence of weak impacts on certain processes of nanotechnology

V A Tupik, V I Margolin and M S Potekhin

Department of Radio Microelectronics and Radio Equipment Technology, Electrotechnical University "LETI", Saint Petersburg 197376, Russia

E-mail: VATupik@etu.ru

Abstract. The article deals with the influence of weak and super weak impacts on certain technological processes in nanotechnology related to the synthesis of nanoscale films and coatings. We also touch upon the impacts of weak diffraction fields of complex shape on the formation of fractal films and coatings.

1. Introduction

The phenomenon of the weak impact of physical factors on physical, chemical and biological systems has long been a subject of scientific interest. Weak and super weak impacts have been registered and proven in numerous aspects of modern technology, but have not been fully explained so far. Of special interest are impacts on the process of synthesis of weak diffraction field nanoscale surface coatings generated by diffraction grates of complex shape, consisting of curved closed circuits [1].

The phenomenon is often explained in the light of resonance theories: resonance surface waves, dissipating resonance, stochastic resonance and electromagnetic acoustic conversion [2]. Active media with internal energy sources can have nonlinear processes of wave interactions. An electromagnetic field can have a necessary distribution of such characteristics as phase and amplitude on surface of a substrate due to the phenomenon of quantum teleportation, which will contribute to the creation of conditions for microcrystals formation and stimulate nanostructure synthesis on the molecular level [3].

2. The influence of weak impacts of physical fields on various physical objects

The scientific school of Gulyaev is currently engaged in studying the effect of highly ordered (coherent) electromagnetic fields of low intensity on biological and physical objects. Since not only electromagnetic oscillations are generated in this frequency range in physical and biological systems, but also acoustic and acoustoelectric oscillations and waves, and all these types of oscillations are transformed into each other, the term "extremely high-frequency (EHF) radiation and oscillations" was introduced [4].

The physical concept of the effect of EHF radio waves is based on the conclusion about their information (not energy) impact. The use of microwave radiometry allowed us to substantiate the idea of the dissipation of the EHF power introduced into the object and its transformation into the noise energy of the intrinsic radio emission of the object, i.e. its radio response, in a wide frequency range, from low frequencies to the microwave range. This is the fundamental difference between EHF/microwave radiospectroscopy and other types of radiospectroscopy [5]. Registering the internal fluctuation thermal electromagnetic field in a condensed medium, which is influenced by a fluctuation of charged particles of the medium and concentrated inside it, turned out to be fruitful [5].

The author of the work cited formulates the following fundamental thesis: in the studies of condensed media physical mechanisms of the impact of weak fields at the very primary level are not taken into account, which leads to contradictions between experiments and theoretical physics. Therefore, the solution of the problem must be based on understanding the properties of the fluctuation thermal electromagnetic field of the medium.

He also introduces the notion of organized external fields as composite fields that, in contrast to deterministic ones, can contain external noise MF, EF, GF and other fields [5].

The mechanism of the action of weak fields on particles of condensed matter is still unclear. Why does a diamagnetic medium change its behavior, for example, in a magnetic field, although the potential energy of interaction of particles with a magnetic field is 6 to 12 orders of magnitude less than the average energy of their thermal motion, determined by the value of kT. Thermal energy must dilute all the results of the impact, [6] which is a generally accepted theoretical fact, but not corresponding to reality. In [7, 8] it is postulated that the basis for considering the influence of multiple fields on the particles of the medium are internal fluctuation of thermal electromagnetic fields and thermal magnetic fields – an integral part of any condensed macroscopic medium, both homogeneous and heterogeneous. The concept of the fluctuation field is introduced as a random inhomogeneous isotropic field that is distributed throughout the volume of the medium and includes all chaotic forces. It is the result of fluctuations, in time, of charged particles of the medium: electrons, ions, molecules, molecular structures, clusters, etc., caused by thermal motion. Only due to fluctuation thermal electromagnetic fields and thermal magnetic fields can weak influences of external fields appear on diamagnetic and paramagnetic media.

3. Electromagnetic field as an agent of low impact

Due to the fact that the body can be viewed, in a certain sense, as a wave structure, then in order not to assist it, it must react with a wave structure having similar characteristics, selecting, for this, the appropriate physical agent. Such a universal physical agent is an electromagnetic field (EMF), covering a range from hundreds of meters (radio waves) to fractions of angstrom (hard gamma quanta). The electromagnetic field can have the necessary distribution of its characteristics such as amplitude and phase on the surface of the material, which allows creating conditions for the formation of microcrystals and stimulate the synthesis of nanostructures at the molecular level. However, the usual irradiation of any EMF system, with a resonant frequency, will not lead to self-consistency and self-correction of the structure, since the field needs to be structured informatively.

There are experimental studies of fractal structures of silicon carbide, which showed that the fractal structures of this material absorb electromagnetic radiation 15 to 30 times more intensely than single crystal silicon carbide [10], and the longitudinal electric field, in combination with weak EMF, strongly influences the formation of ultrathin films, because it promotes the polarization of growing embryos and islands. In this case, the equilibrium shifts toward two-dimensional growth. Thus, there is an orienting effect on the migration of embryos and sets the direction of growth [11]. Resonance interaction, frequency and time structuring of the EM field does not allow achieving the desired result, therefore, it remains possible for the spatial structure of the EM field. A conventional diffraction grating, which is a system of plane-parallel lines or a combination thereof, will not achieve the desired result, since it covers only an extremely limited frequency range. Even a linear combination (such as the worlds used in metrology) will not give a result for the same reasons. The output can be found only on the way of development and application of complex curvilinear diffraction gratings, which are a regular and ordered combination of closed curvilinear structures. Only in this case, there is a real possibility to cover a significant part of the EM field range, since in closed curvilinear structures there is always a section of the lattice multiple of a certain wavelength [12]. The transformation of electromagnetic fields, by means of such lattices, is carried out by the scientific school of V A Soifer, who laid the foundations of a new science – computer optics [13]. The use of such diffraction gratings of a complex configuration, for the conversion of incoherent EMF, makes it possible to create, in space above the lattice, the substrate – the external medium, the interference field of a complex structure that act as additional technological factors.

At this boundary, the deposition processes of the material, in the production of nano-sized thin films, the formation and subsequent growth of the nuclei, occur under special conditions, since the nonequilibrium state arising from the interaction of the deposited material with the light flux initiates the processes of self-organization in the resulting films [14, 15].

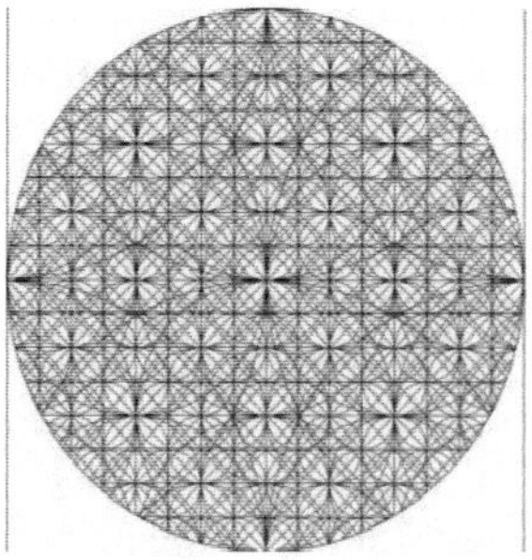

Figure 1. The structure of a diffraction grid.

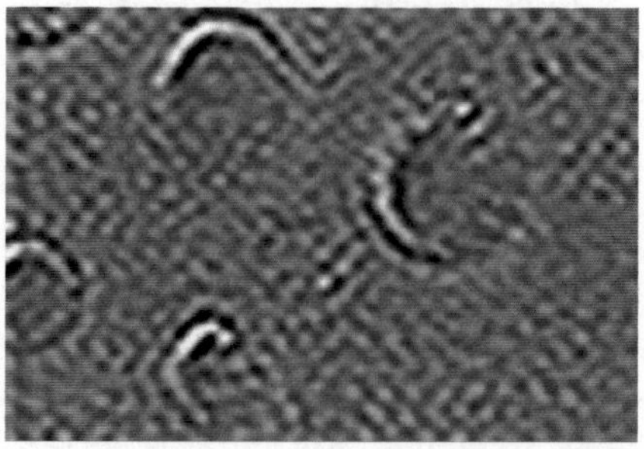

Figure 2. Latent ordered film microstructure.

To create an interference field, diffraction gratings of various configurations, combinations of closed curvilinear contours, made in the form of strokes of various metals, can be used, either in the form of grooves in a silicon or glass substrate [16] (the structure of such a lattice is shown in figure 1) or performed according to the golden section algorithm or Fibonacci series [17–19]. Interestingly, when etching such a structure in a silicon substrate, the process of etching proceeds to a depth of ~3–4 μm, after which it stops abruptly. It is impossible to etch deeper under any conditions. Such a diffraction grating has a complex structure, in which by crossing the lines forming it, a large number of fragments are obtained, which can be regarded as individual diffracting elements. The number of them can reach 400 000, and although they are all self-similar, they have different configurations and sizes [20].

The experimental investigations carried out on the interaction of diffraction gratings with electromagnetic radiation in the optical range have shown that when an visible light spectrum is irradiated from a non-point nonmonochromatic radiation source in space, an interference field is formed in space, which is a system of interference maxima and minima, located in a volume of space a few millimeters above the surface of the substrate and having its own complex internal structure. Such an interference field will in one way or another influence all physical processes taking place in the zone of its localization. In the films synthesized under the influence of the interference field diffraction grating, of great interest were the areas of the smooth film that completely repeated the relief of the substrate and, as it seemed at first glance, corresponded to films synthesized under ordinary conditions i.e. not having their own structure. The investigations were carried out by means of scanning electron microscopy. The available information obtained from the SEM sensors was subjected to additional analysis, using specially developed software. On the final micrograph of figure 2 is the previously revealed latent ordered microstructure of the film.

Similar structures such as diffraction gratings of complex curvilinear shape were used in the experiences on electrophoresis in the laboratories of the RPA "Mezon". A special fluoroplastic bath filled with a solution of salts and tantalum powder and two electrodes with variable EMF from 10 to 120 V were used. A powder of about 1 mm thick settles on the negative electrode. The result of the changes is visible because the appearance of lines of force is found on the applied surface. The effect is that the powder thickens along the lines, these lines helping in the further forming. They act like capillaries and the forming takes place on the whole surface at a time, rather than gradually, as before, and these lines pass through the volume, and not along the plane. Also, a change in the structure of the deposited substance and a change in the field strength vector on the surface were observed.

4. Conclusions

The use of interference fields, initiated by diffraction gratings of complex shape, to affect various processes of nanotechnology and synthesis of nanoscale films with their own explicit or latent structure seems highly practical.

Differences of fractal nanoscale structures, from their usual nanosized analogs, can lie in the field of structure, surface morphology, electrophysical properties, optical properties, and so on. This opens great prospects for the use of fractal nanoscale structures, in particular, fractal nanosized films in engineering and technology.

The fractal structures obtained in nonequilibrium conditions, after the termination of the technological process, tend to an equilibrium state, as a result of which the process of structural evolution (relaxation) continues. Their condition can be compared to a sharp frost and a further slow evolutionary process of restructuring their structure.

Acknowledgments

The authors would like to acknowledge the financial support of this work by the Ministry of the Education and Science of the Russian Federation (basic state assignment No. 8.7130.2017).

References

[1] Grachev V I, Zhabrev V A, Margolin V I and Tupik V A 2015 *Slabye i sverhslabye vozdeystviya na razlichnye struktury v nanotehnologiyah* (Saint Petersburg: Izd-vo Saint-Petersburg Electrotechnical University “LETI”)
[2] Kolesnikov A A and Zarembo V I 2011 *International Scientific Journal for Alternative Energy and Ecology (ISJAEE)* **4** 103–8
[3] Ilyushin G D 2002 *Tez. Dokl 4-go Mezhd. Sem. “Nelineynye process i problem samoorganizacii v sovremennom materialovedenii” (Astrahan)* pp 85–6
[4] Devyatkov N D 1985 *Sb. Statey* (Moscow: IRE AN SSSR) 21–36
[5] Popov I V 2011 *Principy fiziki slabyh vozdeystviy na kondensirovannye sredy* (Saint Petersburg: Izd-vo Politehn. un-ta)
[6] Zeldovich Ya B, Buchachenko A L and Frankevich E L 1988 *Physics-Uspekhi (Advances in Physical Sciences)* **1** 3–45
[7] Novikov V V, Shvecov Yu P and Fesenko E E 1997 *Biophysics* **3** 746–50
[8] Novikov V V and Lisicyn A S 1997 *Biophysics* **5** 1003–7
[9] Zolotuhin I V, Gribanov S A, Popov A A, Zheleznyj V S and Ievlev V M 2000 *Tez. dokl. 3-go Vseross. sem. “Nelineynye process I problem samoorganizacii v sovremennom materialovedenii” (Voronezh)* pp 121–2
[10] Bochkarev V F and Bunin E Yu 2009 *Russian Microelectronics* **3** 181
[11] Serov I N, Zhabrev V A and Margolin V I 2003 *Glass Physics and Chemistry* **4** 241–55
[12] Soyfer V A 2000 *Metody kompyuternoy optiki* (Moscow: Fizmatlit)
[13] Grachev V I, Margolin V I and Tupik V A 2016 *Slabye I sverhslabye vozdeystviya v nanotehnologiyah, biologii i medicine* (Izhevsk: Izd-vo “Udmurtiya”)
[14] Zhabrev V A, Lukyanov G N, Margolin V I, Tupik V A et al. *Proc. of SPIE* **6251** 338–46
[15] Grushina N V, Korolenko P V and Markova S N 2008 *Vestnik Moskovskogo universiteta. Seriya 3. Fizika. Astronomiya* **2** 40–3
[16] Zotov A M, Korolenko P V and Mishin A M 2010 *Crystallography* **8** 965–71
[17] Serov I N and Margolin V I 2002 *Patent RU 2212375 Method for obtaining thin films with a fractal structure*

Research of structural properties of zinc oxide nanopowders obtained by high-energy mechanical milling (attritor) using Raman spectroscopy

N D Yakushova[1], I A Pronin[1,2], I A Averin[1], G V Vishnevskaya[1], M M Sychov[3], V S Levitsky[4], A A Karmanov[1] and V A Moshnikov[1,2]

[1] Department of Nano- and Microelectronics, Penza State University, Penza, 440026, Russia
[2] Department of Micro- and Nanoelectronics, Saint-Petersburg Electrotechnical University "LETI", St. Petersburg, 197376, Russia
[3] Department of Theoretical Foundations of Materials, Saint-Petersburg State Technological Institute, St. Petersburg, 190013, Russia
[4] Ioffe Institute RAS, Saint Petersburg 194021, Russia

E-mail: pronin_i90@mail.ru

Abstract. In the paper the evolution characteristics of structural properties of zinc oxide powders produced by mechanical milling in a high-energy attritor using Raman spectroscopy are studied. Milling time was 1, 3 and 5 h. It is found that the main spectral change observed in the Raman spectra of the samples, is a monotonic intensity increase of E_1(LO) mode associated with increasing concentrations of defects in the material.

1. Introduction

Due to its electrical, optical, and structural properties, pure and alloyed zinc oxide (ZnO) has found its application as a catalyst, a phosphor material for gas sensors and photocatalysts, in varistors [1, 2]. The aim of this paper is to study the characteristics of the structural properties of zinc oxide nanopowders produced out of commercial powders by high-energy mechanical milling (attritor) using Raman spectroscopy. Raman spectroscopy can provide important information on the crystal structure of nanomaterials, lattice distortion, presence of defects, as well as explore the vibrational properties.

2. Experiment

Samples of zinc oxide nanopowders were produced by high-energy mechanical milling out of ZnO commercial powder (pure for analysis, CJSC Vekton – figure 1) in the attritor (Type: BATCH-LAB, Model: HD/01, Union Process, Inc., USA) under the following conditions: the rotational speed of the drum axis is 450 rpm; milling time is 1, 3 and 5 h; the ratio of milling bodies mass (ZrO_2 balls of 3 mm diameter) and the powder is 10:1; milling medium is isopropyl alcohol absolutized. All samples have been dried for 24 h at a temperature of 70 °C after the milling process.

Raman spectra were recorded with LabRamHR800 spectrometer in backscattering geometry at room temperature. An excitation source is the second harmonic of Nd:YAG Laser (λ = 532 nm), the laser beam is focused on the surface of the powder into a spot of 1–2 μm diameter.

Figure 1. SEM images of ZnO commercial powder.

3. Results and discussion

Figure 2 shows Raman spectra of all samples. The wurtzite structure of ZnO belongs to C_{6v}^4 space group, and it has a primitive cell that contains two formula units. Group theory predicts the existence of 9 optical phonon branches of the Γ-point of the Brillouin zone: $\Gamma = A_1 + 2B_1 + E_1 + 2E_2$.

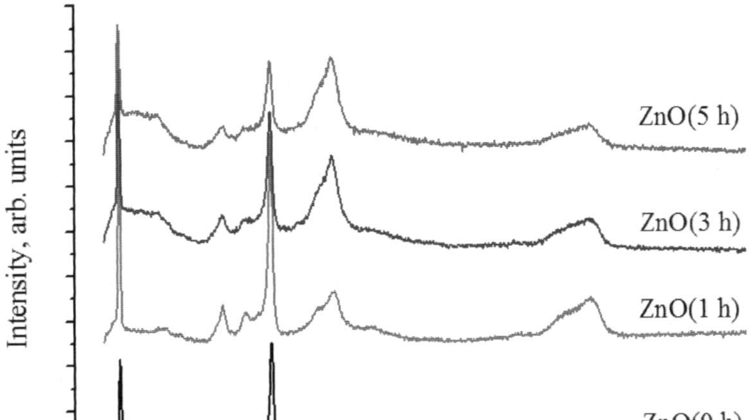

Figure 2. Raman spectra of the samples.

The main spectral change observed in the Raman spectra of the samples, is a monotonic intensity increase of E_1(LO) mode (580 cm^{-1}). According to [3–5], this phenomenon may be due to the presence of impurities or point defects in the material. Therefore, the result is in agreement with the fact that along with milling time increasing, the concentration of defects in zinc oxide will increase.

4. Conclusions

Thus, the evolution of structural properties of zinc oxide powders under a high-energy mechanical milling in the attritor has been studied by the Raman spectroscopy method.

Acknowledgments
The results were obtained within the framework of the government task of the Ministry of Education and Science of the Russian Federation (project No. 16.897.2017/4.6).

References
[1] Pronin I A, Dimitrov D Tz, Krasteva L K, Papazova K I, Averin I A, Chanachev A S, Bojinova A S, Georgieva A Ts, Yakushova N D and Moshnikov V A 2014 *Sens. Actuat. A* **206** 88
[2] Dimitrov D Tz, Nikolaev N K, Papazova K I, Krasteva L K, Pronin I A, Averin I A, Bojinova A S, Georgieva A Ts, Yakushova N D, Peshkova T V, Karmanov A A, Kaneva N V and Moshnikov V A 2017 *Appl. Surf. Sci.* **392** 95
[3] Aggelopoulos C A, Dimitropoulos M, Govatsi A, Sygellou L, Tsakiroglou C D and Yannopoulos S N 2017 *Applied Catalysis B: Environmental* **205** 292
[4] Peleš A, Pavlović V P, Filipović S, Obradović N, Mančić L, Krstić J, Mitrić M, Vlahović B, Rašić G, Kosanović D and Pavlović V B 2015 *J. Alloys Compounds* **648** 971
[5] Kahouli M, Barhoumi A, Bouzid A, Al-Hajry A and Guermazi S 2015 *Superlattices and Microstructures* **85** 7

Tribological behavior and frictional fatigue damage peculiarities in thin *a*-C and *a*-C:N coatings

A Yu Albagachiev[1], D A Tokarev[1], M M Khrushchov[1], E A Marchenko[1] and O S Antonova[1,2]

[1] Institute of Machine Studies, Russian Academy of Sciences, Moscow, 101990, Russia
[2] Institute of Metallurgy and Materials Sciences, Russian Academy of Sciences, Moscow, 119991, Russia

E-mail: michel_x@mail.ru

Abstract. The mechanical and tribological properties of thin vacuum-arc deposited coatings of amorphous carbon in an unalloyed state and alloyed with nitrogen have been investigated especially the peculiarities of their frictional fatigue failure.

1. Introduction

Alloying of diamond-like carbon (DLC) coatings is widely used to improve their properties, to decrease internal stress level and to prevent the film parameters from temporal degradation [1, 2]. Numerous plasma-assisted PVD and CVD technologies may be applied to alloy DLC films with numerous chemical elements, nitrogen being one of them. The interest to N-DLC systems has been stimulated by theoretical predictions [3] of a hypothetic β-C_3N_4-phase with a predicted strength higher than the diamond one. The most of N-DLC films being investigated are hydrogenised (*a*-C:H-based), i.e. have been reactively obtained from the mixtures of gaseous or vaporized hydrocarbons with nitrogen. The high antifriction properties of these *a*-C:H:N coatings may be associated with graphitization of amorphous carbon structure alloyed with nitrogen – the DLC structure becomes more "graphite-like" due to sp^3/sp^2-ratio decrease with the nitrogen concentration increase. Alloying with nitrogen also decreases residual stresses as compared with unalloyed *a*-C:H. The maximum positive effect on adhesion, antifriction and wear properties of *a*-C:H:N has been observed for nitrogen contents ~8...12 at. % (see, for example [4]).

Hydrogen-free amorphous carbon coatings alloyed with nitrogen (*a*-C:N) are in general less studied than *a*-C:H:N PVD films. Nevertheless the investigations of micromechanical and tribological properties of *a*-C and *a*-C:N coatings deposited by ion-beam assisted deposition (IBAD) of graphitic target [5] or closed-field unbalanced magnetron (CFUBM) sputtering of graphite [6] have demonstrated the possibility to improve the film characteristics by nitrogen alloying.

The aim of the present work was to study micromechanical and tribological properties of thin *a*-C and *a*-C:N coatings deposited by cathode-arc evaporation in vacuum and the peculiarities of their frictional fatigue failure.

2. Experimental

The coatings of amorphous carbon have been deposited using a cathode-arc impulse plasma source with graphitic cathode. The vacuum-arc deposition unit (CreepService SARL, Switzerland) used by the authors and the details of deposition are described in [7]. The 12Kh18N10T (AISI 321 type) stainless steel disc-shaped substrates (35 mm in diameter, 1.5 mm thick), polished (R_a = 0.06 μm) and ultrasonically cleaned have been used for deposition. Their surfaces were subjected to ion-cleaning by argon and titanium ions bombardment. A titanium transition layer (~10–20 nm) was deposited to improve adhesion. The graphite cathode was used for *a*-C deposition. The cathode arc on it has been initiated by pulsed laser impulses. The alloying of *a*-C was reached by nitrogen injection into the deposition chamber simultaneously with graphite cathode evaporation. The thickness of coatings was ~150 nm (for unalloyed *a*-C films) and ~500 nm (for films alloyed with nitrogen).

The resulting coatings have been subjected to a comprehensive investigation of their composition, structure, chemical bonding, micromechanical properties, coefficient of friction and tribological performance. To investigate microstructure, wear track morphologies and chemical composition of coatings the scanning electron microscope (JSM 6610 LV, JEOL, Japan) with energy-dispersive X-ray microanalysis system (EDXA) has been used. The presence of hydrogen in coatings was ignored as the deposition technology doesn't presume the presence of hydrocarbon precursor gases.

To determine the prevailing carbon allotropic type in coatings Raman spectroscopy has been used. The nanohardness and elastic modules have been measured with "NanoScan-3D" (NT-MDT, Russia) scanning nanohardness tester by the Oliver-Pharr method.

A ball-on-disc tribometer has been used to estimate tribological parameters at sliding on air without lubricant. This type of tribometers is usually used for frictional contact fatigue studies of structural materials [8]. The ball – spherical silicon nitride indenter 6 mm in diameter was motionless and the disc sample with coating rotated with a constant rotation frequency (100 rev/min). The tests at each load *P* were conducted over 1 h. The results were the average over a series of three tests. The coating performance was estimated as an integer number *N* of the disc rotations preceding the abrupt friction force variation. Traditionally this abrupt change is assumed to be associated with moment of the coating's fracture initiation.

3. Results and their discussion

The data on the chemical composition of *a*-C and *a*-C:N coatings are presented on figure 1. They demonstrate that the light element's concentration ratio [N]/[C]+[N] in *a*-C:N is about 0.11, i.e. [N] and [C] are about 10 and 90 at. % respectively.

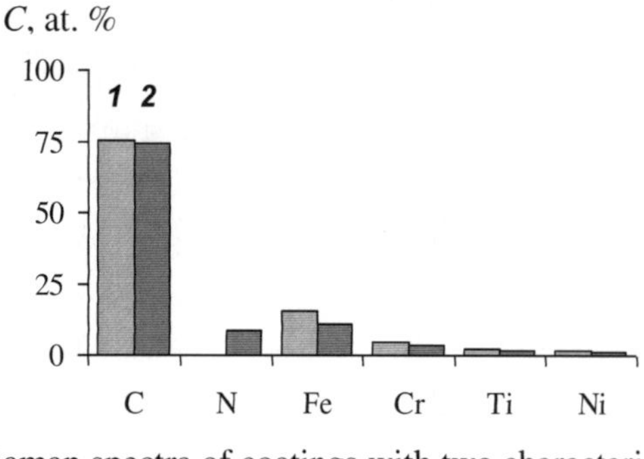

Figure 1. Chemical composition of *a*-C (*1*) and *a*-C:N (*2*) coatings deposited on 12Kh18N10T steel substrates as determined by EDXA. The presence of small amounts of metals from the substrate and titanium transition layer may be explained by the film's small thickness.

The Raman spectra of coatings with two characteristic features known as D- and G-bands are presented on figure 2. The G-band is associated with optical vibrations of carbon atoms in the basal plane of graphite crystal lattice. If some degree of structural disorder is introduced in the structure of graphite phase the G-band becomes broader and shifts in direction of smaller wave vector values. The D-band is characteristic to all disordered sp^2-type structures. The position of this band with sp^2-disordering changes is stable but its intensity may change strongly. At low degrees of disorder the D-band may have the form of a "shoulder" at the left side of the graphitic G-band or may take form of an independent spectral maximum if the disorder increases. The position and the width of the D-band as well as the intensity ratio I_D/I_G and the G-band width may to the opinion of some authors be used to estimate the "graphitization index" – the degree of DLC coating structure graphitization.

For the vacuum-arc technology used in this investigation the I_D/I_G ratio is significantly, ~5 times, higher for coatings alloyed with nitrogen (see table 1 where the characteristics of other nitrogen-doped typically hydrogenised films are also presented for comparison). Thus nitrogen leads to significant changes of the coating's Raman spectrum (figure 2).

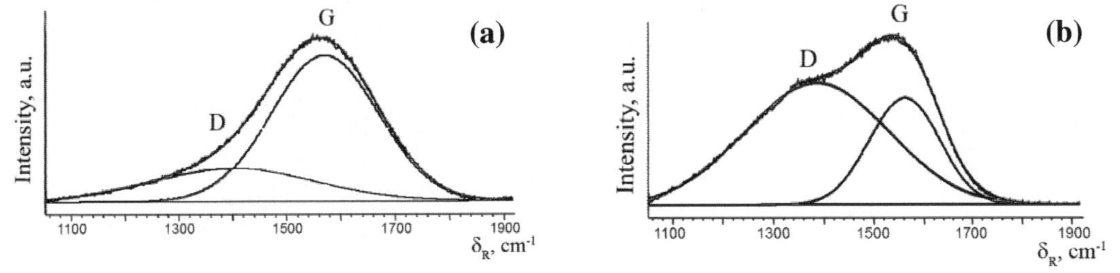

Figure 2. Raman spectra $I_R(\delta_R)$ of *a*-C (**a**) and *a*-C:N (**b**) coatings.

The data collected in table 1 demonstrate that alloying with nitrogen leads not only to DLC structure graphitization but also significant changes of mechanical characteristics are observed. The trend of these changes may be different depending on the type of deposition technology used by the authors.

Table 1. Composition, structural and micromechanical characteristics of coatings.

Coating type	[N], at. %	Thickness, nm	I_D/I_G	*H*, GPa	*E*, GPa	*H/E*	Carbon source (technology)	Issue
a-C	0	150	0.21	28	178	0.157	Graphitic cathode (vacuum cathode arc)	Present work
a-C:N	10	500	1.14	35	146	0.24		
a-C:H	0	1500	0.97	14.4	105	0.137	$C_2H_2 + CH_4$ (CFUBM)	[3]
a-C:H:N	12	1900	0.98	8.6	87	0.099		
a-C:H:N	29	2300	2.51	4.2	41	0.102		
a-C:H	0.1	–	1.5	14.3	125.5	0.114	CH_4 (PA CVD)	[8]
a-C:H:N	3.8	–	2.5	13	109.5	0.119		
a-C:H	0	160–276	–	38	460	0.083	Graphitic target + C_2H_2 (PVD)	[9]
a-C:H:N	max		–	21.5	330	0.065		

It may be concluded for example that using graphite targets or cathodes it becomes possible to obtain DLC coatings whose hardness exceeds 30 GPa, these values being more typical for the coatings of tetrahedral amorphous carbon *ta*-C [2]. At the same time the hardness of *a*-C:H and *a*-C:H:N films obtained by reactive deposition of hydrocarbon gases is typically less than 15 GPa. The type of technology may also influence the "index of plasticity" *H/E* values. $H/E < 0.1$ in combination with high hardness may certify the DLC coating's ability to be crack resistant [9]. Thus it may be concluded that the process of friction-induced damage of vacuum-arc deposited coatings with $H/E > 0.15$ have to be brittle.

The results of tribological tests (figure 3) have demonstrated the low antifriction properties of unalloyed *a*-C film (f = 0.3…0.7, and the f value gradually grows with increasing P). Alloying with nitrogen improves the antifriction behavior reducing f to about 0.2…0.3 for all the loads.

The performance of *a*-C was extremely low, less than 100 cycles for all P values, and the alloying with nitrogen significantly changed the situation: the durability of *a*-C:N period of friction coefficient stability increased to $N > 6000$ for $P = 0.02$ and 0.05 N and was $N \approx 1000$ for $P = 0.1$ N.

The nature of surface friction and wear induced damage the coatings have been studied with SEM. It was established that that the surface of DLC coatings both in unalloyed and nitrogen-alloyed state inherits the grain structure morphology of the 12Kh18N10T steel substrate. The secondary-electron SEM images presented on figure 4 demonstrate that the local defects on the wear track surfaces leading to further fracture damage develop along the grain boundaries of the substrate and at ternary junctions of these boundaries. The friction damage-induced fracture of both types of coatings was brittle.

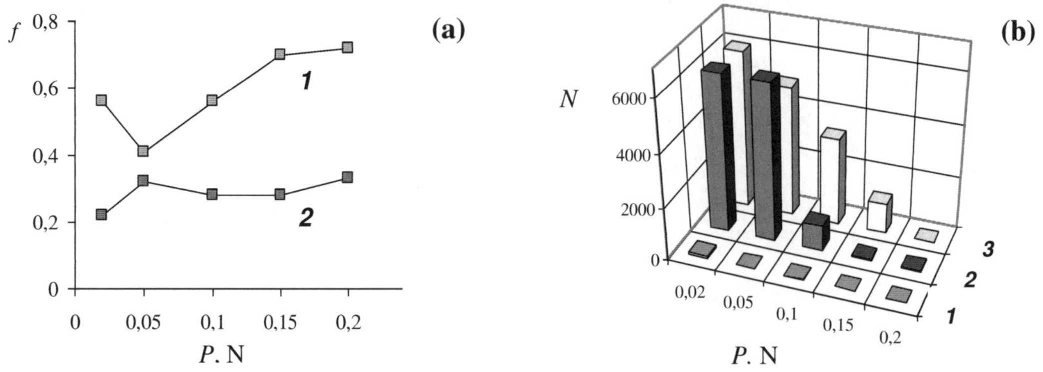

Figure 3. Load dependence of coefficient of friction *f* (**a**) and of *N* – number of cycles preceding the of friction fatigue-induced damage initiation (**b**) for coatings: *a*-C (*1*); *a*-C:N (*2*) and for substrate material (12Kh18N10T stainless steel) (*3*).

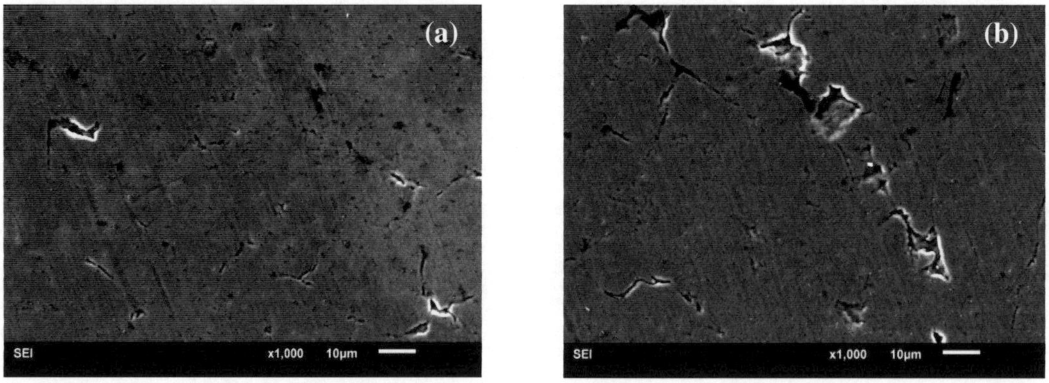

Figure 4. Surface damage evolution with load in *a*-C:N film: wear tracks at P = 0.10 (**a**) and at P = 0.15 N (**b**).

4. Conclusions

The mechanical and tribological properties of thin vacuum-arc deposited hydrogen-free amorphous carbon *a*-C and *a*-C:N coatings have been studied. Their extremely high micromechanical properties have allowed us to suppose that the amorphous carbon in both unalloyed and nitrogen-alloyed coatings of this type has tetrahedral amorphous structure of *ta*-C. The tribological tests have demonstrated that alloying with nitrogen significantly improves of performance of *a*-C:N coatings as compared with *a*-C.

The investigations of coating surfaces with scanning electron microscopy have shown that their morphology is inherited from the steel substrate and that the processes of the coating's frictional fatigue-induced damage and cracking initiates in the vicinity of these "substrate-determined" boundaries and of their ternary junctions.

Acknowledgement

This work has been financially supported by the Russian Foundation for Basic Research project no. 15-08-05264.

References

[1] Sánchez-López J C and Fernández A 2008 *Tribology of Diamond-Like Carbon Films. Fundamentals and Applications* (New York: Springer) pp 311–38
[2] Khrushchov M M 2013 *Modern Techniques of Machine Parts Surfaces Modification* (Moscow: Lenand) pp 78–113
[3] Liu A Y and Cohen M L 1990 *Phys. Rev. B.* **41** 10727–34
[4] Kao W H, Su I L, Yao S H and Huang Y C 2010 *Surf. Coat. Technol.* **204**. 1277–87
[5] Wang D F, Kato K and Umehara N 2000 *Surf. Coat. Technol.* **123** 177–84
[6] Park Y S and Hong B S 2009 *Appl. Surf. Sci.* 255 3913–7
[7] Bashkov V M, Beliaeva A O, Tokarev D A and Sedykh N S 2013 *Nanoinzheneria* **11** 35–8
[8] Kombalov V S 2008 *Methods and Tools for Tests of Friction and Wear of Structural Materials and Lubricants: Handbook* (Moscow: Mashinostroenie)
[9] Rabbani F, Escobar Galindo R, Amoldbik W M, Zwaag S V D, Veen A V and Schut H 2004 *Diamond Rel. Mat.* **13** 1645–57
[10] Dwivedi N, Kumar S, Rauthan C M S and Panwar O S 2011 *Appl. Phys. A.* **102** 225–30

IOP Conf. Series: Journal of Physics: Conf. Series **872** (2017) 012034 doi:10.1088/1742-6596/872/1/012034

Chromium-nanodiamond coatings obtained by magnetron sputtering and their tribological properties

M V Atamanov[1], M M Khrushchov[2], E A Marchenko[2], N V Shevchenko[3], I S Levin[2,4], M I Petrzhik[5], V I Miroshnichenko[1] and M D Relianu[4]

[1] INACOTEC Centre Ltd., Moscow, 117325, Russia
[2] Institute of Machine Studies, Russian Academy of Sciences, Moscow, 101990, Russia
[3] Petrovsky Scientific Centre "FOUGAS", JSC, Moscow, 117447, Russia
[4] Department of Physics, Moscow State University, Moscow, 119991, Russia
[5] National Research Technological University "MISiS", Moscow, 119049, Russia

E-mail: michel_x@mail.ru

Abstract. Peculiarities of structure, chemical and phase composition, micromechanical and tribological properties of chromium-based coatings obtained by magnetron-sputtering of composite and/or compacted chromium-nanodiamond targets have been investigated.

1. Introduction
It is well known that the composite coatings obtained by electrolytic codeposition of nanodiamonds with metals typically posses high wear resistance and hardness in combination with low porosity, high corrosion resistance and low friction, and also demonstrate good adhesion and cohesion [1, 2]. The time of life of the units with these coatings is typically sufficiently (2 to 10 times) higher than that of uncoated ones and their wear rate may be significantly decreased by increasing the nanodiamond phase content in them. It has been demonstrated that the chrome-nanodiamond coatings obtained by ion-assisted and PVD methods have nanocomposite structure with grain size typically of about 30–70 nm and that the nanodiamond phase segregates at the boundaries of these metallic nanosized grains [3], i.e. the structure of such chromium-nanodiamond (Cr-ND) coatings must significantly differ from that of metal-alloyed nanocomposite diamond-like carbon (DLC) coatings where the nanostructured metallic and carbide phase inclusions are randomly distributed in an amorphous carbon matrix [4, 5].

The investigation of the micromechanical, tribological and strength properties of such Cr-ND coatings, the comparison of their structure and phase composition with that of alloyed Cr-DLC coatings is of high interest both from the practical point of view and as a scientific base for non-equilibrium synthesis of nanocomposite structural materials with high mechanical properties and strength.

The analysis of peculiarities of the structural state, phase composition and functional properties of magnetron-sputtered chromium-nanodiamond coatings has been the aim of the present investigation.

2. Coating deposition and characterization techniques
The deposition unit equipped with an ion-source for ion-cleaning and with two magnetron sputtering units has been used for deposition of Cr-ND coatings in argon and/or argon-reactive gas mixtures. The polished substrates of AISI 430 have been used to deposit Cr-ND coatings 6–10 μm thick. The

magnetron technology was similar to that used for *a*-C:H:Cr:N coatings deposition [6, 7]. The bulk chromium targets with blind holes symmetrically arranged in the zone of maximal target erosion and filled with detonation nanodiamond powder to form the "inclusions" of ND material have been used as well as the Cr-ND targets obtained by powder metallurgy methods. The bulk targets with the nanodiamond "inclusions" will be mentioned below as "composite" (sometimes also called "mosaic"). For comparison with them the coatings obtained by sputtering pure chromium targets have been also investigated.

Structure and phase composition of coatings have been studied by X-ray powder diffractometry technique. The details of X-ray experiment and the methods of XRD patterns interpretation have been analogous to those used earlier [6, 7]. The chemical composition of coatings has been investigated by energy-dispersive X-ray microanalysis (EDXA) on a JSM 6610 LV (JEOL) scanning electron microscope. Nanohardness and elastic modules have been measured by instrumental nanoindentation with an NHT nanohardness tester (CSM International) by the standard Oliver-Pharr method. Tribological tests have been performed on a ball-on-disc tribometer with the scheme used to estimate frictional fatigue of materials [8]. The ball of silicon nitride ceramics 6 mm in diameter has been used as counterpart. The tests have been performed on air during 1 h (disc rotation speed – 100 rev/min, the maximum number N of disc revolutions ("cycles") being 6000) at loads P = 0.02–0.2 N (contact pressures of about 0.2–0.45 GPa). The number of cycles preceding an abrupt change of the coefficient of friction f that is usually associated with the beginning of coating fracture has been registered.

3. Results and discussion

The data on target type, sputtering conditions, technology parameters, coating chemical and phase compositions are presented in table 1. The results of nanohardness tests are presented on figure 1 and those of tribological tests – on figure 2.

Table 1. Coatings obtained by magnetron sputtering of chromium-nanodiamonds (ND) targets, their chemical and phase composition.

No., target type	Technological parameters	Chemical composition, at. % Cr	C	N	O	Phase composition
1. Cr	Ar	90.7	9.3	–	–	Cr
2. Cr+ND (composite)	Ar	91.3	8.7	–	–	Cr
3. Cr	Ar + N_2 (5:1)	78.2	7.5	12.2	2.0	Cr + Cr_2N
4. Cr+ND (composite)	Ar + N_2 (5:1)	75.9	7.6	14.3	2.2	Cr + Cr_2N
5. Cr+ND (composite)	Ar + N_2 (7:1)	72.5	16.4	8.1	3.1	Cr + $Cr_2N_{0.39}C_{0.61}$
6. Cr+ND (composite)	Ar + N_2(7:1), U_b = –60V	78.6	8.0	11.6	1.8	Cr + Cr_2N
7. Cr+ND (composite)	Ar + air	82.5	0	14.6	2.9	Cr + CrN
8. Cr+ND (sintered)	Ar	58.5	40.6	–	0.9	Cr + Cr_7C_3

Note. If not otherwise stated, the negative bias voltage $U_{b.}$ is zero.

From the analysis of these data it may be concluded that:

1. The carbon content [C] is less than 10 at. % in coatings Nos. 1–4;
2. The structure of coatings sputtered in pure argon (Nos. 1 and 2) is bcc;
3. The nitrogen content [N] in coatings Nos. 3 and 4 deposited in the presence of nitrogen is about 12–14 at. % and their structure is bcc Cr + Cr_2N chromium nitride.

The tribological tests of coatings Nos. 1–4 have demonstrated that they failed practically at the beginning even at loads being less than 0.1 N (figure 2). The tribological parameters of coatings Nos. 5–7 have been better. In particular at P = 0.1 N their behavior has been stable. The noticeable friction coefficient fluctuations due to friction fatigue damage being observed at P = 0.15 N (figure 2(b)). This is possibly due to the nitrogen carbide phases formed in nitrogen-argon gas mixture.

The coating No. 5 with [C] ~ 15 at. % and [N] ~ 10 at. % contained a carbonitride phase $Cr_2N_{0.39}C_{0.61}$. In coating No. 6 the nitrogen content was ~ 11 at. %, the same as in coatings Nos. 3 and

4, but their tribological properties were better probably as a result of ion bombardment of growing film due to non-zero negative bias potential $U_b = -60$ V being applied during deposition. The fact of coating densification and porosity decrease in the presence of ion bombardment stimulated by the negative bias applied to the substrate is well known from the literature [9].

The coating No. 7 deposited in argon-air gas mixture and having [N] > 14 at. % was decarburized. Despite of this decarburization the tribological properties of the coating were close to those of coatings Nos. 5 and 6 that may be explained by the presence of chromium nitrides. The absence of carbon in No. 7 may be due to the oxygen present in air and to carbon monoxide CO formation during reactive magnetron sputtering in nitrogen-air mixtures [7].

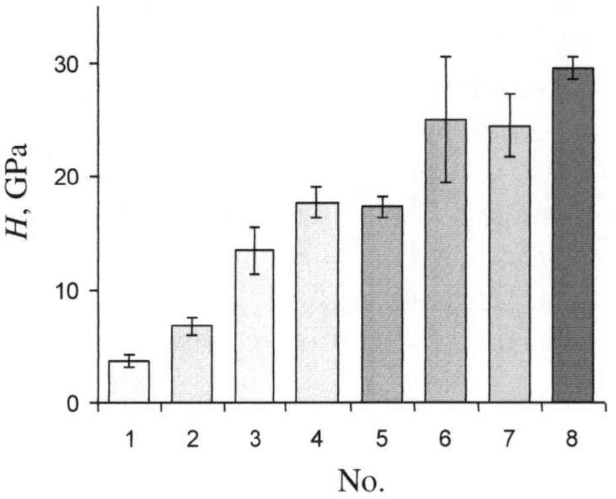

Figure 1. Nanohardness of coatings (the coating numbers are the same as in table 1).

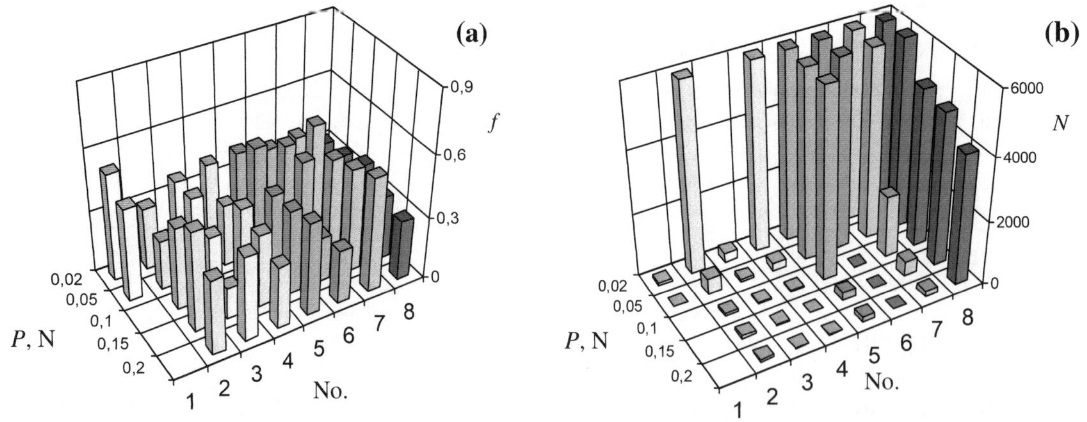

Figure 2. The dependence of the coefficient of friction f (**a**) and the number of cycles N (**b**) before the coating fracture initiation on load P (the coating numbers are the same as in table 1).

The results of coating Nos. 1–7 performance tests at high contact pressures (figure 2(b)) may be compared with the results of nanohardness measurements (figure 1). It may be noticed that the most of the abovementioned coatings posses nanohardness that doesn't exceed 17 GPa. In particular the hardness of chromium coatings Nos. 1 and 2 is close to 4 and 7 GPa respectively. It may be concluded that the low micromechanical characteristics may be a sign of poor tribological performance at high contact pressures. In this connection one can notice that the performance of coatings Nos. 6 and 7 with nanohardness close to ~25 GPa is higher than that of other coatings deposited by magnetron sputtering of composite targets. The fact may indicate that there is a close correspondence between high micromechanical properties of Cr-ND coatings and their high tribological performance. In particular it may be supposed that the hardness of Cr-ND coatings with high tribological efficiency must be close to 30 GPa or better.

The carbon content in Cr-ND films deposited by magnetron sputtering of composite targets is not higher than 17 at. % that is the problem is in obtaining coatings that contain more carbon and the way to improve their characteristics is to fabricate targets with higher nanodiamond component contents. It is also desirable to obtain a target with an even distribution of chromium and nanodiamond components. To obtain this composition uniformity the methods of powder metallurgy have to be used for Cr-ND target fabrication. The targets of two types have been prepared: one was compacted by hot-pressing (type I) and the other – by sintering (type II).

The hot-pressed targets of type I possessed a rather heterogeneous loose structure with strong porosity. Their structure was composed of small faceted granules ~5…20 μm in size and of separate randomly distributed large formations of irregular shape (figure 3). The size of some of these formations reached ~50…70 μm and they were 100 % carbon. This was in contrast with small faceted granules rather rich in chromium (according to EDXA the [Cr] content in them varied from 40 to 80 at. %). Unfortunately the mechanical strength of the type I targets was low and they crumbled easily. It was impossible to use then as a source of sputtered material for Cr-ND coatings deposition.

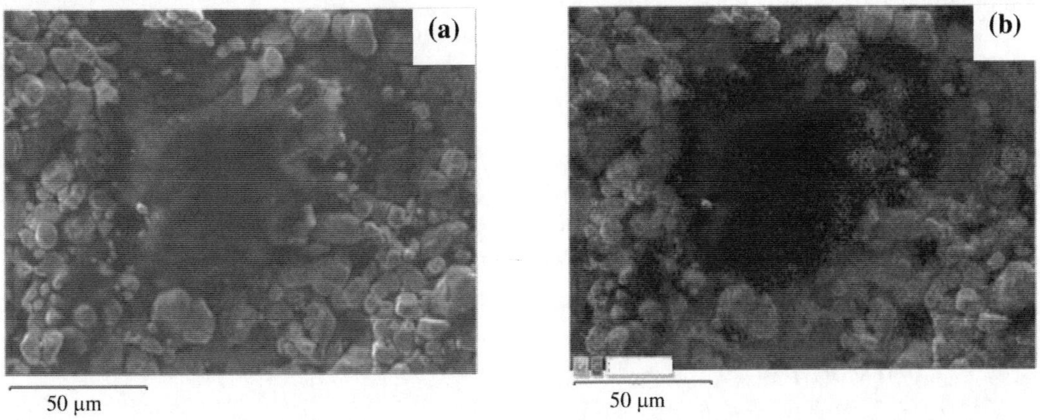

Figure 3. The surface of a type I (hot-pressed) target with a carbon formation: images obtained in secondary electrons (**a**) and in characteristic X-ray radiations of chromium and carbon (in on-line version – superimposed areas colored in green and red respectively) (**b**).

The type II targets fabricated by sintering were more suitable for magnetron sputtering – they were less porous with characteristic granules size less than 5 μm and stronger than those of type I. Their structure was more uniform and suitable for magnetron sputtering. Their chemical composition and the composition of appropriate Cr-ND coatings No. 8 are presented in table 2.

Table 2. Compositions of a type II sintered Cr-ND-target and of coatings deposited by their sputtering.

	Chemical composition, at. %		
	Cr	C	O
Target	59.3	23.5	15.6
Coating	58.5	40.8	–

One can see that the target in as fabricated state contains oxygen (about 16 at. %). This oxygen is eliminated from the final coatings probably due to the effects of preferential sputtering and as a result they contain only chromium and carbon. The carbon content in this type of coatings is large (about 40 at. %). This is close to values observed in Cr-DLC films reactively sputtered in argon-acetylene mixtures [5, 7] that have demonstrated a good working performance and antifriction behavior.

The results of tribological tests of coatings deposited from the type II sintered targets (No. 8 in figures 1 and 2) also were good. The coating No. 8 in spite of its rather high coefficient of friction ~0.3 has kept functionality (N > 4000 cycles for P = 0.1…0.2 N). This is much higher as compared with other Cr-ND coatings (Nos. 1–7). To the author's opinion this may confirm that the magnetron sputtering of targets fabricated by sintering is a promising technology that allows deposition of nanocomposite chromium-nanodiamond coatings for tribological applications. Further investigations devoted to study the Cr-ND nanocomposite film microstructure and their tribological behavior are desirable.

4. Conclusions

Chemical and phase composition, micromechanical and tribological properties of coatings deposited by magnetron sputtering of chromium-nanodiamond "composite" (mosaic) or powder-metallurgy fabricated targets have been investigated.

These coatings posses composite structure consisting of chromium with inclusions of chromium nitride, oxide or carbide phases formed during deposition as a result of chromium interaction with carbon or with reactive gases added to argon.

The highest tribological characteristics and performance in conditions of heavily loaded frictional contact have been demonstrated by the coatings deposited by non-reactive sputtering of targets fabricated by sintering of chromium and nanodiamond powders (type II).

Acknowledgements

The authors appreciate the financial support from the Russian Foundation for Basic Research, project no. 15-08-05264 and from the "MISiS Increase Competitiveness Program" of the Ministry of Education and Science of the Russian Federation (in part of nanohardness measurements).

References

[1] Dolmatov V Y 2001 *Russ. Chem. Rev.* **70** 607–26
[2] Dolmatov V Y 2007 *Russ. Chem. Rev.* **76** 339–60
[3] Vityaz P A 2004 *Phys. Sol. St.* **46** 606–10
[4] Andreev M A, Chekan V A, Markova L V and Kuznetsova T A 2000 *Proc. of 4th Belarusian Seminar on Scanning Probe Microscopy* (*BelSPM-4*) (Gomel: MPRI) pp 60–3
[5] Gassner G, Mayerhofer P H, Mitterer C and Kiefer J 2005 *Surf. Coat. Technol.* **200** 1147– 50
[6] Khrushchov M M 2013 *Modern Techniques of Machine Parts Surfaces Modification* (Moscow: Lenand) pp 78–113
[7] Khrushchov M M, Marchenko E A, Levin I S, Atamanov M V and Tarelkin Y A 2016 *J. Frict. Wear* **37** 407–14
[8] Khrushchov M M, Marchenko E A, Levin I S, Atamanov M V and Petrzhik M I 2014 *Bull. Russ. Acad. Sci. Phys.* **78** 1007–14
[9] Kombalov V S 2008 *Methods and Tools for Tests of Friction and Wear of Structural Materials and Lubricants: Handbook* (Moscow: Mashinostroenie)
[10] Semenov A P, Kovsh I B, Petrova I M et al. 1992 *Methods and Tools of Machine Parts Surface Hardening by Concentrated Flows of Energy* (Moscow: Nauka)

Mechanical properties of titanium nitride films obtained by reactively sputtering with hot target

R V Babinova[1], V V Smirnov[1], A S Useenov[2], K S Kravchuk[2], E V Gladkikh[2], V I Shapovalov[1] and I L Mylnikov[1]

[1] Department of physical electronics and technology , Saint-Petersburg State Electrotechnical University "LETI", 197376, Saint Petersburg, Russia
[2] Technological Institute for Superhard and Novel Carbon Materials, 142190, Troitsk, Moscow, Russia

E-mail: vishapovalov@mail.ru

Abstract. The mechanical properties and morphology of the surface of titanium nitride films deposited by reactive sputtering with a hot target have been investigated. It was founded that films synthesized at low nitrogen flow rate and at high discharge current density have the highest hardness (up to 30 GPa) and Young's modulus (up to 300 GPa). The analysis of films surface images obtained by atomic force microscopy is carried out. The typical grain size and its shape were determined. Films with the greatest hardness have the largest grains.

Titanium nitride (TiN) retains leading position among transition metals nitrides that can be synthesized as thin films. TiN films have high electrical conductivity, adhesion, hardness and chemical inertness therefore they are used as decorative, wear-resistant and corrosion-resistant coatings, for modifying the metal surface against fretting corrosion. Also they are applied in silicon technology as diffusion barriers for aluminum and copper metallization, when creating MOS transistors, ohmic and rectifying contacts. In the recent time high power reactive magnetron sputtering (sometimes with hot target) has become the most popular method of obtaining titanium nitride thin films.

In the present work research of titanium nitride films deposited by DC reactive sputtering of a hot target in a mixture of argon and nitrogen gas have been carry out. The films were made in a vacuum chamber with a volume of $7.8\cdot10^{-2}$ m^3 equipped with a planar magnetron with a titanium target 130 mm in diameter. The hot target regime is provided by fixation a 1 mm thick titanium disk with a gap of 1 mm on a 4 mm thick water-cooled copper plate. The residual pressure in the chamber did not exceed 10^{-2} mTorr. Sputtering was carried out in Ar/N_2 at a pressure p_{Ar} = 2 mTorr, nitrogen flow rate 2–6 cm^3/min, and discharge current density 14–33 mA/cm^2.

The mechanical properties (hardness and modulus of normal elasticity) were measured by the nanoindentation using the NanoScan-4D nanohardness measuring [1–3] (FGBN TSUM, Russia). This method allows studying evolution processes of both elastic and plastic deformation in small volumes. The Berkovich indenter in the shape of triangular diamond pyramid with an apex angle of about 142° was used for measurements. In nanoindentation method the indentor is pressed into the surface of the sample at a set rate with registration of the applied load and the corresponding displacement of the indenter.

The study of surface morphology was performed by atomic-force microscopy (AFM) using the INTEGRA Prima atomic force microscope (NT-MDT, Russia) in the semi-contact mode using silicon cantilevers with a needle radius of 10 nm (NSG01 series) (figure 1). The scanning speed did not exceed 0.5 μm/s. Image analysis is performed using the Nova software (NT-MDT).

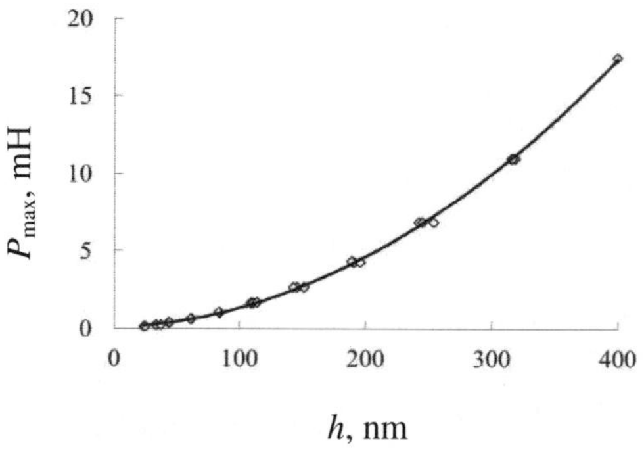

Figure 1. Loading diagram of the sample.

Figure 1 shows a typical loading diagram of the samples in our studies. The diagram is shown in the coordinate system "load *P* vs. depth of the indenter penetration *h*" with a maximum load of about 20 mN. Based on experimental results, for each sample nano-hardness and Young's modulus were calculated. The results of the calculations are shown in figure 2.

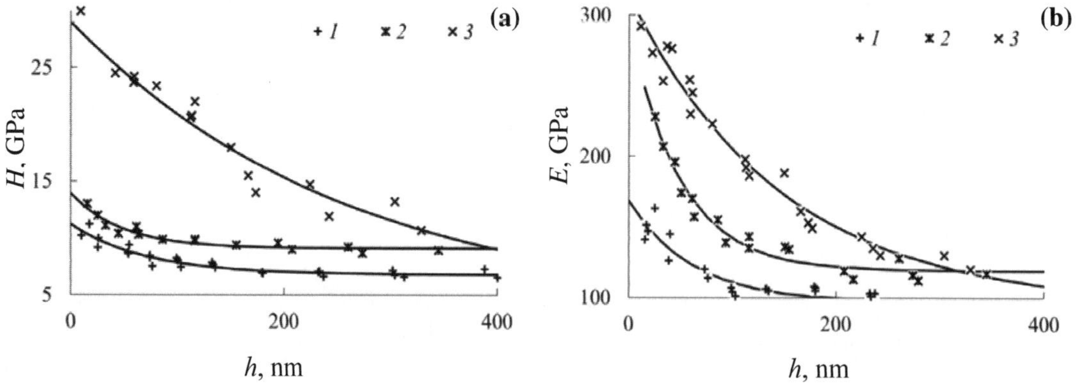

Figure 2. Nanohardness **(a)** and elastic modulus **(b)** of films produces with nitrogen gas flow of 2 cm^3/min and current density (mA/cm^2) of: *1* – 14; *2* – 22; *3* – 33. Dots are for experimental results of indentation with penetration depth h, lines are approximations.

It can be seen from figure 2 that the highest hardness (around 30 GPa) and Young's modulus (around 300 GPa) are possessed by films deposited with the highest current density. The ones deposited with the lowest current density have surface properties that insignificantly vary from the substrate surface properties. For glass substrates, the values of *H* and *F* are equal to 7 and 80 GPa respectively.

The surface morphology studies were performed by AFM. For results processing we used a specialized Nova software that allows accurate surface visualization and perform a film characterization by various statistical methods.

Figure 3 shows typical 3D images of films deposited with constant nitrogen inlet flux. Images were modified by using a line-by-line filtering for elimination of inclined plane effect occurred due to impossibility of placing a sample exactly perpendicular to atomic-force microscope cantilever.

A root mean square (RMS) roughness was chosen as a parameter for characterizing film surface. By using Nova software tool "Roughness analysis" it was found that with increasing of gas flow from 2 to 6 cm^3/min the RMS roughness decreases from 1.5 to 1.3 nm at j = 13.7 mA/cm^2 and from 6.1 to

3.3 nm at j = 32.9 mA/cm^2. This decreasing of RMS roughness with increasing of gas inlet is in good correlation with data for titanium nitride films obtained by reactive magnetron sputtering studied by other authors [4].

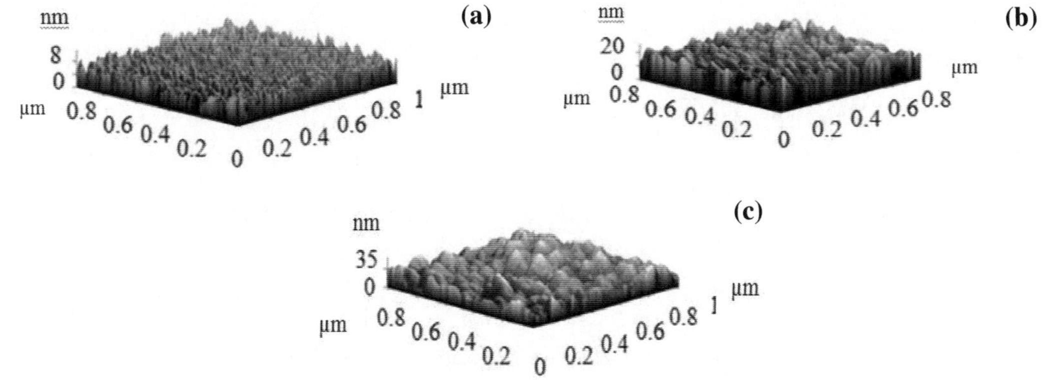

Figure 3. AFM images of films produced with current density (mA/cm^2): **(a)** –14; **(b)** – 22; **(c)** – 33.

Analysis of film grains was performed using "Grain analysis" tool. The principle of the method is as follows: user cut the sample with a plane parallel to the substrate and placed on a given distance from it. It was defined that at constant gas flow rate of 2 cm^3/min with increasing of current density from 14 to 33 mA/cm^2, a typical size of grains increases from 50 to 70 nm. At the same time, a shape of grains becomes flat-topped. At the constant current density of 14 mA/cm^2, increasing of gas flow rate from 2 to 6 cm^3/min cause not only decreasing of a typical grain size to 20 nm but also a more needle-like form of grains.

The proposed method can be used in the complex study of surface morphology as it per-form to make a conclusion about a main mechanism of crystallite growth in a film. The titanium nitride films have been deposited by DC reactive sputtering of a hot target in a mixture of argon and nitrogen gas.

Comparing the results of film morphology and its micro-hardness studies it can be said that the increasing of crystallites typical size correlates with the increasing of film micro-hardness. The sample having the largest crystallites exhibit the most hardness as well. This sample was obtained under conditions of the most current density and the least gas flow rate i.e. the most deposition rate.

Acknowledgments
In the part of films fabrication, AFM results analysis the research and preparation of the text of the article was funded by RSF (grant number 15-19-00076). All measurements were made at the Institute of Superhard and New Carbon Materials.

References
[1] Zhuikov V, Bonartsev A, Bagrov D et al. 2017 *Solid State Phenom.* **258** 354–7
[2] Kulevoy T, Oks E, Chalykh B et al. 2016 *Rev. Sci. Instr.* **87** 02C102
[3] Useinov A S, Kravchuk K S, Rusakov A A et al. 2015 *Phys. Procedia* **72** 194–8
[4] Ponon N K, Appleby D J R, Arac E et al. 2015 *Thin Solid Films* **578** 31–7

IOP Conf. Series: Journal of Physics: Conf. Series **872** (2017) 012036 doi:10.1088/1742-6596/872/1/012036

About possibility of detecting micron-size defects in layered structures using the method of microfocus tomography

V B Bessonov[1], A V Obodovskiy[1], A Y Gryaznov[1], V V Klonov[1], I A Larionov[1] and V M Osokin[2]

[1] Saint Petersburg Electrotechnical University "LETI", 197022, Saint-Petersburg, Russia
[2] Perm National Research Polytechnic University, 614990, Perm, Russia

E-mail: vbbessonov@yandex.ru

Abstract. The paper describes the experience of detecting defects in layered structures using microfocus tomography. The basic features of the equipment for microfocus tomography are given. The sizes of defects (pores and inclusions) and their location in the object are analyzed.

1. Introduction

Quality control is an important part of the manufacturing process in any industrial sphere, because quality is the most significant criterion of product reliability, and therefore its safety and trouble-free operation. Currently, when controlling finished products, it is customary to use several types of non-destructive testing:

- visual-measuring control;
- acoustic (usually ultrasonic) control;
- capillary methods of nondestructive inspection;
- magnetic methods;
- X-ray inspection.

At the same time, taking into account the latest trends in the development of technologies and equipment for non-destructive testing, a special place among the main methods take ones, based on the use of precisely X-ray units [1, 2]. This fact is due to the possibility of determining the defects with the smallest, in comparison with other methods, dimensions. Perhaps, the top disadvantage is the summative form of the output data obtained (X-ray images). Eliminate this drawback – the use of X-ray tomography.

When using tomography, the study of the internal structure and detection of defects in the volume of the controlled article is carried out by visual analysis of images of individual flat sections (tomograms) of the restored spatial structure. Thus, it is possible to control in detail the geometric structure and the character of the volume distribution of density and elemental composition without destroying the product.

2. The object of study

In recent years, substantial interest research and development in the field of additive manufacturing and composite materials. Among the latter are separately allocated polymeric composite materials used in the aircraft engine building. Their manufacture and study of properties is given considerable attention. In this case, special requirements are imposed on the means of control and diagnostics [3].

In general, a product made from polymeric composite materials is a layered structure with alternating carbonaceous fibers (in some cases glass fiber is also used) and a binder that are pressed together. Obviously, such a technology making possible the presence of defects between the layers (pores, inclusions, folds, etc.). As the object for the research was the blade of aircraft engine, having a fairly complex form (figure 1). The specified product is in the process of operation under high loads and presence of defects depends on both its reliability and service life.

Figure 1. The object under investigation – blade made of composite materials.

3. Equipment and methods of conducting research

The work was carried out in the laboratory of X-ray systems at the department of electronics and devices in Saint Petersburg Electrotechnical University "LETI". The study was performed on the prototype of microfocus X-ray computer tomograph (figure 2).

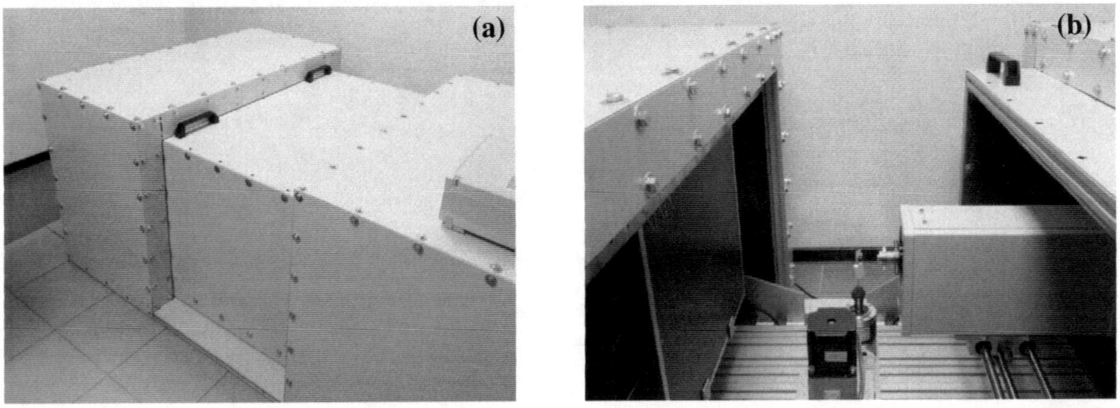

Figure 2. Microfocus X-ray computer tomograph: **(a)** – appearance; **(b)** – the source and receiver of radiation with the movement system.

As the radiation source used is specially designed microfocus X-ray apparatus based on sealed type X-ray tube with grid control and external anode [4]. This allows to achieve the minimum focal length is 1 mm and the dimensions of the focal spot at the level of a few micrometers. For the registration of the projection data uses a digital flat panel detector based on the matrix of thin-film transistors with a phosphor of cesium iodide. The use of this equipment allows to obtain projection data with magnification up to 200 times and, given the size of the pixel X-ray detector, to achieve a resolution of 2–3 μm.

4. Experiment

The study sample was mounted on a special movable object table. In the process of research table rotated by a specified angle. Taking into account the chemical composition, size, geometry and spectral sensitivity of the X-ray detector, the values of anode voltage and anode current were 90 kV and 35 μA, respectively. The number of projection data was 800. Projected increase was based on assumptions about the size of possible defects of the internal structure and made up 7 times. This has resulted in a voxel size of the reconstructed image at the level of 15–16 μm.

Based on the obtained projection data were produced layer-by-layer reconstruction of the internal structure of investigated object by the method of inverse projection with filtering by convolution. After the specified operation, for ease of presentation the results of the study were performed three-dimensional reconstruction (figure 3).

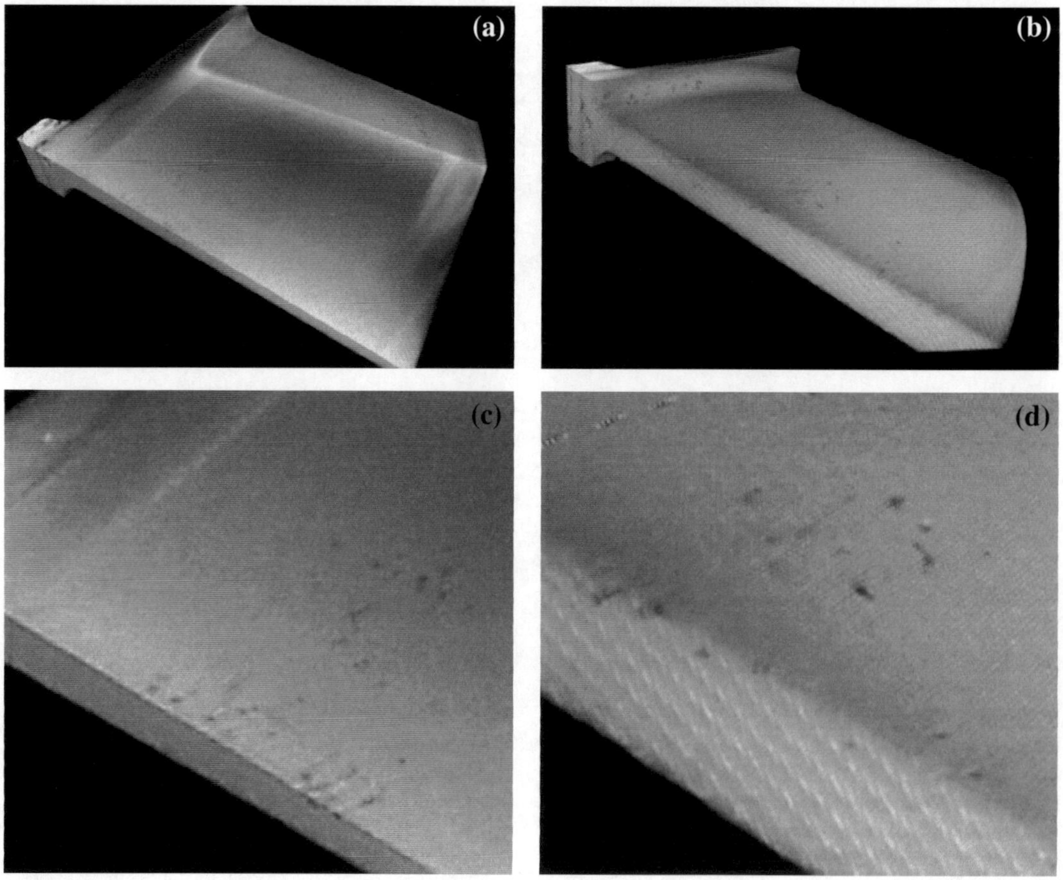

Figure 3. Three-dimensional models: **(a)**, **(b)** – appearance; **(c)**, **(d)** – the presence of defects such as pores.

5. Conclusions

During the analysis of the data revealed the following defects:

– the pores in the surface and deep layers;

– the inclusion of foreign materials.

The dimensions of these pores and inclusions range from 30 μm to 0.15 mm. The distribution of defects within the object is uneven. Pores of small size are arranged in sections according to their geometry is close to flat. Large defects are located at corners and kinks. It is the latter pose the greatest danger, since they are in areas with the highest loads.

The obtained results should be used in the testing of technological modes of production details, as well as the establishment of numerical criteria for the acceptable number of pores and inclusions in products from polymer composite materials. Studies have shown the possibility of detecting small size defects in layered structures with high detail.

Acknowledgments
The work was supported by the Russian Science Foundation in the framework of the project «Creating portable systems for microfocus radiography with the purpose of operational control of the microstructure, physicochemical properties and determining the residual life of aircraft parts and assemblies made of composite materials». Number of project 15-19-00259.

References

[1] Fayzrakhmanov R A, Murzakayev R T, Bakunov R R and Mekhonoshin A S 2016 *Information-measuring and control systems* **9** 12–6
[2] Anoshkin A N, Kochurov V I and Osokin V M 2016 *Parallel computing (PaVT'2016) proceedings of an international scientific conference* p 754
[3] Potrakhov N N, Zhamova K K, Bessonov V B, Gryaznov A Yu and Obodovskiy A V 2015 *Vestnik PNIPU. Aerospace engineering* **43** 97–115
[4] Potrakhov N N, Podymskiy A A, Bessonov V B, Obodovskiy A V, Gryaznov A Yu, Potrakhov Ye N and Zhamova K K 2017 *Defectoscopy* **3** 69–73

IOP Conf. Series: Journal of Physics: Conf. Series **872** (2017) 012037 doi:10.1088/1742-6596/872/1/012037

Modeling the electromagnetic processes in a technological device for producing ultradispersed particles in pulsed arc discharges

V D Goncharov[1], R V Yashkardin[1], K S Sorokin[1] and E M Fiskin[2]

[1] Saint Petersburg Electrotechnical University "LETI", Laboratory of "Electric impulsive technologies", 197376, Saint Petersburg, Russia
[2] Irkutsk state technical university, Russia

E-mail: vdgoncharov@rambler.ru

Abstract. The article contains a brief description of the operation principles of the device for creating ultrafine metal powders under the action of electrically charged flows with a high power density. The results of creating a mathematical model of electromagnetic processes in the installation are presented. The results of solving the problem of current flow in the installation during its operation are presented. The dependence of the magnetic field strength in the section of the plasma bunch during the operation of the apparatus is illustrated.

1. Introduction

Nowadays the ultradispersed particles (nanoparticles) and materials and their unique properties are used in many scientific and technological fields. The most perspective of which for today are medicine, chemical industry and automobile manufacturing [1–3].

There're many principally different ways of producing different types of ultradispersed particles [4–6]. And in the Laboratory of "Electric Impulse Technologies" the new method of producing the ultradispersed particles in pulsed arc discharge [7–9], moving along the extended electrodes is being developed [10]. The principles of the developed device are schematically presented in figure 1.

2. Model description

The main arc discharge *2* is being ignited by the additional spark in gap between the main *1* and secondary *11* electrodes. The main discharge is supported by the energy accumulated in the capacitive batteries *10*. The main discharge exists in the gap between the main electrodes, which constructively are two extended parallel metal bars. The main electrodes are connected to the capacitors with a co-axial cable *13*. In between the main electrodes the dielectric inset is placed *3*.

The interaction of the magnetic fields in the main discharge and the main electrodes (rails) causes the motion of the discharge along the surface of the electrodes. The impact of the main discharge *6* on the electrodes cause the forming *7* and dispersing *8* the particles. Ultradispersed particles condence *9* on the surface of the product *4*, where they are supposed to be used.

In order to evaluate the parameters of the device in the paper [11] the mathematical model was suggested, which was based on presenting the electromagnetic processes as the number of substitution circuits, connected in a certain way. The parameters (R, L, C) needed for the circuit were determined experimentally, using the immitance measuring device MNIPI E7-20 in a wide dynamic range.

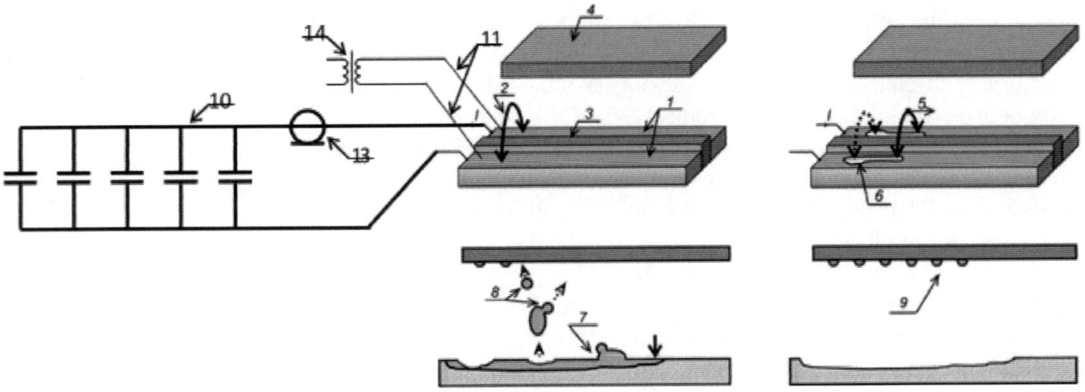

Figure 1. The schematic presentation of processes in device.

Such an approach gives a detailed and correct image of the processes. Though, in order to define them numerically, a lot of effects must be taken into consideration, which couldn't be determined using simple measurements. Moreover, the presented model could only describe completed experiments. Particularly, it can't predict changes in the electromagnetic processes while changing the geometry of the arc gap. So, the solution required much more detailed description of the electromagnetic processes in the device.

Creation of such a mathematical model, capable for describing transient electromagnetic processes is only possible using Maxwell equations. Due to model being principally three-dimensional, (apart from the ones, presented in [12, 13]) only numerical solution is possible, which considered the properties of geometry of the device. So, the very paper is dedicated to a mathematical model development.

In case when the radiation energy losses are not needed to consider, the description of the processes requires only two first of Maxwell equations without a derivative vector flux member. The numerical solution was processed in ASNYS program environment. It is comfortable to solve three-dimensional transient models in it. The Cartesian system was used for modeling. The borders of the volume where the model was built (figure 2) were chosen in the way not to affect the processes while defining zero electrical and magnetic fields boundaries.

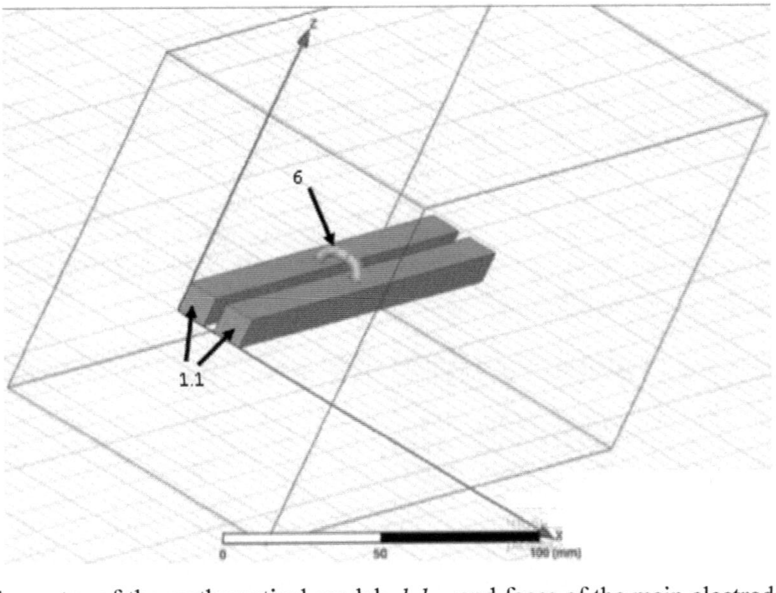

Figure 2. Geometry of the mathematical model: *1.1* – end faces of the main electrodes; *6* – main discharge (same as one in figure 1). The resistance of the main electrodes $R = 11 \cdot 10^{-4}$ Ohm.

Boundaries at the end face of the electrodes *1.1* were assigned after solving the system of equations for the rest of the substitution circuit (the capacitor and co-axial cable). Such boundaries allow to simulate the capacitor discharge process considering its own parameters and the parameters of the matching circuit between it and the main electrodes.

To check the adequacy of the developed mathematical model, the results received during the solution were compared to experimental ones, measured in the device. During the experiments, the capacity changed from 100 up to 1000 uF, the initial capacitor voltage changed from 1500 up to 3500 V, the height of semi-conductive insert from 0.4 to 3 mm.

3. Solution analysis

We only measured and compared the integral parameters (the voltage at the end faces of the electrodes) as the transient parameter of the electromagnetic field is extremely complicated to measure. The voltage dynamics at the end faces of the rails was measured using the oscilloscope Tektronix TDS 3012C. The variation between experimental and calculated results never exceed by 10 %. Particularly, in figure 3 the typical results of the comparison are presented.

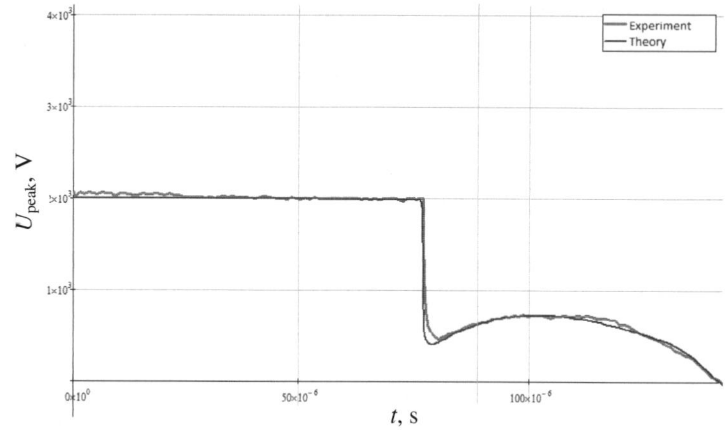

Figure 3. Experimental and numerically calculated voltage dynamics at the end faces of the main electrodes (voltage in the capacitors = 2000 V, the capacitance = 500 uF, the height of the semi-conductive insert = 2 mm).

Current density distribution across the conductor at different time moments is presented in figure 4. It's easy to see that the most of the current is positioned only in a narrow "skin" area of the conductor. It depends on time. And the most of the current is at the side of the electrodes where a semi-conductive insert belongs. Such a current distribution explained by the surface effect and the short-range effect. According to their effects the resistance of the conductors changes in time, which previously was considered to be constant [10]. Such transient changes are very difficult to measure experimentally.

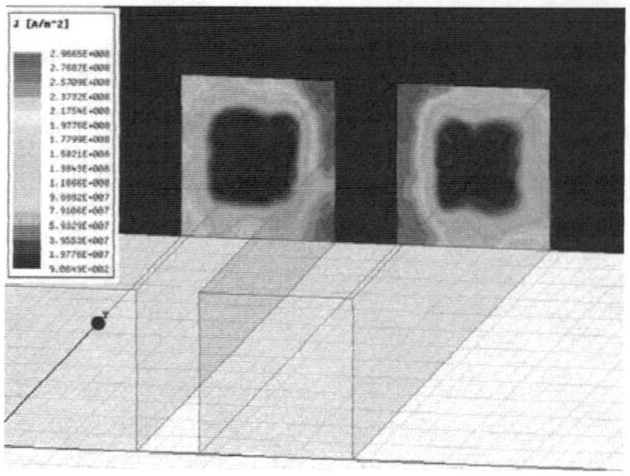

Figure 4. Current density distribution in the main electrodes after 20 us after the ignition

Received results would improve the precision of the calculations dramatically. Particularly, in the presented case, the rails resistance, calculated considering the experimental frequency characteristics, appeared to be three times less that in numerically calculated one.

Current flow and density dynamics allow to figure the magnetic field distribution. The gradient of the square of the absolute value of it shows the direction and the value of the force affecting the plasma model. Such a distribution is presented in figure 5. The gradient direction of the magnetic field distribution is angled to the rails surface. Such force direction is characteristic for the initial discharge moments conditions. In steady-state conditions this direction is supposed to be parallel to the main electrodes.

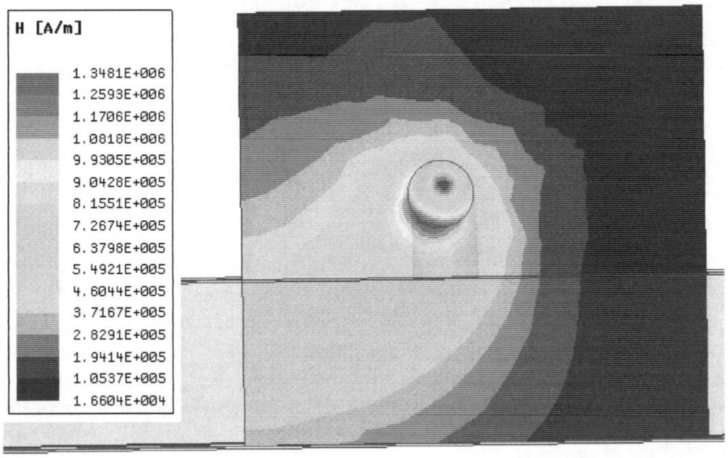

Figure 5. Magnetic field distribution in the cross-section at the middle of the main discharge. The conditions are; Voltage in the capacitors = 2000 V, the capacitance = 500 uF, the height of the semi-conductive insert = 2 mm.

4. Conclusions

1. Developed mathematical model simulated the electromagnetic processes in the device.

2. Achieved results allow to learn the electromagnetic processes in the device and particularly: transient current density parameters and power in rails and discharge plasma; figure the direction and the value of the forces acting on the electrodes and discharge plasma; improve the substitutional circuit parameters.

3. To predict the parameters of the device when changing the discharge gap.

References

[1] Moshnikov V A, Terukov E I et al. 2010 *Fundamentals of Hydrogen Energy* (Saint Petersburg: Publishing house SPbETU "LETI")

[2] Antonova I N, Goncharov V D, Kopchuk A V and Bobrova E A 2014 *Morphology* **5** 52–6

[3] Gubin S P, Koksharov Yu A, Khomutov G B and Yurkov G Yu 2005 *Uspekhi Khimii* **6** 539–74

[4] Chernik G, Fokina E, Budim N et al. 2008 *Nanoindustry* **5** 304–9

[5] Goncharov V D and Novik A A 2013 *Izvestiya SPbETU "LETI"* **5** 98–103

[6] Gorokhov M V, Kozhevin V M, Yavsin D A et al. 2012 *Letters to JTP* **6** 135–41

[7] Bystrov Yu A, Vetrov N Z, Lisenkov A A and Kostrin D K 2014 *Vakuum in Forschung und Praxis* **5** 19–23

[8] Bystrov Yu A, Kostrin D K, Lisenkov A A and Vetrov N Z 2015 *Vakuum in Forschung und Praxis* **2** 22–5

[9] Kostrin D K and Lisenkov A A 2016 *Materials Science Forum* **843** 278–83

[10] Goncharov V D and Samsonov D S 2015 *Journal of Technical Physics* **5** 37–42

[11] Yaskardin R V and Sorokin K S 2016 *IEEE NW Russia Young Researchers in Electrical and Electronic Engineering Conference (2016 ElConRusNW)* pp 736–9

[12] Glinov A P et al. 2004 *Thermal Physics of High Temperatures* **4** 635–9

[13] Lelevkin V M et al. 2002 *Vestnik of the KRSU* **22** 25–34

IOP Conf. Series: Journal of Physics: Conf. Series **872** (2017) 012038 doi:10.1088/1742-6596/872/1/012038

X-ray inspection of arduous welds in rocket and space technology with the use of a microfocus X-ray apparatus

N N Potrakhov[1], E N Potrakhov[2], E Y Usachev[3], M M Gnedin[4], D V Voroschuk[5] and I M Shavrina[5]

[1] Saint Petersburg Electrotechnical University "LETI", 197022, Saint-Petersburg, Russia
[2] Electronic Technique-Medicine, 197376, Saint-Petersburg, Russia
[3] MTU MIREA, 119454, Moscow, Russia
[4] LLC Diagnostika-M, 111024, Moscow, Russia
[5] JSC RCC Progress, Moscow, Russia

E-mail: vbbessonov@yandex.ru

Abstract. The article presents the design of a specialized X-ray machine to perform circumferential weld inspections various structural elements of air and space technology. Shows the main specifications of the device and describes a particular application of the apparatus. The results of the use of the device in conditions of real production on one of the local engine companies in the aircraft and space industry.

1. Introduction

Welds, which used in the products of rocket and space technology (RST), characterized by not only the variety of designs, brands of materials, methods of welding, but also, in some cases, is difficult access in the welding and control zone. Examples of such areas are circumferential welds of pipes mounted inside all sorts of tanks (figure 1). As it is known, to assess the quality of welded connections used radiation monitoring techniques, and in the first place, X-ray radiography [1]. However, the wide range of materials to be welded, a large variation of geometric dimensions and complex form of structural elements in many cases do not allow to be used for their control mass-produced X-ray machines based on conventional X-ray tubes. In this case, advantageously may be used a specially designed technical equipment of X-ray diagnostics.

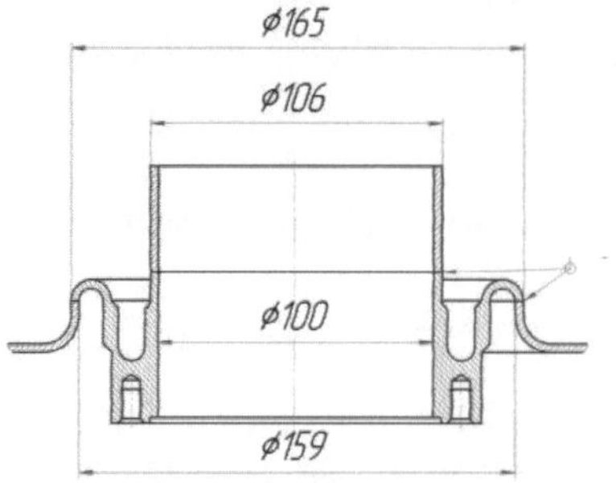

Figure 1. Circular welded joint.

Typical specifications of known circular welded joints are:

– the type of welding – argon-arc unconsumable electrode;
– material – aluminum alloy;
– the type of welded joint – butt joint;
– radiation thickness of one wall, mm – from 5 to 10;
– the control sensitivity corresponds to the first class according to GOST 75.12 – 82;
– the inner diameter of the weld joint, mm – 100…170;
– the depth of the weld plane relative to the outer surface of the tank, mm – 80;
– the volume of non-destructive control – 100 % of the weld length.

Since the tank design does not allow to place X-ray machine in its cavity for radiography "two walls" of the welded joint (figure 2(a)), it is advisable to use a scheme of panoramic X-ray radiography (figure 2(b)).

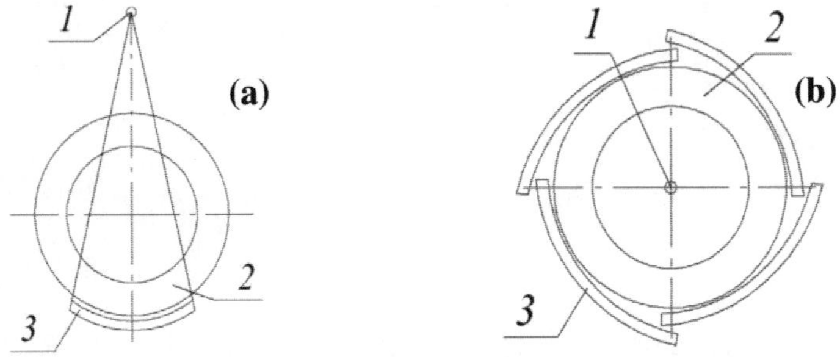

Figure 2. The scheme X-ray radiography of circular welded joint: **(a)** – "two walls"; **(b)** – panoramic X-ray radiography.

As is well known, section 3.5 GOST 75.12 – 82 defines the requirements for the size of the focal spot of X-ray tube used for the implementation panorama scheme of shooting.

$$Ø \le Kd/2(D - d) \text{ [mm]}, \qquad (1)$$

where Ø – focal spot size of X-ray tubes; K – sensitivity control; d – inner diameter of the controlled joint; D – outer diameter of the controlled joint.

In accordance with expression (1) at the required value of contrast sensitivity: $K \le 0.1$; $d = 100$; $D = 110$; focal spot size must not exceed 0.5 mm. In view of these requirements by the specialists of JSC "Svetlana-X-ray" was developed a specialized X-ray tube [2] with the imposition target of the vacuum balloon and "reverse" output panoramic beam 0.144BPV33-120.

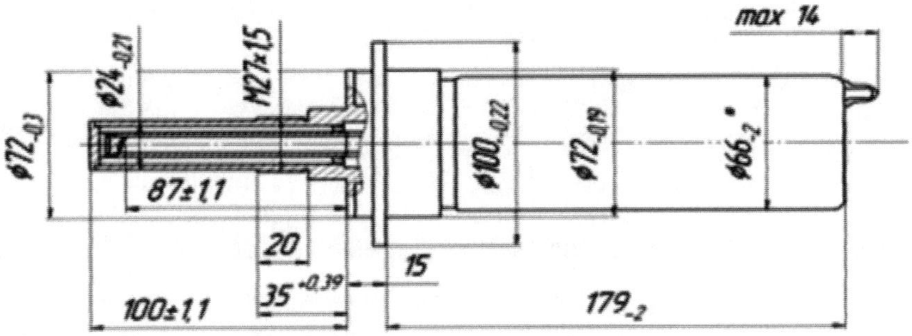

Figure 3. X-ray tube 0.144BPV33-120.

The maximum operating voltage of the tube is 120 kV, the maximum current – 1 mA, the minimum focal spot diameter – 0.1 mm. The tube has a three-electrode cathode assembly, so the electric circuit of the high-voltage power of the tube additionally contains the source of grid (control) voltage. In contrast to the traditionally used in conventional panoramic tubes conical target in the described tube used flat tungsten target on the aluminum anode tube outside of the vacuum balloon. This target provides the necessary uniformity of the intensity of the generated radiation beam around the corner of his solution – 360×40 degrees and therefore, the entire length of the controlled weld.

2. Experiment

On the basis of this X-ray tube in JSC "ELTECH-Med" was developed and mastered in serial production monoblock type X-ray apparatus RAP-120PM (figure 4).

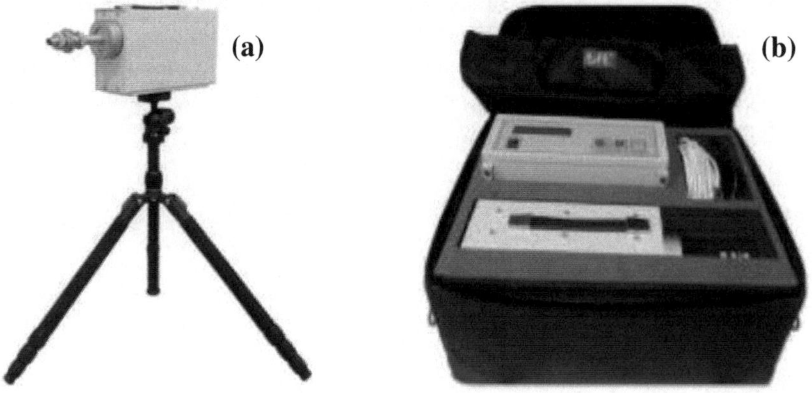

Figure 4. Monoblock of X-ray apparatus RAP-120PM on a tripod **(a)** and a complete set of the apparatus in the transport bag **(b)**.

The experimentally obtained radiation diagram, which generated by the X-ray apparatus shown in figure 5. Zero count on the horizontal axis in figure 5 is the outer end of the X-ray tube target.

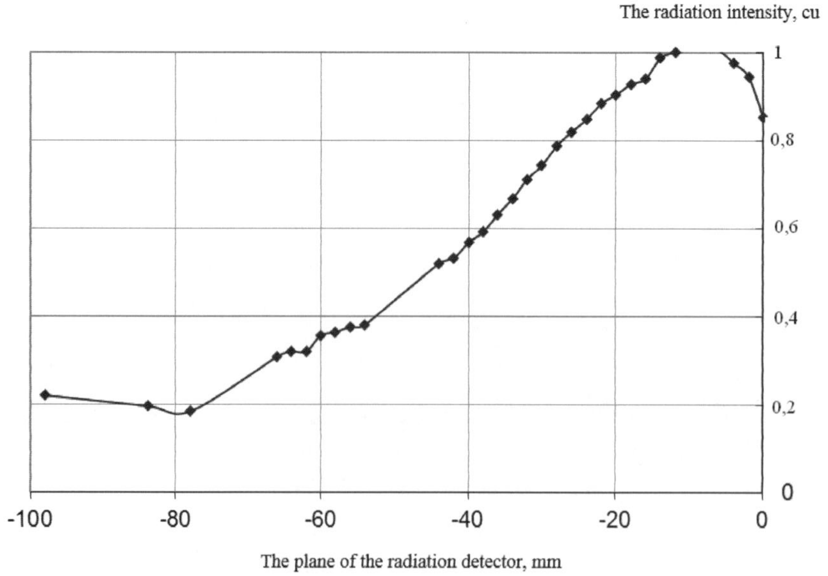

Figure 5. Diagram directivity radiation intensity of the X-ray apparatus RAP-120PM.

The apparatus RAP-120PM was successfully tested at one of the domestic enterprises of the aerospace industry. As an example, figure 6 is an X-ray image of the weld of an aluminum component with a diameter of 140 mm. The total thickness of aluminum walls in the weld region is 14 mm. The picture was taken on a roll X-ray film D7 and "digitized". Shooting modes: voltage – 70 kV, current – 0.1 mA, exposure time – 90 s.

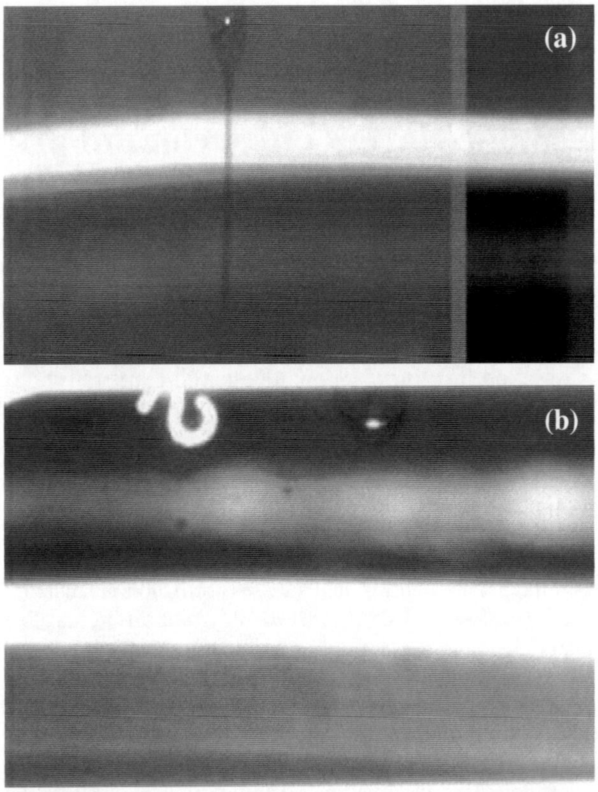

Figure 6. Fragments of an X-ray image of an annular welded seam of an aluminum body part with the image of a standard **(a)** and welding defects **(b)**.

To assess the radiographic sensitivity, a standard № 1 was used. With a standard technique for analyzing X-ray images using a negatoscope, the sixth groove of the standard (the groove dimensions 0.1×0.1 mm) differs confidently on the image of the annular welded seam.

3. Conclusions

The technical and technological solutions tested during the development of the X-ray tube 0.144BPV33-120 and the apparatus RAP-120PM allow satisfying any "reasonable" requirements for voltage, current, focal spot size, and also dimensions and weight of an X-ray source intended for monitoring circular welded seams of rocket and space technology products according to a panoramic scheme of shooting.

Acknowledgments

The work was carried out with the financial support of the Russian Science Foundation in the framework of the project "Robotechnic system of multi-angle microfocus X-ray radiography of aviation parts and assemblies of polymer composite materials under complex influence". Project number 16-19-00155.

References

[1] Anoshkin A N, Osokin V M, Tretyakov A A, Potrakhov N N and Bessonov V B 2017 *Journal of Physics: Conference Series* **808** 012003
[2] Podimsky A A and Potrakhov N N 2017 *Journal of Physics: Conference Series* **808** 012011

IOP Conf. Series: Journal of Physics: Conf. Series **872** (2017) 012039 doi:10.1088/1742-6596/872/1/012039

Structure of the surface layers on zirconium products after dynamic impact of the microbodies

S V Ivanova[1], S Ya Betsofen[2], A A Lozovan[2], D E Molostov[2] and M I Pikus[3]

[1] National Research Nuclear University, MEPhI, 115409, Moscow, Russia
[2] Moscow Aviation Institute (National Research University), 105080, Moscow, Russia
[3] Saint-Petersburg State Electrotechnical University "LETI", S.-Petersburg, Russia

E-mail: s.betsofen@gmail.com

Abstract. The effect of modifying the surface of products made of zirconium alloy E110 by magnetic abrasive treatment (MAT), abrasive blasting treatment (ABT) and cold jet treatment (Cold Jet) on the texture and residual stresses formation is studied. Cold Jet does not change the basal texture of the initial sheet of E110 alloy. The ABT leads to a significant weakening of the basal texture in the initial sheet, the basal texture fraction decreases from 85 down to 44 %, and the proportion of the prismatic texture increases from 4 up to 27 %. Surface modification by MAT results in weak texture changes which are manifested only in the increase of the pole density for non-basal orientations. After the ABT compressive stresses of 400–450 MPa are formed on the surface of the alloy; as the distance from the surface increases, the value of compressive stresses decreases and at a depth of 80 μm the stresses become tensile. The depth of the modified layer after MAT is ~15–20 μm. The value of residual compressive stresses on the surface of the alloy after MAT is 200–250 MPa.

1. Introduction

To improve corrosion resistance of zirconium products surface cleaning operation is required. In this case the use of grinding operation results in tensile residual stresses formation and various types of etching in hydrofluoric solutions lead to an increase in the amount of hydrides in the surface layers. In this work, we investigated the effect of magnetic abrasive treatment (MAT), abrasive blasting treatment (ABT) and Cold Jet thermo kinetic treatment on the structural state of the surface in zirconium products made of the E110 alloy.

The MAT process is the exposure of the surface to magnetic abrasive powders which are elastically fixed in a magnetic field existing in the working gap between magnetic poles of a magnetic abrasive machine. The ABT process consists in the impact-kinetic effect on the surface by working microbodies. The Cold Jet process consists in thermo kinetic surface treatment using dry ice pellets (solid fraction of CO_2) being sprayed by air stream under pressure; its effect on the surface combines thermo shock occurring due to the difference in temperatures of the dry ice (–79 °C) and the treated surface with the kinetic impact of beating granules. Cast iron chipped shot (CICHS) and steel chipped shot (SCHS) were used as magnetic abrasive powders for MAT, and microspheres from glass and oxidized zirconium served as microbodies for ABT.

In this work, we investigated the effect of surface modification on the crystallographic texture of surface layers and distribution of residual stresses in surface layers. These characteristics are chosen because it is known that crystallographic texture influences not only performance of zirconium

IOP Conf. Series: Journal of Physics: Conf. Series **872** (2017) 012039 doi:10.1088/1742-6596/872/1/012039

products [1–6], but also the orientation of monoclinic and tetragonal oxide modifications and their ratio in the oxide layer [7–9]. It is also known that morphology and texture of hydrides depend on texture of the substrate and residual stresses [10, 11]. These effects are conditioned by the presence of orientation relationships between the lattices of zirconium with oxides and hydrides which can change under the influence of residual stresses.

2. Texture and residual stresses after surface modification

Figure 1 shows the results of the study on the effect of treatment type for zirconium components surface on texture characteristics. The texture of the initial material in the form of a sheet with the 2 mm thickness was characterized by a pronounced basal texture with the fraction of 85 % (figure 1(a)), while the prismatic texture fraction was only 4 % (figure 1(b)). Modification of the surface is accompanied by a significant weakening of the basal texture for the ABT process, which is characterized by a reduction in the basal texture fraction to 44 % and an increase in the prismatic texture fraction from 4 to 27 %. Modifying the surface by the MAT leads to slight changes in the texture which are manifested only in an increase in the pole density of non-basal orientations. This is most clearly manifested in a certain strengthening of the prismatic texture fraction (from 4 to 7–8 %), which is clearly seen in figure 1(b).

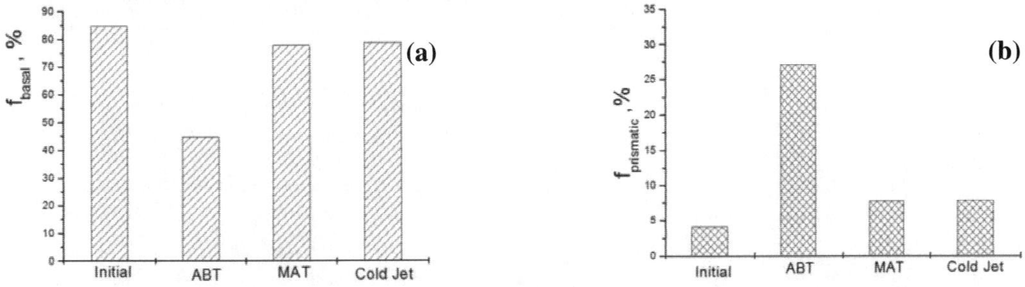

Figure 1. Effect of the surface treatment type for the fuel cladding made of the E110 alloy on texture characteristics: **(a)** the basal texture fraction; **(b)** the prismatic texture fraction.

Figure 2 shows the results of a layer-by-layer distribution of residual stresses after processing by the ABT mode which results in the most noticeable changes in the texture on the surface. Compressive stresses of 400–450 MPa are formed on the surface.

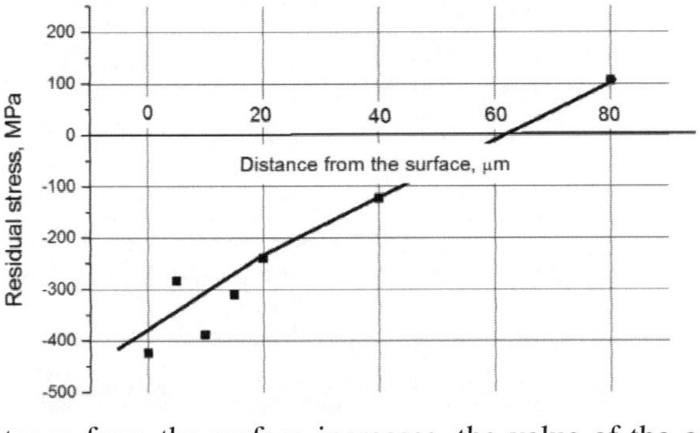

Figure 2. Distribution of residual macro-stresses over the cross section in E110 alloy sheet of 2 mm after ABT.

As the distance from the surface increases, the value of the compressive stresses decreases and at a depth of 80 μm the stresses become tensile. The depth of the modified layer after the MAT by CICHS and SCHS does not differ and equals ~15–20 μm, and the treatment with Cold Jet does not lead to appreciable changes in the texture. The values of the compressive residual stresses in the fuel cladding after the MAT by CICHS and SCHS are 200–250 MPa.

3. Investigation of the oxide layers structure on a modified surface

The formation of oxide layers on the surface of alloy samples differs significantly for different modification variants with a clear correlation between the structure of the metal substrate and the morphology of the oxide layer. The ABT (figure 3(d), (e)) is characterized by the multiple nature of oxidation sites which reflect a strong hardening of the surface accompanied by fragmentation of the initial grains and un-texturing, which leads to an increase in the number of nuclei of oxide grains. After the MAT (figure 3(c), (d)) and Cold Jet (figure 3(a), (b)) the oxide replicates structure of the metal substrate while, as it is clearly seen in the 3D image, the rate of the oxide growth at the grains boundary dominates, which results in discontinuities formation, and in the case of Cold Jet treatment in reduction of protective properties of the surface oxide.

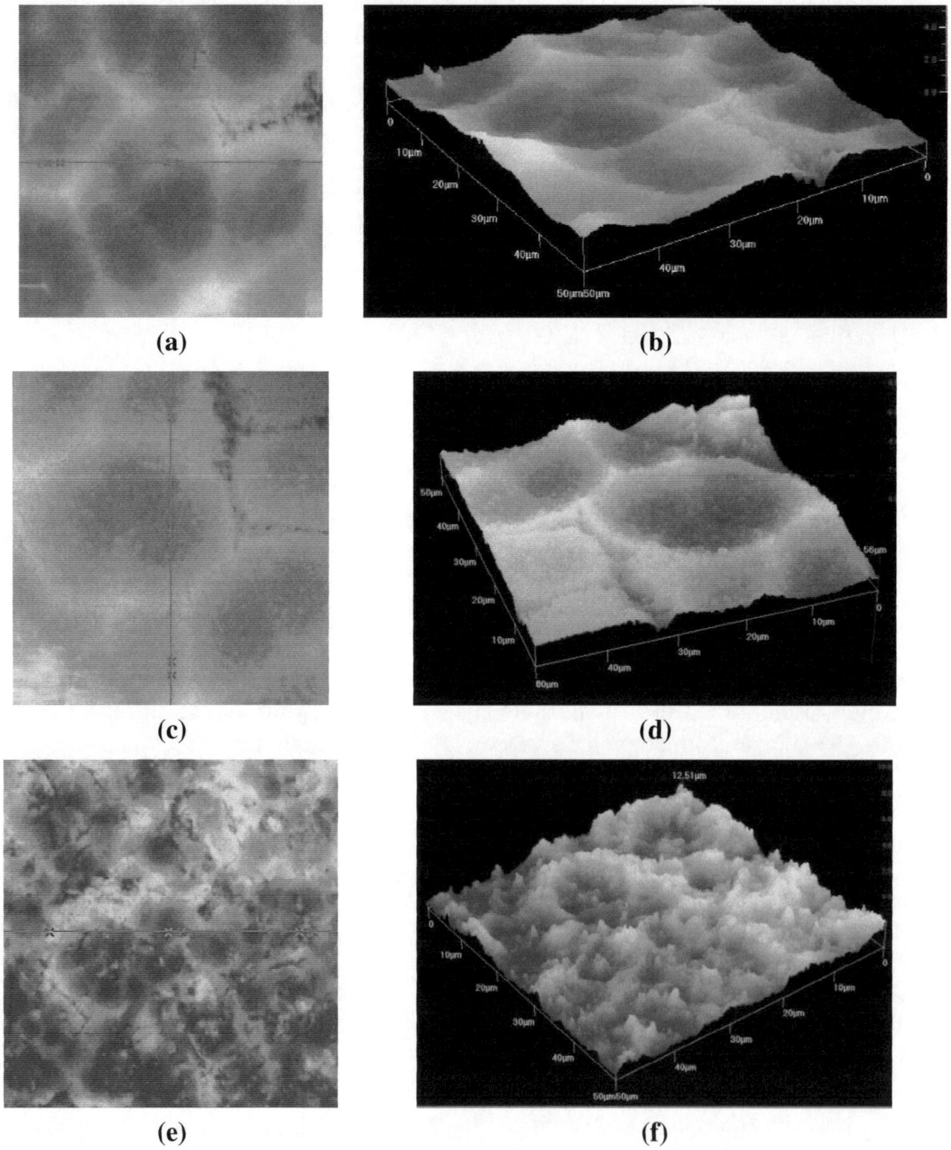

Figure 3. 2D **(a)**, **(c)**, **(e)** and 3D **(b)**, **(d)**, **(f)** images of the oxides surface on the modified surface of the E110 alloy after exposure for 7.2 ks at 800 °C: **(a)**, **(b)** Cold Jet treatment; **(c)**, **(d)** MAT by SCHS; **(e)**, **(f)** ABT.

However, the oxide having been formed on the surface of zirconium after the MAT has higher protective properties. Naturally, in the case of the ABT which corresponds to a much larger fraction of the oxide grains boundaries and, correspondingly, a larger fraction of discontinuities (figure 3(d), (e)), protective properties of the oxide (table 1) are substantially lower than for other modification modes.

Table 1. Effect of the surface treatment method for the E110 alloy on the properties of oxide films being formed during corrosion tests for 30 days at the temperature of 400 °C and the pressure of 20.3 MPa.

Treatment method	Microhardness (HV), kgf/mm^2	Barrier voltage, V	Protective properties index (1/CW), cm^4/mg·μF
Etching	257	250	3.8
Grinding	420	250	2.9
MAT by CICHS	683	300	13.3
MAT by SCHS	695	275	12.2
ABT	880	150	1.8
Cold Jet	280	250	4.5

Acknowledgments
The research was carried out at the expense of a grant from the Russian Science Foundation (project No. 15-19-00241).

References

[1] Kearns J J 2001 *Journal of Nuclear Materials* **299** 2171–4
[2] Murty K L and Charit I 2006 *Progress in Nuclear Energy* **48** 325–59
[3] Romero J, Preuss M and Quinta da Fonseca J 2009 *Acta Materialia* **57** 5501–11
[4] Gruber J A, Brown S A and Lucadamo G A 2011 *Journal of Nuclear Materials* **408** 176–82
[5] Gey N, Humbert M, Gautier E, Béchade JL 2004 *Journal of Nuclear Materials* **328** 137–45
[6] Daymond M R, Holt R A, Cai S, Mosbrucker P and Vogel S C 2010 *Acta Materialia* **58** 4053–66
[7] Szpunar J A, Qin W, Li H and Kumar K 2012 *Journal of Nuclear Materials* **427** 343–9
[8] Lin J, Li H, Nam C and Szpunar J A 2004 *Journal of Nuclear Materials* **334** 200–6
[9] Akhiani H and Szpunar J A 2013 *Applied Surface Science* **285** 832–9
[10] Santisteban J R, Vicente-Alvarez M A, Vizcaíno P, Banchik A D and Almer J D 2010 *Acta Materialia* **58** 6609–18
[11] Vicente Alvarez M A, Santisteban J R, Domizzi G and Almer J 2011 *Acta Materialia* **59** 2210–20

IOP Conf. Series: Journal of Physics: Conf. Series **872** (2017) 012040 doi:10.1088/1742-6596/872/1/012040

Studies of the Cr-CrN coating characteristics formed by means of the magnetron sputtering method from bulk target

G V Kachalin, A F Mednikov, A B Tkhabisimov and S V Sidorov

Moscow Power Engineering Institute, 111250, Moscow, Russian Federation

E-mail: kachalingv@mpei.ru

Abstract. The paper presents the study's results of ion-plasma chromium based coating characteristics produced on blade steel samples 12Kh13 and EI961 by means of the magnetron sputtering method from the bulk "hot" target. A set of metallographic studies and erosion tests of coatings were carried out using the research equipment URI (unique research installation) "Hydroshock rig Erosion-M" of NRU "MPEI". Cr-CrN based coatings have a layered structure; thickness of intermediate Cr layers ranges from 0.7 to 1.7 μm, thickness of nitride layers CrN ranges from 1.5 to 4 μm, while the overall coating thickness is 17.0–21.5 μm; coating microhardness is 1830–1880 HV0.05. The resulting coatings are found to increase 1.5 times the incubation period duration of erosion wear for steels 12Kh13 and EI961; they reduce the maximum erosion rate 1.3 times, and the steady erosion rate – 1.5 times.

1. Introduction

Rotating blades of the high capacity turbine units last stages are particularly susceptible to erosion wear [1]; they have a complex geometry, and are relatively long. In view of that, the production process of building erosion resistant of ion-plasma coatings on these products is quite difficult, and characterized by a combination of subsequent operations [2]. The choice of a particular type of ionic-plasma coating and its data, such as thickness, composition, application method, etc., depends strongly on the material, weight, geometry and operating conditions of specific products. The formation process of such coatings is developed as a result of multi-optional studies in order to create a coating with specified characteristics on experimental samples.

Within the scope of this study, characteristics of the Cr-CrN coating on samples made of blade steels 12Kh13 and EI961 were studied by means of the magnetron sputtering method from the bulk "hot" target. The advantage of using this process lies in increasing the deposition rate of coatings and reducing their cost, compared to cooled targets [3, 4]. Application of the "hot" target ensures high process efficiency and stability. Besides, the possibility of using bulk targets instead of expensive solid target would decrease considerably the cost of the production cycle.

2. Formation methods of ion-plasma coating, methods of metallographic studies and erosion tests

Within the scope of this study, formation of the Cr-CrN coating on samples made of blade steels 12Kh13 and EI961 was carried out by means of the vacuum apparatus "Gefest-HIPIMS" [5] equipped with the "hot" bulk target (figure 1).

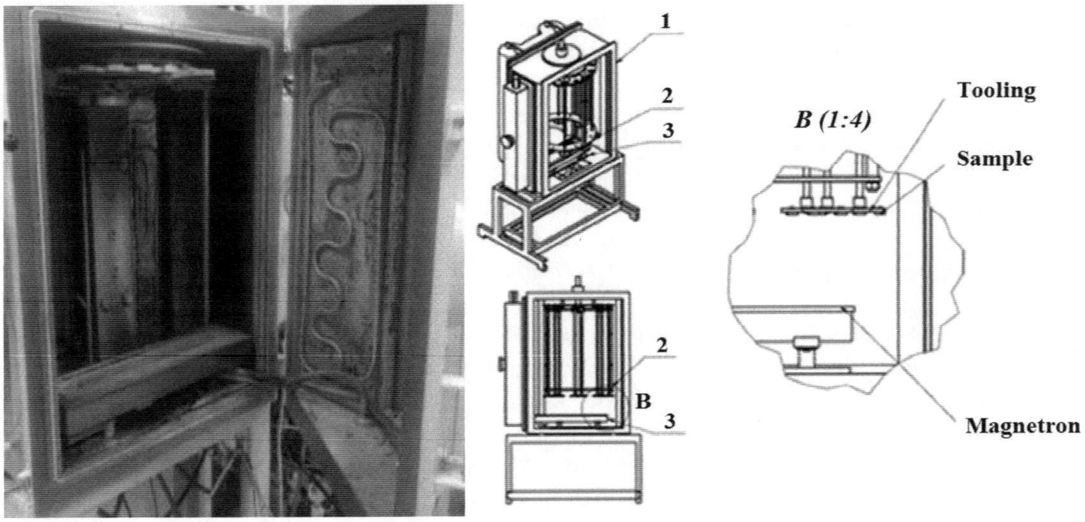

(a) (b)

Figure 1. Photo **(a)** and layout **(b)** of the vacuum apparatus configured for sputtering from the "hot" bulk target, where: *1* – vacuum chamber; *2* – process tooling; *3* – magnetron in the tooling.

The operational sequence of the Cr-CrN coating formation included the following stages:

1) preliminary preparation of sample surfaces (plasma electrolytic polishing, surface cleaning and degreasing);

2) loading of samples into the vacuum chamber (the distance between the processed surface and the source was 100 mm);

3) air pumping from the vacuum chamber to reach the high vacuum condition (the max. pressure was $8\cdot10^{-3}$ Pa) after preheating it to 150 °C to intensify the degassing process of the chamber and the samples;

4) supply of the inert plasma forming gas (argon) to reach the pressure of 1.2 Pa and ignition of glowing discharge between the samples and the vacuum chamber walls;

5) sample surface cleaning and activation by bombing them with the inert gas ions in order to remove the oxide film and activate the surface (the product was exposed to a negative voltage ranging from 800 to 1150 V during 10 min);

6) coating formation on the sample surface using the bulk target (discharge power – 6 kW, offset voltage – 115 V; during the coating deposition process, tooling with the samples was not rotated with relation to the source);

7) process interruption and cooling of the samples in vacuum;

8) removal of the samples from the vacuum chamber.

Characteristics of formed coatings were measured using the research complex URI "Hydroshock rig Erosion-M" of NRU "MPEI" capable of investigating the composition, structure, thickness, microhardness and erosion resistance of protective coatings. Coating structure and thickness were studied using the scanning electronic microscope TESCAN MIRA 3 LMU equipped with the EMF spectrometer X Max 50 (Oxford Instruments). Metallographic sections were made using a set of sample preparation equipment including: abrasive cut-off wheel with movable linear cutting system POWERMET 3000 (Buehler GmbH), electrohydraulic press SIMPLIMET 1000 (Buehler GmbH) and section polishing machine BETA/1 (Buehler GmbH). Coating microhardness as well as microhardness of uncoated samples was measured by means of the hardness meter DuraScan 20 (Emco-Test). Measurements were carried out according to the Vickers method, under load of 0.05 kgf (0.491 N). Testing of the samples for resistance to high-speed impact of liquid drops was carried out on the erosion test bench "Erosion-M" [6]. The test bench "Erosion-M" allows simulating of different interfacing conditions between liquid drops and the surface of test samples: impact speed, size (diameter) of liquid drops, number of drop flows and incidence angle. For the purposes of these studies, the speed of samples during the erosion tests was 300 m/s, and the diameter of liquid drops (distilled water) was 800 μm. In order to perform comparative analysis of the erosion tests results, diagrams were built showing correlation between the average depth of erosion wear $E(m)$ and the weight of the liquid falling on the eroded surface – m (kg/m^2). Based on the correlations obtained, relative erosion resistances of uncoated and coated samples were compares in terms of the incubation period duration, maximum and steady rates of the erosion wear process.

3. Results and discussion

Figure 2 shows characteristic images of cross sections of blade steel samples 12Kh13 and EI961 with Cr-CrN coating. After obtaining the images of cross sections and investigation of coating structure as well as measuring its thickness, the elemental composition was analyzed by means of the energy-dispersive microanalysis. Table 1 shows elemental analysis results for Cr-CrN coatings. Results of formed coatings microhardness measurements as well as of uncoated samples are shown in table 2. Relative erosion resistance comparison of uncoated and coated samples based on the incubation period duration, maximum and steady rates of the erosion wear process is shown in figures 3 and 4.

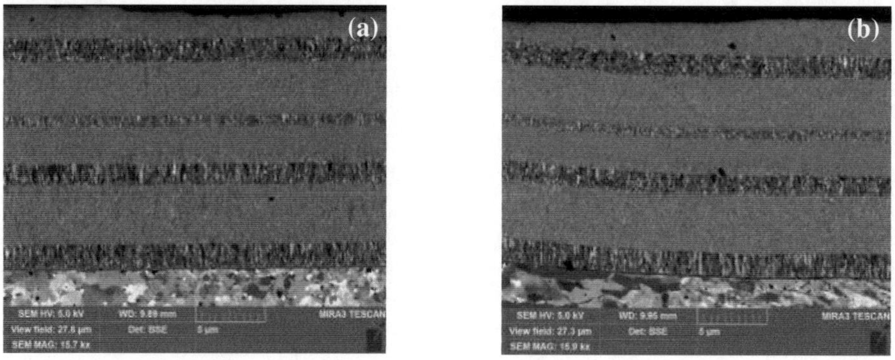

Figure 2. Image of microstructure of Cr-CrN coating on samples made of blade steels 12Kh13 **(a)** and EI961 **(b)**.

Table 1. Results of coatings elemental composition analysis.

Sample	Average elemental contents in the coating, % by weight				
	N	O	Si	Cr	Fe
Cr-CrN coating on steel EI961	0.9 ± 0.4	0.93 ± 0.04	0.12 ± 0.02	97.9 ± 0.6	0.53 ± 0.08
Cr-CrN coating on steel 12Kh13	1.2 ± 0.9	0.77 ± 0.09	0.10 ± 0.02	98.0 ± 0.9	0.58 ± 0.20

Table 2. Results of mechanical properties studies of uncoated and coated steels 12Kh13 and EI961 samples.

Item No.	Sample description	Microhardness acc. to Vickers HV0.05
1	12Kh13, uncoated	271 ± 5
2	EI961, uncoated	280 ± 7
3	12Kh13 with Cr-CrN coating	1830 ± 30
4	EI961 with Cr-CrN coating	1880 ± 40

As a properties comprehensive studies result of the Cr-CrN coating formed on experimental samples of blade steels EI961 and 12Kh13, using magnetron sputtering from the bulk "hot" target, findings were as follows:

– chromium content is 98 %, nitrogen content is 1 %; also, 0.8–0.9 % of oxygen as well as low additives of silicone and iron are observed;

– coatings have a layered structure; thickness of intermediate Cr layers ranges from 0.7 to 1.7 μm, thickness of nitride layers CrN – from 1.5 to 4 μm, while the overall coating thickness is 17.0–21.5 μm;

– coating microhardness is 1830–1880 HV0.05;

– coatings increase the incubation period duration for steels 12Kh13 and EI961 in 1.5 times; they reduce the maximum erosion rate in 1.3 times, and the steady erosion rate in 1.5 times.

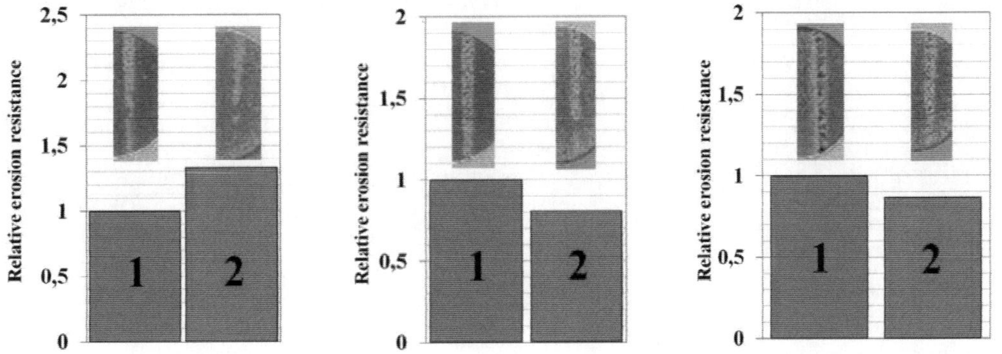

(a) (b) (c)

Figure 3. Histogram of relative erosion resistance based on the incubation period duration **(a)** and on the maximum **(b)** and steady **(c)** erosion rates of uncoated steel 12Kh13 (*1*) and steel 12Kh13 with Cr-CrN coating (*2*).

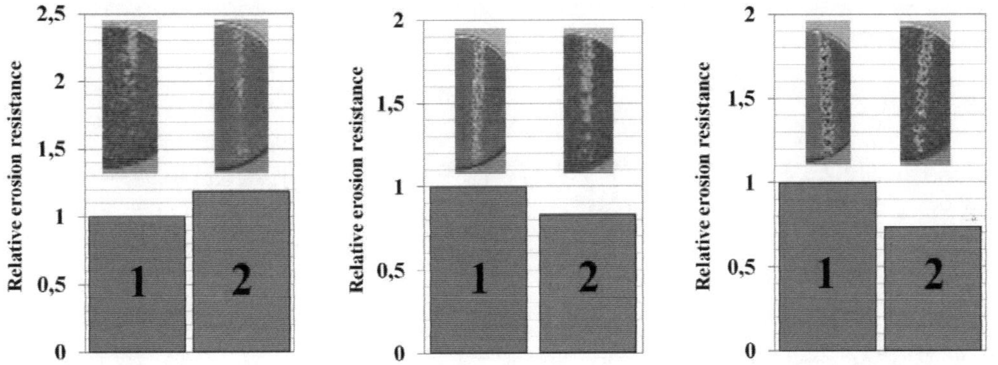

(a) (b) (c)

Figure 4. Histogram of relative erosion resistance based on the incubation period duration **(a)** and on the maximum **(b)** and steady **(c)** erosion rates of uncoated steel EI961 (*1*) and steel EI961 with Cr-CrN coating (*2*).

4. Conclusions

The results obtained within the scope of this study prove that application of ion-plasma coatings produced by means of the magnetron sputtering method from the bulk “hot” target is an advanced method for surface protection of the power equipment from various wear types, including erosion wear. Application of this method also helps to reduce the prime cost of the production cycle compared to widely used cooled targets thanks to enhancing the efficiency due to higher deposition rate of coatings and cost reduction of consumables (target cathodes).

Acknowledgements

This study was financially supported by the Ministry of Education and Science of the Russian Federation under agreement № 14.577.21.0180 dated October 27, 2015 (unique identification number RFMEFI57715X0180).

IOP Conf. Series: Journal of Physics: Conf. Series **872** (2017) 012040 doi:10.1088/1742-6596/872/1/012040

References

[1] Kachalin G V, Mednikov A F, Tkhabisimov A B and Zhukova E A 2016 *Journal of Pharmaceutical, Biological and Chemical Sciences* **7(5)** 955–63
[2] Kachalin G V, Mednikov A F and Tkhabisimov A B 2014 *WIT Transactions on Ecology and the Environment* **186** 729–37
[3] Daninlin B S and Syrchin K 1982 *Magnetron sputtering systems* (Moscow: Radio i Svyaz)
[4] *Patent US 6599405 B2 Recessed sputter targets 29.07.2003*
[5] Kachalin G V, Ryzhenkov A V, Medvedev K S, Bychkov A I and Parfenenok M A 2014 *Nadezhnost i Bezopasnost Energetiki* **2(25)** 8–12
[6] Ryzhenkov V A, Seleznev L I, Mednikov A F and Tkhabisimov A B 2014 *Thermal Engineering* **61** 594–9

X-ray photoelectron spectroscopy method for obtaining a $CuIn_{0.95}Ga_{0.05}Se_2$ thin film: development of technology and a phase composition studying

Kh Kh Kalazhokov[1], Z Kh Kalazhokov[1], T M Gadjiev[2], M A Aliev[2], A M Ismailov[2], R M Gadjieva[2], A Sh Asvarov[2], R K Arslanov[2], E N Kozyrev[3], V I Filonenko[3] and R O Askerov[3]

[1] Kabardino-Balkarian State University (KBSU), 360000, KBR, Nalchik, ul. Gorky, 5, Russia
[2] Amirkhanov Institute of Physics, Daghestan Scientific Center, Russian Academy of Sciences (RAS), 367003 Makhachkala, Russia
[3] North Caucasian Institute of Mining and Metallurgy (State Technological University), NCIMM (STU), 362021, North Ossetia-Alania, Vladikavkaz, ul. Nikolaeva 44, Russia

E-mail: asker2005@rambler.ru

Abstract. A technology of two stage selenezation method in the carrier gas (nitrogen) stream of the reaction component (selenium) has been suggested for the $CuIn_{0.95}Ga_{0.05}Se_2$ thin film growing. Morphology and structure of the $CuIn_{0.95}Ga_{0.05}Se_2$ film have been studied by SEM and XRD methods. Phase compound of film material for the $CuIn_{0.95}Ga_{0.05}Se_2$ semiconductor solution has been studied by X-ray photoelectron spectroscopy method.

Thin films of $CuInGaSe_2$ with chalcopyrite structure are promising materials from the viewpoint of their application as absorbing layer in solar cells. However, depending on the technology method for obtaining these thin films, there is a dispersion of the electrophysical and photoelectric parameters of the photoconverters [1]. This is due to the phase composition and the structure of these films. A selenization method is a promising route in production of not only $CuInGaSe_2$ thin films, but also many semiconductor materials containing a volatile component in their composition [2, 3].

The aim of this work is a development of technology for obtaining and studying the phase composition of a $CuIn_{0.95}Ga_{0.05}Se_2$ semiconducting solution film using X-ray photoelectron spectroscopy.

For deposition of preliminary intermetallic $CuIn_{0.95}Ga_{0.05}$ films on the glass substrates a dc magnetron sputtering method was employed by using the "Watt AMK-MI" sputtering setup. The vacuum system of the sputtering setup consists from oil-free pumping tools (Anest Iwata ISP-500C dry spiral pre-vacuum pump, cryogenic high-vacuum pump – Cryogenics CryoTorr 8). For vacuum measurement a wide-range SS-100 vacuum gauge was used. An ion source is placed in the vacuum chamber that allows to carry out the final cleaning of the substrate before deposition and ionic assisting of the deposited layers. The target was a conductive disc of 40 mm in diameter and 4 mm in thickness. The target was synthesized by melting of initial elements placed in a quasi-hermetic container with using a high-vacuum furnace of SNVE 1.3-1/16. The process of the target synthesis was carried out at a pressure of 10 Pa and a temperature $T = 1100$ °C for 0.5 h.

IOP Conf. Series: Journal of Physics: Conf. Series **872** (2017) 012041 doi:10.1088/1742-6596/872/1/012041

Thin films of $CuIn_{0.95}Ga_{0.05}Se_2$ were obtained by two-step method of controlled selenization of copper-indium gallium layers, in a two-zone thermal diffusion device with the participation of the carrier gas (nitrogen) of the reaction component (selenium).

The surface morphology and the chemical composition of the film $CuIn_{0.95}Ga_{0.05}Se_2$ were studied using a scanning electron microscope LEO-1450 equipped with EDX. Chemical analysis of thin films showed that the film composition is quasi-stoichiometric and the inhomogeneity of the distribution of the components along the film surface exceed 3 %. Figure 1 shows a micrograph of the film surface with a thickness of 1.3 μm. It is established that the film consists from micrograins, with the merging of micrograins into conglomerates.

Figure 1. Microphotographs of $CuIn_{0.95}Ga_{0.05}Se_2$ films.

X-ray diffraction studies were performed on the Empyrean Series 2 diffractometer (PANalytical, Netherlands) using Cu-K_α radiation (1.5405 Å) within scanning angles of 10–140 deg, using β-filter. Photography scheme was performed by Bregg-Brentano, geometry – 2θ. The phase identification was carried out by comparing experimentally obtained values of interplanar spacings (*d*) with the data of Joint Committee on Powder Diffraction Standard (JCPDS) and calculation methods.

Typical X-ray patterns for $CuIn_{0.95}Ga_{0.05}Se_2$ thin-films are presented in figure 2. There are replicable diffraction lines characteristic of chalcopyrite: (112), (013), (121) (220/024) (116/132) (040/008) (136/332) (224/228), and the line (112) has a maximum of intensity. According to data for interplanar distances a unit cell parameters for the synthesized films are calculated, in good agreement with JCPDS card.

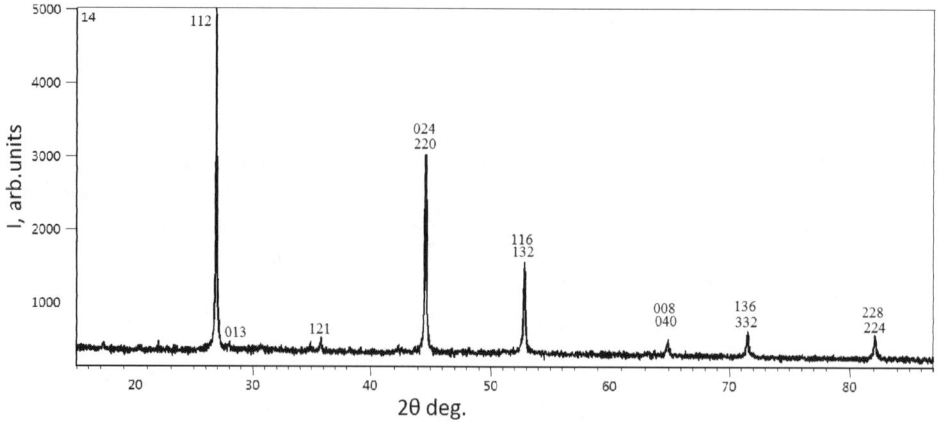

Figure 2. The X-ray diffraction pattern of $CuIn_{0.95}Ga_{0.05}Se_2$ films.

The chemical composition was studied by XPS method on the X-ray photoelectron spectrometer of the K-Alpha system at the Center for Surface Physics, Nanosystems and Nanostructure Technology of the Kabardino-Balkar State University. During the experiments, a vacuum in the analyzer chamber was maintained at least of $4.5 \cdot 10^{-9}$ mbar. To compensate the charge displacement of the photoelectron peak positions, a low-energy electron gun was used and the spectra were calibrated by the position of the C1*s* carbon line with a binding energy of 284.7 eV.

Before studying the composition, the surface was etched with argon ions until completely purified from atoms of chemisorbed oxygen and carbonaceous contaminants. Figure 3 shows overview spectrum for synthesized sample of $CuIn_{0.95}Ga_{0.05}Se_2$, which mainly indicates next elements lines: doublet of copper Cu2*p*, indium In3*d*, selenium Se3*d* and peak of Ga2*p*, as well as intense peaks of carbon C1*s* and oxygen O1*s*.

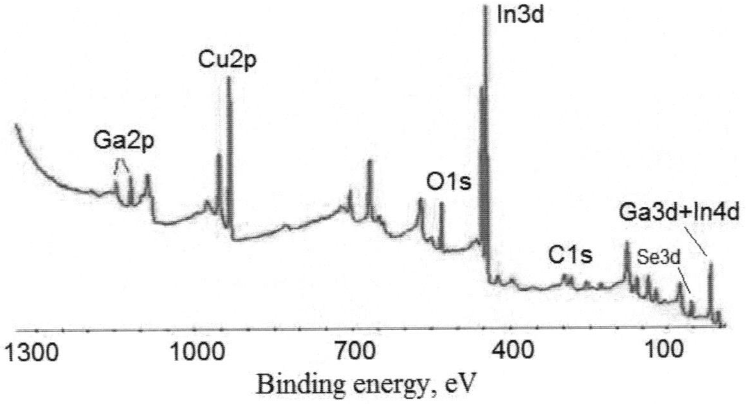

Figure 3. Overview spectra of the Cu(In, Ga)Se_2 film surface.

The atomic concentrations of components obtained from analysis of the overview spectrum and high-resolution spectra are correlated with the chemical analysis data. To analyze the chemical states of elements forming the film material, the chemical shifts of the peaks Se3*d*, Cu2*p*, In3*d* and Ga3*d* were considered. Significant changes in the positions of the peaks that are not related to their surroundings can be observed due to the effect of charging the surface under X-rays. Therefore, Auger parameters have been calculated for these components, which are the sum of the kinetic energy of the Auger peak and the binding energy of an electron for given line. The Auger parameters obtained from the analysis are given in table 1.

Table 1. The values of the binding energies of electrons and Auger parameters for material components of the Cu(In, Ga)Se_2 film.

Elements	PE-peak (eV)	Auger peak (eV)	α′ (eV)	α′ (eV) from [1]	Compounds
Se	3*d*: 54.1	LMM: 1307.4	1361.5	1361.5	$CuSe_2$
Cu	2*p*: 932.7	LMM: 917.18	1849.8	1849.8	$CuSe_2$
In	3*d*: 444.65	MNN: 407.46	852.1	852.4	In_2Se_3
Ga	3*d*: 19.2	LMM: 1063	1082.2	1082.8	Ga_2O_3

The Auger parameters for Se and Cu elements are exactly the same as given in the [4] for $CuSe_2$ phase. For indium, the Auger parameter practically coincides with the values recommended in the [4] and presumably, indium can form the In_2Se_3 compound.

In the case of gallium a partial overlap of Ga3*d* and In4*d* peaks occurs. Thus, the resulting photoelectron peak becomes asymmetrical. This problem is resolved by expanding the Gaussian function for this peak and obtaining the binding energy for Ga3*d* line that is necessary to determine the Auger parameter (figure 4). Apparently, all gallium is connected to oxygen lattice in the form of Ga_2O_3 [5].

A technology of two stage selenezation method in the carrier gas (nitrogen) stream of the reaction component (selenium) has been suggested for the $CuIn_{0.95}Ga_{0.05}Se_2$ thin film growing.

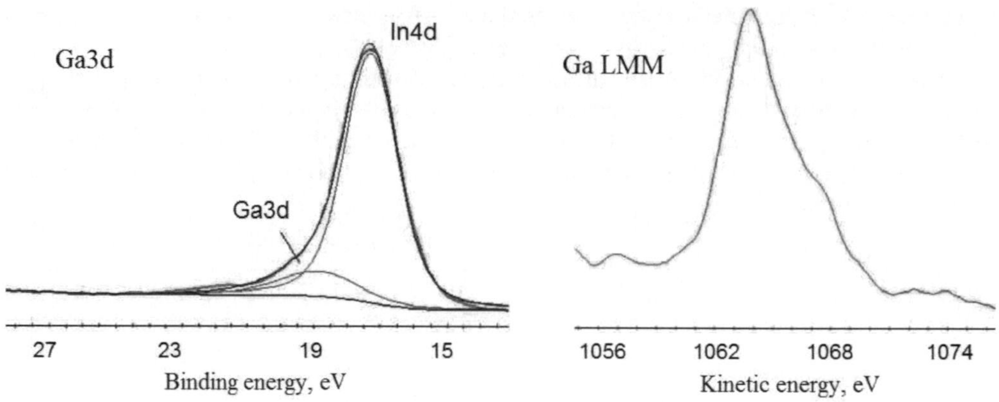

Figure 4. The high-resolution spectrum of the Ga3*d* line and the Auger line Ga LMM.

Morphology and structure of the $CuIn_{0.95}Ga_{0.05}Se_2$ film have been studied by SEM and XRD methods.

Phase compound of film material for the $CuIn_{0.95}Ga_{0.05}Se_2$ semiconductor solution has been studied by X-ray photoelectron spectroscopy method.

Acknowledgments
The results of the work were obtained with the financial support of the Ministry of Education and Science of the Russian Federation, the unique identifier of the work RFMEFI57715X0196.

References

[1] Jackson P, Hariskos D, Wuerz R, Kiowski O, Bauer A, Magorian Friedlmeier T and Powalla M 2015 *Phys. Status Solidi RRL* **1** 28–31
[2] Bilalov B A, Gadjiev T M and Sapharaliev G K *RU Patent № 2354006 27.04.2009*
[3] Gadjiev T M, Gadjieva R M, Arslanov R K, Kurutsov M G and Zubailov I G *RU Patent № 116614 27.05.2012*
[4] Niemi E and Stolt L 1990 *Surf. Interface Anal.* **15** 422
[5] Mizokawa Y, Iwasaki H, Nishitani R and Nakamura S 1978 *J. Electron Spectrosc. Relat. Phenom.* **14** 129

IOP Conf. Series: Journal of Physics: Conf. Series **872** (2017) 012042 doi:10.1088/1742-6596/872/1/012042

The method of controlling the thickness of the deposited film on the basis of the surface plasmon resonance effect

A E Komlev[1], R V Dyukin[2] and E S Shutova[1]

[1] Department of Physical Electronics and Technology, St. Petersburg State Electrotechnical University, Prof. Popov St. 5, St. Petersburg, Russia
[2] JSC LOMO, Chugunnaya St. 20, St. Petersburg, Russia

E-mail: a.komlev@mail.ru

Abstract. New method, based on surface plasmon resonance, for thickness control of thin films deposition in vacuum is offered by authors. This method allows to determine with high accuracy thickness of nanoscale optical coatings. Numerical simulation of surface plasmon resonance conditions in Kretchman geometry and it's frequency changes during deposition of different thickness Ta_2O_5 films are presented. Modeling was given in WinSpall program.

1. Introduction

Different methods are used nowadays for control of thin film thikness during deposition in vacuum. Widely used two methods – quartz frequency change (mass change method) and photometric measurements (reflectance/transmission coefficient change). Both of them have limitations of usage in nanoscale film thikness control. This limitations are especially relevant for multilayer optical coatings task.

Authors offer method for film thikness detection prevented imperfection of quartz and photometric systems. Excitation of surface electromagnetic wave (SEW) in Kretchman geometry was taken as a base.

2. Surface electromagnetic waves

Surface electromagnetic waves or surface polaritons are special kind of electromagnetic waves, propagating along boundary matter surface. Field of this waves localized in near surface layer with dimention, as usual, about wave length. As a result of this localization, even insignificant dielectric permittivity changes take an effect on SEW excitation conditions [1–3].

Different dielectric permittivity sign of bordering matters is an indispensable condition for SEW exitation. Surface wave energy produced by electromagnetic field of falling light and a component linked to resonance properties of matter. It can be lattice mechanical oscillations (phonons), electron oscillations in metals (plasmon), excitons in semiconductors. Independently of quasi particle, taking part in SEW excitation – surface localization is typical for all of them [4].

Conversion of incident light in SEW by prism method is based on total internal reflection effect at falling of *p*-polorized light on sarface active layer (SAL) from matter with higher optical density. Propagation length of SEW along surface is strongly sensitive to processes in absorption layer, surface state and it's changes, adsorption layers presence, films, surface roughness and other. It provides possibility to analyse surfaces and matters boundary with SEW mechanism.

Angle of incidence is chosen by well-known in optics condition:

$$\theta > \left[\arcsin\left(\frac{\varepsilon_2}{\varepsilon_3} \right) \right]^{1/2}; \ \varepsilon_2 < \varepsilon_3, \tag{1}$$

were ε_2 – dielectric permittivity of air; ε_3 – dielectric permittivity of prism material. Surface plasmon exitation method exist in two modifications: Otto (figure 1(a)) and Kretchman (figure 1(b)) geometry. Generally lasers are light sources for SEW exitation.

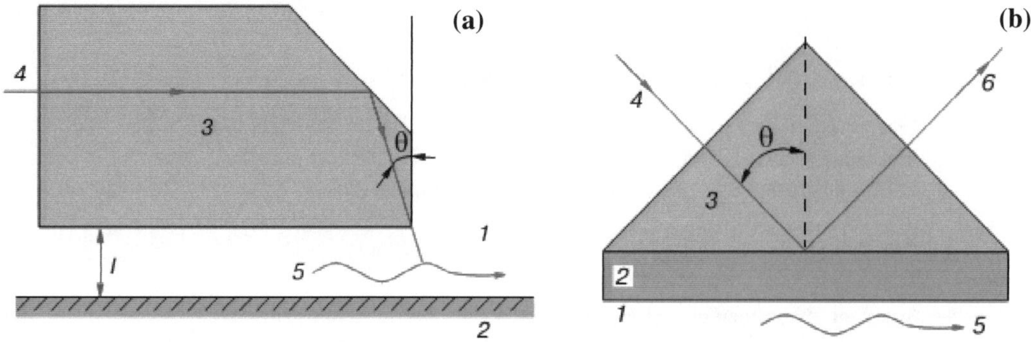

Figure 1. Exitation of SEW by light: **(a)** – prism method in Otto geometry; **(b)** – Kretchman geometry; *1* – dielectric (air); *2* – surface active layer; *3* – prism; *4* – incident light; *5* – surface electromagnetic wave; *6* – mirrored light.

Convertion effectiveness of incident light into SEW in Otto scheme, usualy applied for infrared radiation region, is about one tenth, while in Kretchan scheme, for visible region, reaches up to 1. But in last case taking out SEW from prism is embarrassed and SAL usualy deposite on prism facet as a thin film [4].

3. Film thikness control method

Thin film thickness detection method, proposed by authors, prevent from disadvantages of photometric and quartz resonator methods. As mentioned above, resonance exciting of electron gas oscillations (plasmons) by electromagnetic wave in thin layer of conductive matter, placed between materials with different refractive index, is called surface plasmon resonance (SPR) [1]. In most cases for SPR effect used system wich consists of opticaly transparent material with high refractive index (fused silica, for example [4]) and thin gold film, deposited on it's surface. High chemical inactivity makes gold mainly used material for SPR applications [4].

Layer of gold deposit on hypotenuse facet of the prism. According to numerical simulation and literature 50 nm film is optimal. At definite incidence angle (behind total reflection) of laser beam on gold surface (from prism side) SEW excites at gold/air boundary. Surface plasmon frequency depends on dielectric permittivity of gold ε_1 and contiguous matter ε_2 by dispersion ratio:

$$\frac{\omega_p}{k_{px}} = \frac{c\sqrt{\varepsilon_1 + \varepsilon_2}}{\sqrt{\varepsilon_1 \varepsilon_2}}, \tag{2}$$

were k_{px} – wave vector projection of p-polarized light on x axis.

If such system mount in vacuum chamber and start material deposition then on gold surface will growth thin film. This leads to ε_2 change and, correspondingly, change of surface plasmon frequency. Now resonance conditions will be corresponding to another incidence angle of laser beam.

Illuminating prism by laser beam with some aperture we will see resonance angle shifting during film deposition. Main advantage of proposed method is that system directly registrate dielectric permittivity change of material deposited on sensor surface. It allows calculate optical thickness of the layer in real time.

4. Modeling results

Results of modeling in WinSpall program are shown below. Light source is *p*-polarized He-Ne laser beam. Wave length is 632.8 nm. Dielectric permittivity of prism material is $\varepsilon_3 = 2.2952$. Dielectric permittivity of gold is $\varepsilon_1 = -12.033 + i1.1634$. Illuminating 50 nm thick gold film from prism side by laser beam with 40° aperture we will see on screen (or CCD) narrow strip of resonance at 43°.

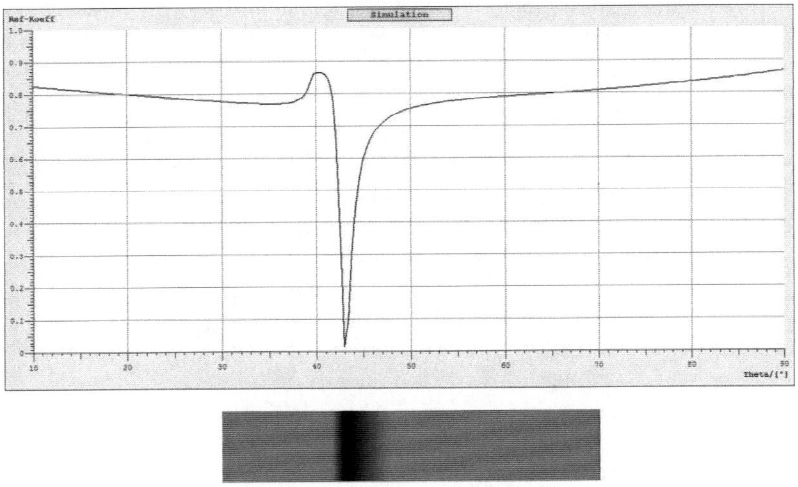

Figure 2. SEW excitation by He-Ne laser. Gold film thickness is 50 nm.

As mentioned above – growing film thickness can be determine by SPR angle shift. To show this, different thickness of Ta_2O_5 film, deposited on gold surface, ware simulated. Dielectric permittivity of Ta_2O_5 is $\varepsilon(Ta_2O_5) = 4.4958 + i0.0041967$. Shifting of signal minimum on screen (or CCD) allows to watch on resonance angle change.

Numerical simulation results for different thickness of Ta_2O_5 film (1; 5; 10; 15; 20; 25 and 30 nm) are shown on figure 3 and in table 1.

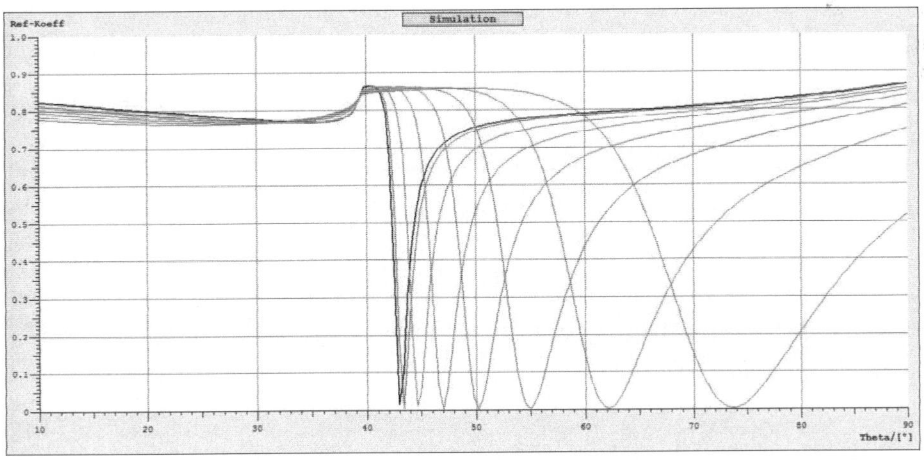

Figure 3. Numerical simulation results of SPR angle shift for different Ta_2O_5 film thickness. The values of thickness and angels are shown in table 1.

Table 1. Modeling results.

Ta_2O_5 thickness (nm)	0	1	5	10	15	20	25	30
Resonance angle (deg)	43.0	43.4	44.7	46.9	50.2	55.0	62.1	73.7

IOP Conf. Series: Journal of Physics: Conf. Series **872** (2017) 012042 doi:10.1088/1742-6596/872/1/012042

As it is seen from presented results – proposed method allows thickness measuring of nanoscale films with high accuracy. Now we are working on measuring system prototype, which will be mounted in vacuum material evaporation plant as a real time thickness detection system.

5. Conclusions

Computer modeling results shows operability of invention such thickness measuring system. Offered by authors method allows to calculate optical thickness of growing nanoscale film in real time, that is relevant task for optical coatings evaporation.

References

[1] Schasfoort R B M and Tudos A J 2008 *Handbook of surface plasmon resonance* (Cambridge: The Royal Society of Chemistry)
[2] Mayer K M and Hafner J H 2011 *Chemical Reviews* **111** 3828–57
[3] Li M, Cushing S K and Wu N 2015 *Analyst* **140** 386–406
[4] Oates T W H, Wormeester H and Arwin H 2011 *Progress in Surface Science* **86** 328–76

Research of mechanical stresses in micromechanical structures based on silicon carbide films produced by magnetron sputtering

O N Mikhailova, A V Korlyakov and A V Lagosh

Saint Petersburg State Electrotechnical University, 197376, Saint Petersburg, Russia

E-mail: novaja.tasamaya@yandex.ru, akorl@yandex.ru, anton.lagosh@gmail.com

Abstract. Investigations of the effect of residual atmosphere in the magnetron chamber on the mechanical stresses and the shape of micromechanical structure based on SiC film are discussed. Measurements of the curvature radius of SiC micromechanical structure deflection are presented.

1. Introduction

Mechanical stress is a film parameter of various materials. It determines the characteristics and shape of micromechanical structures formed on their basis. For example, it was found that an undesirable structure bending occurs during carrying out the micromechanical switch technology. This leads to an increase of the driving voltage value and deterioration of speed and reliability.

The reason of this deflection is the mechanical stress that occurs during the deposition process of a silicon carbide film. Thus, it is obvious that mechanical stresses in silicon carbide films are required further investigations.

2. Deflection and mechanical stresses in micromechanical structures based on silicon carbide films

In this paper, the mechanical stress of released micromechanical structures based on silicon carbide films was investigated. Structures were anchored to the substrate in various ways. These structures were produced by the surface micromachining technology with the help of sacrificial layers. These layers were subsequently etched, and a released micromechanical structure was formed. Silicon carbide films were deposited by magnetron sputtering.

Silicon carbide films bent after releasing from the substrate during the process of sacrificial layers etching. An image of one of these structures is shown in figure 1. It was obtained with the help of a scanning electron microscope.

These deflections lead to malfunction and even a failure of micromechanical switches based on silicon carbide films. The measurements of the deflection were carried out by means of a three-dimensional model created with the help of the optical microscope KH 7700.

The created model was cut by a plane, and the deflection value was estimated from the cross-section profile. The maximum gap was 3.5 μm, and the radius of curvature was 2.4 mm. The three-dimensional model and the cross-sectional profile are shown in figure 2.

Deflection can occur due to the occurrence of high compressive mechanical stresses in silicon carbide films and due to the nonuniform distribution of mechanical stresses along the thickness of the film. Conceivably it is connected to the conditions of the SiC film deposition process.

Figure 1. Image of a micromechanical structure bent under the action of mechanical stresses.

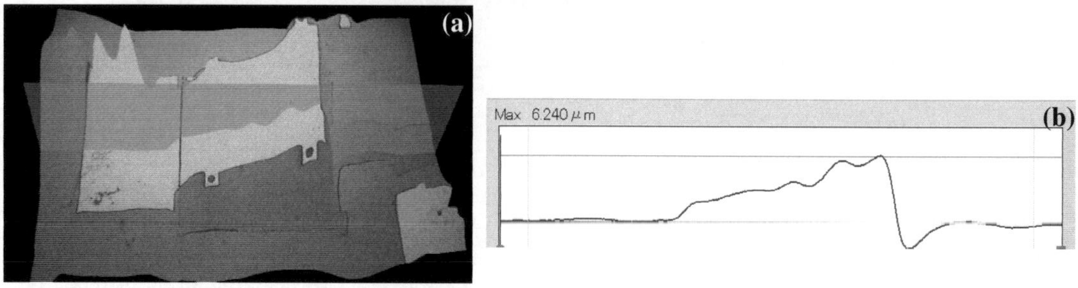

Figure 2. The three-dimensional model **(a)** and the cross-sectional profile of the micromechanical structure based on silicon carbide film **(b)**.

This is also proved by experimental measurements of mechanical stresses in micromechanical membranes based on silicon carbide film [1]. At the thickness of 0.24 µm, the average mechanical stresses in the silicon carbide film are compressive and range from –500 to –100 MPa, depending on the deposition conditions.

As the thickness of the film was increased to 1.2 µm, the mechanical stress becomes tensile and reaches 120 MPa, which may indicate an nonuniform distribution along the thickness and possible influence of the oxidation process at the initial stages of the film deposition process.

3. Technological parameters correction of SiC film deposition process to minimize mechanical stresses and micromechanical structure deflection

High compressive stress in silicon carbide films can occur because of plasma-stimulated oxidation. This process can be nonuniform during the magnetron deposition of the film. It was really noted that the residual pressure at the beginning of the process is almost twice as much as the value at the end of deposition process. This fact indicates that growth conditions of the film varies at different stages of its deposition.

The silicon carbide film was investigated by X-ray spectral microanalysis. It was found that the concentration of oxygen impurities from the bottom surface of the release film are greater than from the upper surface. X-ray characteristic radiation spectra was obtained with the help of a scanning electron microscope and results are shown in figure 3.

So, the chamber was preliminary pumped for a long time. This operation made it possible to reduce the influence of plasma-stimulated oxidation at the initial stages of silicon carbide film growth. The change of the chamber residual pressure during the silicon carbide film deposition was reduced by a factor of over 3. Besides, ehe radius of curvature increased and the maximum gap decreased by a factor of 3.5.

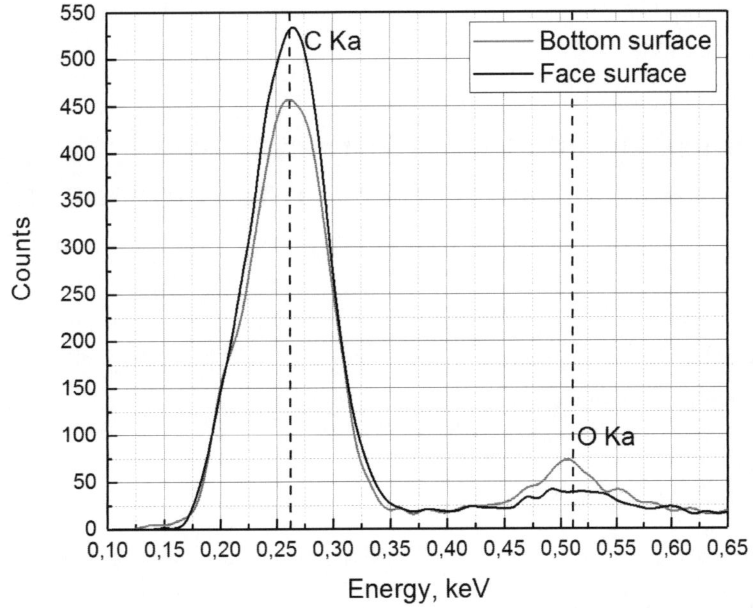

Figure 3. The X-ray spectra from the bottom and face surface of the film.

4. Results and discussion

Thus, one can draw a conclusion that the deflection shape of the released micromechanical structures based on silicon carbide films depends on the parameters of the deposition process namely the pressure change of the residual atmosphere during deposition process of the silicon carbide film by the magnetron sputtering plant.

Nonuniform distribution of mechanical stress along the film thickness occurs because of the time nonuniform process of plasma oxidation. These stresses results in high bending of the film after releasing from the substrate.

5. Conclusions

If one wants to obtain uniform mechanical stresse distribution along the film thickness, it is necessary to preliminary pumped the magnetron chamber for a long time and maintain the same residual pressure during the whole deposition process of the silicon carbide film.

References

[1] Astashenkova O N and Korlyakov A V 2014 *Modern science researches, ideas, results, technologies* **2(15)** 57–61

Improvement of the measurement accuracy of the spectral method for evaluation parameters of the optically transparent thin films

A A Uhov, D K Kostrin, V A Gerasimov, L M Selivanov and V A Simon

Department of electronic instruments and devices, Saint Petersburg Electrotechnical University “LETI”, 197376, Saint Petersburg, Russia

E-mail: aauhov@yandex.ru

Abstract. In this paper the ways of increasing the measurement accuracy of the parameters of the optically transparent thin films using the spectral method are described. The proposed plotting of the envelope curves for the minima and maxima of the interference oscillations in a film and additional filtering of the spectral data, allows improving the accuracy of determining the extrema position and, consequently, increasing the accuracy of the thickness measurements of the deposited coatings.

Thin films and coatings are widely used in industry and household appliances and devices. Control of the parameters of such coatings should be conducted using the non-destructive methods, the most suitable of which is the spectral method [1, 2].

The process of radiation propagation in thin films and related effects are well studied and described. Registered with the spectrometer passed through the film or reflected from the film radiation has oscillations, characteristics of which depend on the thickness and refractive index of the film and the substrate material on which the film is applied [3, 4]. Having information about the film and the substrate materials, and knowing, therefore, their refractive indices it is possible to calculate the film thickness according to the position of the neighboring extrema. However, this method has a significant drawback which is the difficulty of determining the exact position of the extrema of the oscillations in the real noisy spectrum, especially when the control of the film parameters is conducted in the conditions of industrial production.

Industrial manufacture of products with film coating means a rather high reproducibility of their parameters, and, consequently, the known refractive indices of the used materials. This fact allows increasing the calculation accuracy of the measured film parameters through the use of the theoretical calculations of the reflectance or transmittance spectrum shape. For measurements of film parameters in real time can be used a compact spectrometer with fiber-optic input of the radiation and an optical scheme of the experimental installation, allowing to carry out measurements both on reflection and transmission, depending on the type of a substrate [5, 6]. Parameters of the spectrometer must allow measurement of the transmittance or reflectance in a wide wavelength range; meanwhile there are no special requirements on the spectral resolution of the device, since the extrema of the recorded spectrum are obviously separated from each other by more than a few nanometers.

During the passage of monochromatic radiation through a thin film occurs interference, which in a broad spectral range looks like oscillatory modulation of the spectral signal. The theoretical form of

IOP Conf. Series: Journal of Physics: Conf. Series **872** (2017) 012044 doi:10.1088/1742-6596/872/1/012044

the spectrum for the film of a given thickness with known refractive index at normal incidence of radiation can be modeled based on expressions for the transmission T and reflection R coefficients [7]:

$$T = \frac{n_2}{n_0}\left(\frac{t_1^2 t_2^2}{1 + 2r_1r_2\cos 2\delta_1 + r_1^2 r_2^2}\right); \quad (1)$$

$$R = \left(\frac{r_1^2 + 2r_1r_2\cos 2\delta_1 + r_2^2}{1 + 2r_1r_2\cos 2\delta_1 + r_1^2 r_2^2}\right), \quad (2)$$

here

$$\delta_1 = \left(nd\frac{2\pi}{\lambda}\right);\ t_1 = \left(\frac{2n_0}{n_0 + n}\right);\ t_2 = \left(\frac{2n}{n + n_2}\right);\ r_1 = \left(\frac{n_0 - n}{n_0 + n}\right);\ r_2 = \left(\frac{n - n_2}{n + n_2}\right), \quad (3)$$

where d – film thickness; n_0 – refractive index of the air; n – refractive index of the film; n_2 – refractive index of the substrate, provided that the values of refractive indices $n > n_0$; $n > n_2$.

In the general case, the refractive indices of the film and substrate materials have dispersion dependent on the wavelength of the radiation. Therefore, in expressions (3) it is necessary to use not constants, but the values calculated for each wavelength.

For calculation of the film parameters the spectrometer management program ASpect2010 [5] has been modified – was added calculation of the theoretical shape of spectrum using expressions (1)–(3) for a given film thickness with a known refractive index. The program has a database of most common materials used for film deposition, taking into account the dispersion of the refractive index (figure 1).

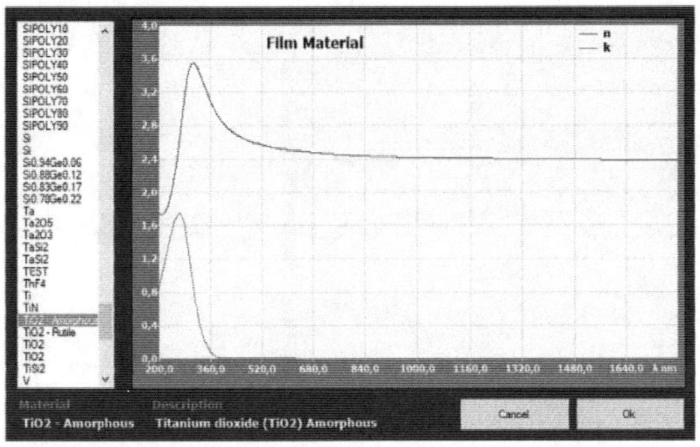

Figure 1. Choice of a film material from the Aspect2010 database.

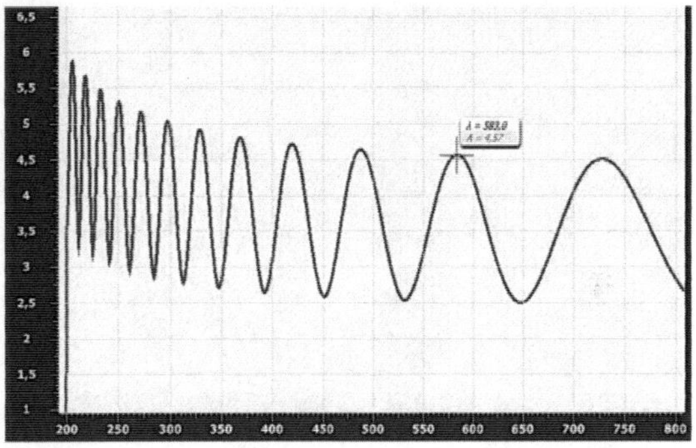

Figure 2. Simulation of the reflectance of SiO_2 film with a thickness of 1 μm.

Step between the points and their number for each material is different, and for correct calculations of T and R for each material approximation is performed for the specified points to obtain the intermediate values. For approximation cubic splines are used because they provide a continuous first and second derivatives, which guarantees the continuity of the plotting of the spectral distribution of T and R. Comparison of simulation results of the reflection coefficient of the SiO_2 films in the program ASpect2010 (figure 2), and using the widely known Filmetrics site [8] gives almost identical results, confirming the correctness of the implemented calculation procedures.

The maxima and minima of T and R are recorded under certain conditions. For example, for T_{min} can be written

$$nd = \frac{\lambda}{4}(2m + 1), \quad (4)$$

where $m = 0, 1, 2, \ldots$ – order of the minimum.

For T_{max}, the expression would be as follows

$$nd = \frac{\lambda}{4}(2m + 2). \quad (5)$$

For calculation of the film thickness it is necessary to transform expressions (1), (2) for the values of the extrema of transmittance and reflectance [9]:

$$T_{\min}=\left(\frac{4n_0n^2n_2}{\left(n_0n_2+n^2\right)^2}\right);\ T_{\max}=\left(\frac{4n_0n_2}{\left(n_0+n^2\right)^2}\right); \tag{6}$$

$$R_{\min}=\left(\frac{n_0-n_2}{n_0+n_2}\right)^2;\ R_{\max}=\left(\frac{n^2-n_0n_2}{n^2+n_0n_2}\right)^2. \tag{7}$$

From (6), (7) can be obtained the expressions for calculation of the refractive index of a film:

$$n=\left(\sqrt{n_0n_2}\left(\frac{1+\sqrt{1-T_{\min}}}{\sqrt{T_{\min}}}\right)\right);\ n=\sqrt{n_0n_2\left(\frac{1+\sqrt{R_{\max}}}{1-\sqrt{R_{\max}}}\right)}. \tag{8}$$

Thus, it is necessary to determine the position of extrema in the spectrum of the reflectance or transmittance of a film. To obtain the data on the actual index of refraction in the whole spectral range in the program are plotted the envelope curves for maxima and minima of the spectra (figure 3).

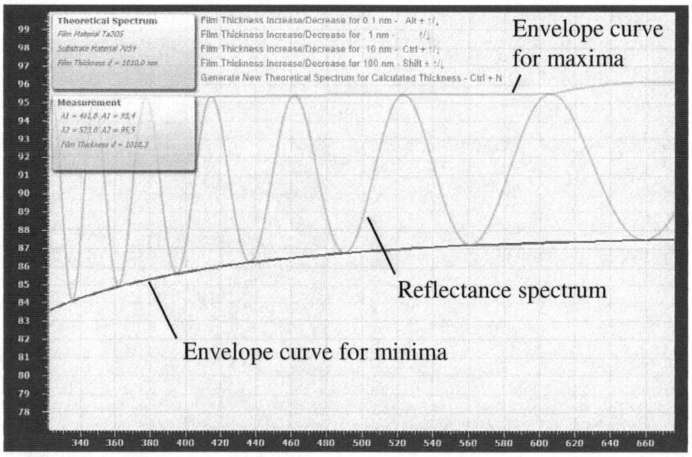

Figure 3. Reflection spectrum of the Ta_2O_5 film with plotted envelope curves for maxima and minima.

When building envelope curves it must be considered that the number of extrema depends on the film thickness and can be small. To plot the envelope curves across the entire spectral range to the found extrema are added the additional counts on the edges of the band. If there is only one extremum the envelope curve is given as a horizontal line, in the presence of two extrema the envelope curve is defined as the line passing through them, if there are three or more extrema the end points are specified using linear interpolation on the nearest to each edge pair of extrema. These values are approximated by a third degree polynomial and are displayed on the screen together with the original spectra, which allows to visually verify the correctness of building envelope curves.

Then using the constructed envelope curves with the help of (8), depending on the type of spectrum is plotted the dependence of the refractive index of a film from the wavelength. Taking into account (4), (5) the expression for calculation of the film thickness takes the form

$$d=\frac{\lambda_{e1}\lambda_{e2}}{2\left(\lambda_{e1}n_{e2}-\lambda_{e2}n_{e1}\right)},$$

where λ_{e1} – wavelength of the first extremum; λ_{e2} – wavelength of the second extremum; n_{e1} – refractive index of the film in the first extremum; n_{e2} – refractive index of the film in the second extremum.

The "classic" method of finding the maximum or minimum in the spectra of reflection or transmission leads to the significant errors due to the presence of noise on the fairly flat extrema. For example, two of the reflection spectrum of the same film of SiO_2 on a glass substrate (figure 4(a)) in search for one of the peaks gave results 723.6 and 730.7 nm. Such a large variation in the measurement of the extremum position leads to the serious errors of measurement of the film thickness.

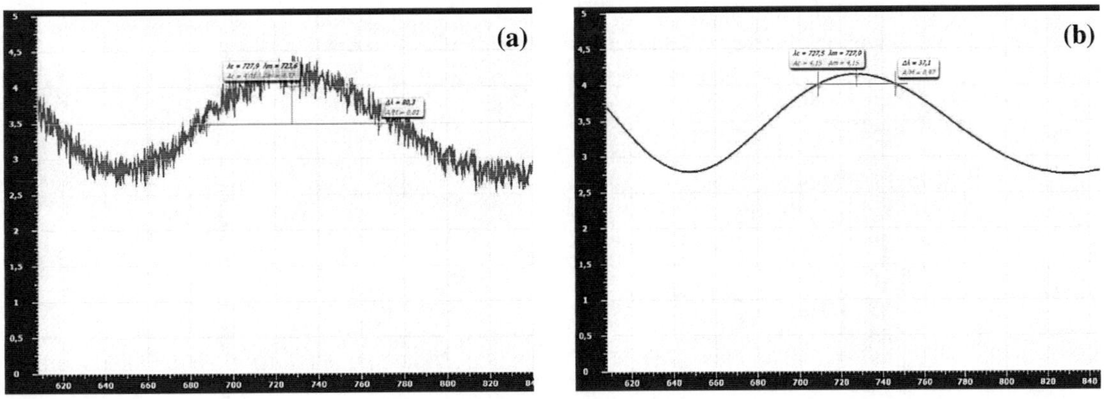

Figure 4. Measuring of the position of one of the peaks on SiO_2 film spectrum with different limits of integration: **(a)** – before filtering of the signal; **(b)** – after filtering of the signal.

For increasing the measurement accuracy by reducing the effect of noise can be used the calculation of the center of gravity for the peak, which is carried out according to the expression

$$\lambda_c = \frac{\sum_i \lambda_i I_i}{\sum_i I_i}, \qquad (9)$$

where i – element number from the array of spectral data from the left to the right border of the peak (are taken into account only the elements of the array greater than the given level); λ_i – wavelength of the i-th element of the spectrum; I_i – amplitude of the i-th element of the spectrum.

The values of the wavelength for the center of gravity of the peak λ_c (figure 4(b)) are 727.9 and 728.2 nm, so the dispersion is about 20 times less than for conventional extremum seeking. However, must be kept in mind the fact that the extremes are not symmetrical. This leads to the changes in results when moving the boundaries of numerical integration according to (9).

For minimization of the additional errors associated with the choice of boundaries their position should be always selected on the same level. When narrowing the boundaries of the integration distance between the maximum value and center of gravity of the extremum λ_c value is reduced to 0.5 nm from the initial distance of 5 nm for the widest possible area of integration. For the experimentally found level of 20 % of the difference between neighboring extrema the received location values of the extrema with the minimum error correspond to the theoretically calculated values. While calculating film thickness as a first approximation can be taken the value of the film thickness, obtained for any two neighboring extrema. To improve the measurement accuracy the filtering of a signal is conducted, however the thickness value will be clearly erroneous, because in the presence of noise it is impossible to accurately determine the position of extrema. Improvement of the measurement accuracy for the calculation of film thickness is achieved at the following stage. The cumulative absolute value of the difference between the samples of the theoretical and real spectra is calculated:

$$D = \sum_{\lambda=180}^{1080} I_{ex} - \sum_{\lambda=180}^{1080} I_{th},$$

where I_{ex} – count values of the experimental spectrum; I_{th} – count values of the theoretical spectrum. The wavelength range 180...1080 nm can be reduced depending on the actual parameters of the device.

The presence of noise in the real spectrum means that the calculated integrated difference with the theoretical spectrum will never be zero, but it can be minimized bringing the shape of the theoretical spectrum closer to the shape of a real spectrum. For this should be built several theoretical spectra for thicknesses that are close to the initial calculated thickness, in a range, obviously exceeding the measurement error. For each of the plotted theoretical spectra it is necessary to calculate the integral difference with the real spectrum and to compare them with each other. As a result a spectrum with a minimum integral difference between the real and theoretical spectrum will give the most accurate result of the film thickness (figure 5).

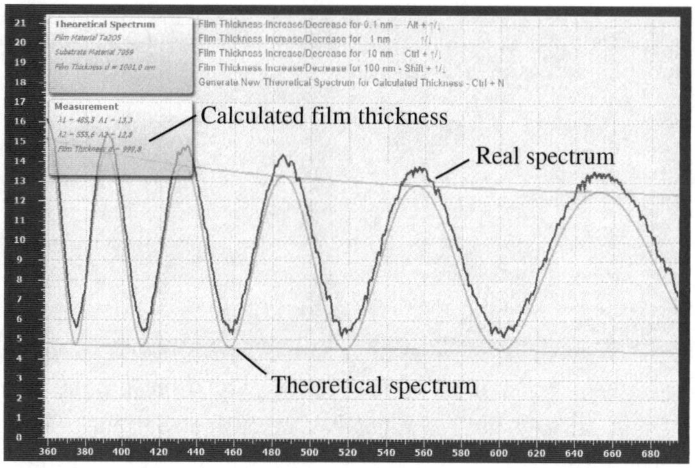

Figure 5. Theoretical spectrum with the minimum difference from the real spectrum, calculation of the film thickness.

Thus, the conducted research and developed methodology allowed to minimize the influence of the real spectrum noise on the measurement results of the film parameters, greatly increasing the accuracy of calculations. However, it should be noted that in the absence of reliable information on the film material and the refractive index shape of the theoretical spectrum will differ significantly from the real one. This circumstance does not allow obtaining the exact value of the film thickness by the proposed method, since the refractive index of the films influences not only the period of the oscillations, but also their amplitude.

Measurement of the thin films parameters by proposed spectral method includes the following steps:

- acquisition of the reflectance or transmittance spectrum;
- normalization of the spectrum on a radiation source (calculation of the reflectance or transmittance coefficient);
- filtering of the obtained reflectance or transmittance coefficient;
- calculation of the refractive index of the film and substrate;
- selection of the extrema for pre-calculating of the film thickness;
- generation of a model of reflectance or transmittance based on the conducted measurements;
- refinement of the film thickness value for obtaining the best fit of the model to the real reflectance or transmittance coefficient.

References

[1] Kostrin D K, Uhov A A and Lisenkov A A 2016 *Vakuum in Forschung und Praxis* **3** 34–7
[2] Kostrin D K, Lisenkov A A and Uhov A A 2016 *Journal of Physics: Conference Series* **735** 012055
[3] Simon V A, Kostrin D K, Uhov A A, Ramazanov A N and Lisenkov A A 2017 *Proceedings of the 2017 IEEE Russia Section Young Researchers in Electrical and Electronic Engineering Conference* 342–5
[4] Kostrin D K and Uhov A A 2013 *Sensors and Systems* **5** 13–5

[5] Uhov A A, Gerasimov V A, Kostrin D K and Selivanov L M 2014 *Journal of Physics: Conference Series* **567** 012039
[6] Kostrin D K and Uhov A A 2013 *Biotechnosphere* **3** 21–5
[7] Ayadi K and Haddaoui N 2000 *Journal of Materials Science: Materials in Electronics* **2** 163–7
[8] Filmetrics Reflectance Calculator. URL: http://www.filmetrics.com/reflectance-calculator
[9] Sreemany M and Sen S 2004 *Materials Chemistry and Physics* **83** 169–77

Effect of the lead oxide content on the microstructure and properties of PZT films obtained by RF magnetron sputtering

N V Mukhin and D A Chigirev

St. Petersburg State Electrotechnical University "LETI", ul. Prof. Popova 5, St. Petersburg, 197376, Russia

E-mail: mukhin.nikolay.v@gmail.com, dachigirev@mail.ru

Abstract. Experimental studies of the influence of lead oxide as well as temperature and time of heat treatment on the microstructure and ferroelectric properties of PZT films obtained by the high-frequency magnetron sputtering method were carried out. It is shown that the change in the ferroelectric properties of polycrystalline PZT films can be explained by their heterophase structure with inclusions of lead oxide. It was found that the presence of a sublayer of lead oxide leads to a self-polarization of the film of the PZT. During the formation of the perovskite structure, the diffusion of lead oxide to the surface of the film occurs more intensively than after its formation.

1. Introduction

Thin-film ferroelectrics are widely used in microelectronics. Ferroelectric films are used in microelectromechanics, various sensor devices based on piezoelectric and pyroelectric effects in the ferroelectric memory devices, microwave devices, sensors, photovoltaics, etc. [1–6]. The solid solutions of lead zirconate titanate (PZT) are among the most widespread materials for these applications and characterized by a broad spectrum of values of the physical parameters depending on the composition [2, 7].

Studying processes of the PZT thin films, managing their structure and properties are widely discussed in the scientific literature [7–9]. It is caused by the complexity of physico-chemical processes that occur during the formation of the perovskite PZT phase and their dependence on a number of factors. A specific interest in this case is the behavior of lead, vapor volatility and migration ability that in the process of application and crystallization of PZT may lead to changes in structure, composition as well as in the nature of electrically active defects and can cause an influence on the electrical properties of the films [7, 10].

Due to that, a study of the effect of the lead oxide content on the kinetics formation and electrophysical properties of the polycrystalline PZT films as well as development of model representations of the lead diffusion processes in polycrystalline PZT films for predicting their properties are an urgent task both from scientific and practical point of view.

According to modern concepts in [11], a high-temperature annealing in an oxygen containing environment of thin polycrystalline PZT films leads to diffusion of lead on the grain periphery. Similar phenomena were studied in other lead-containing materials where the lead diffusion leads to the formation of the lead oxide phase at the grain boundaries of the polycrystalline films [12]. The impact of forming conditions and defectiveness of PZT films on their ferroelectric polarization were studied in [13].

The aim of this work is study of the effect of lead oxide content and time-temperature regimes of heat treatment on the microstructure and properties of polycrystalline $Pb(Zr, Ti)O_3$ (PZT) films obtained by RF magnetron sputtering.

2. Experimental

PZT films with thickness of 0.3 to 1.5 μm of composition near the morphotropic phase boundary were obtained by reactive magnetron RF sputtering method with subsequent heat treatment in an oxygen containing environment to crystallize the films into the perovskite structure.

Films of lead zirconate titanate were obtained by a two-stage technology (ex-situ). In the first stage, PZT films were deposited by reactive magnetron RF sputtering of a ceramic target with a composition of $Pb(Zr_{0.58}Ti_{0.42})O_3$ on the "cold" (≈130 °C) platinized substrates. At the second stage, PZT films heat treatment in a muffle furnace at air (from 530 to 630 °C within 30...180 min) was performed at various time-temperature conditions. Films rate of heating in a furnace was 10 °C/min, and their cooling was carried out with the furnace. The upper platinum electrodes with an area of 0.18 mm^2 were precipitated by ion-plasma sputtering method through a free mask.

For investigation of the effect of lead oxide on properties of PZT and thin film structures based on them, an additional series of samples was made where an additional thin (10…20 nm) layer of lead oxide between the ferroelectric film and the lower structure-directing Pt electrode was formed.

Thus, two series of film structures samples were made:

- capacitor structures with polycrystalline films of lead zirconate titanate of stoichiometric composition: $Pt/PZT/Pt/Ti/SiO_2/Si$;
- multilayer structures with PZT films and with thin underlayer of lead oxide between the ferroelectric film and the lower structure-directing electrode: $Pt/PZT/PbO/Pt/Ti/SiO_2/Si$.

The values of the coercive fields and residual polarization of PZT films were determined from the dielectric hysteresis loops that were obtained by the method of Sawyer–Tower [14]. The values of the dielectric constant and dielectric loss tangent were determined using immittance meter E7-12 at a frequency of 1 MHz. The study of layers morphology surface was conducted on the atomic force microscope (AFM) "Integra Thermo" with a resolution of 10 nm in the horizontal plane of the investigated film and better than 1 nm in the vertical plane. Microstructure layers study was conducted by raster electron microscopy (REM) with the help of the technological complex Helios Nanolab D449 FEI Company. Studies of the elemental composition were performed on Auger spectrometer ESO-3. The main technical parameters of the instrument are: vacuum in the measuring chamber is less then10^{-9} Torr, energy of the primary electrons is 3 keV, electron probe diameter is 5 μm, electron probe current is $5\cdot10^{-7}$A, resolution is 250. The study of the samples phase composition was carried out by X-ray diffraction on Shimadzu XRD 6000 diffractometer. The phase identification was performed using radiometric file cabinets Powder Diffraction File (PDF-2).

3. Results and discussion

Conducted researches have shown a significant influence of temperature – time regimes of heat treatment on the structure and properties of PZT films. The films that are annealed at 550 °C have small values of dielectric permittivity (ε), residual polarization and management factor. This can be explained by heterophase structure of the films i.e. inclusions of pyrochlore phase together with crystallites of perovskite phase are stored at interfaces. With increasing temperature, the proportion of pyrochlore phase is reduced, which leads to an increase of permittivity and residual polarization at 580 °C. With further increase of the annealing temperature, the observed decrease in dielectric constant values can be associated with the formation of thin layers of lead oxide, which occurs in the intergrain space and at the interface with platinum electrodes.

The presence of heterophase PZT film structure is confirmed by research results that were carried out by X-ray diffraction the beam moving method. After the heat treatment, PZT films have a polycrystalline structure with a preferential orientation in the direction of (110). With the rise of annealing temperature to 580 °C, the degree of texturing films increases. With further increase in temperature or time of heat treatment, the appearance of lead oxide phase occurs, which is accompanied by a decrease of intensities of all reflections from the perovskite phase. Presumably, a formation of lead oxide occurs near interfaces and in the periphery of PZT crystallites. It is also reasonable to assume that the structural changes associated with a reduction in the degree of ordering caused by the advent of lead oxide, should lead to a modification of the film surface.

The study results of surface morphology of PZT films showed that the film that has passed heat treatment at 530 °C has a lower surface roughness in comparison with films formed at higher temperatures. Moreover, PZT films formed at a temperature of 530 °C are characterized by low values of dielectric constant. This indicates that the films are in the pyrochlore phase, which begins its formation at temperatures above 450 °C.

With the increase of annealing temperature, the surface of films becomes more developed due to the crystallization process of the perovskite phase and the formation of the columnar structure. The surface of formed PZT films is characterized by developed morphology at temperature of 580 °C during 70 min, due to the presence of formations from an individual crystallites with sizes up to 1 μm. The crystallites have a diameter of about 150…250 nm. With increasing temperature, the formations become smaller and larger crystallites become more clearly with an approximate diameter of 200…300 nm. A longer duration of heat treatment at temperature of 580 °C leads to reduction in sizes of formations and decrease of the surface roughness. A heat treatment at 600 °C leads to the disintegration of formations into individual crystallites with a diameter of 250…350 nm.

The maximum value of the dielectric constant was 770 among investigated series of samples had films that were annealed at 580 °C. Obviously, such low values of dielectric constant can be explained by violation of films stoichiometry, i.e. the formation of PZT layers occurred without compensation of the lead oxide loss.

Reduction of ε while increasing temperature or time of heat treatment higher than 580 °C is connected with a pinning process of polarization charges located in the formed interlayers of lead oxide. It is assumed that the formation of lead oxide interlayers occurs by lead atoms located in a perovskite crystal lattice. Their move to the intergrain and interface boundaries of section leads to decrease of switching ferroelectric volume and reduce ε. Apparently, during longer heat treatment, the thickness of interlayers of PbO and quantity of pinning charges at the interfaces increase, which lead to the cessation of polarization of the ferroelectric volume. This is confirmed by the increase of the switching fields for films that were annealed at 600 °C. Studies of the effect of lead oxide sublayer on the properties of PZT and thin film structures based on them were performed (figures 1–3). Figure 1 shows that the PZT film that was formed at 600 °C is characterized by an average grain size of 90 nm without a sublayer and has a surface roughness of 50 nm.

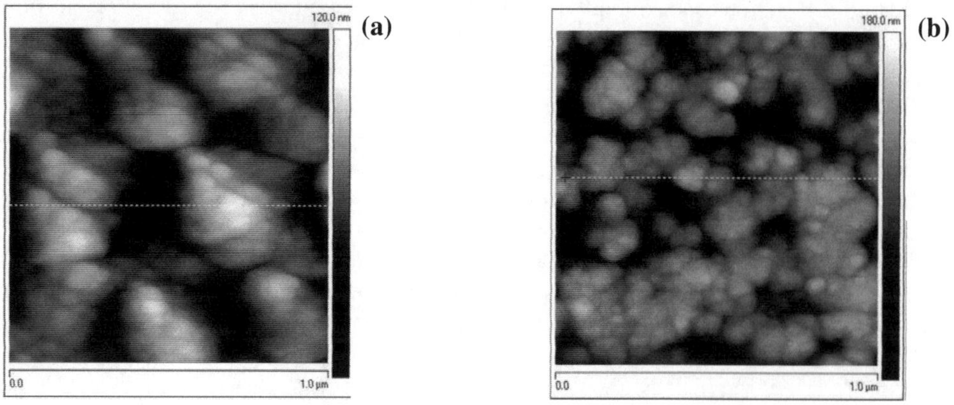

Figure 1. AFM images of the surface of PZT film without a PbO sublayer **(a)** and the surface of PZT film with a 10 nm thick PbO sublayer **(b)**.

The film formed at 600 °C with a sublayer of 10 nm has a grain size of 70 nm and a surface roughness of 60 nm; with a sublayer of 20 nm has the grain size of 30 nm and the surface roughness of 10 nm. Therefore, the presence of the lead oxide underlayer leads to a more dense formation of crystallization centers of the perovskite phase that results in a decrease of the grain size and increase of their number. This prevents the formation of grain conglomerates that leads to a significant decrease in surface roughness of the PZT film.

Studies of electrical characteristics showed that the permittivity of the sample without a sublayer is significantly higher (ε = 760) than in the sample with the sublayer (ε = 410). PZT films without a sublayer and with PbO are characterized by the residual polarization in the 13 and 11 μC/cm^2, the value of the coercive field is 110 and 185 kV/cm, respectively. It is worth noting that the dielectric hysteresis loop of PZT-PbO sample is shifted to the left that indicates the presence of a built-in electric field of about 15 kV/cm (figure 2).

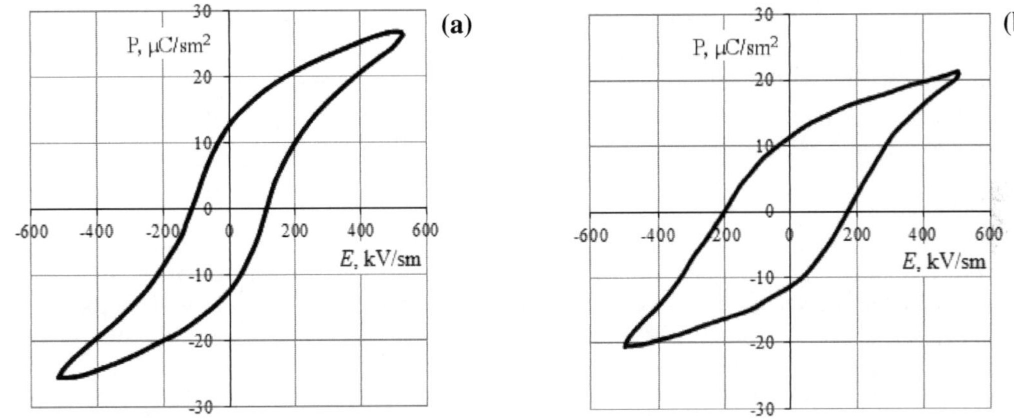

Figure 2. Dielectric hysteresis loops of the capacitor structures based on PZT film without a PbO sublayer **(a)** and with a 10 nm thick PbO sublayer **(b)**.

It is also confirmed by the shift of the current-voltage characteristics of the capacitor structures based on the PZT-PbO film (figure 3).

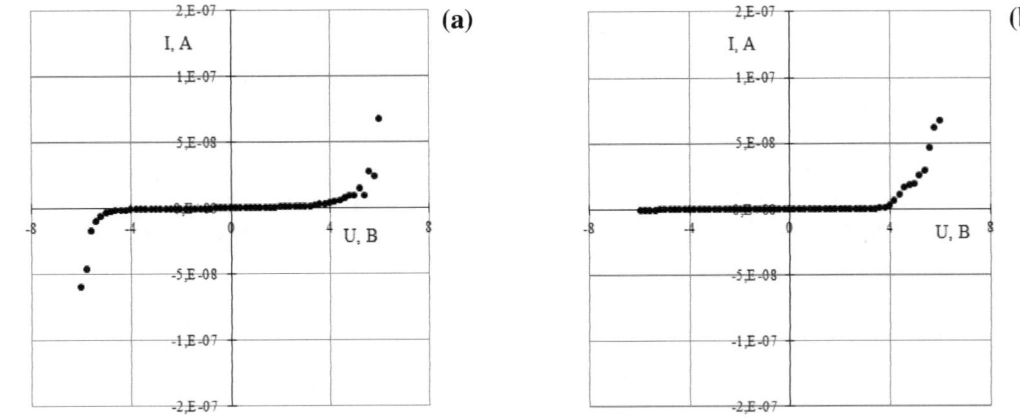

Figure 3. Current-voltage characteristics of the capacitor structures based on PZT film without a PbO sublayer **(a)** and with a 10 nm thick PbO sublayer **(b)**.

Results of researches testify that high-temperature heat treatment of PZT films leads to loss of lead oxide. In this paper, the attempt to experimentally determine such losses was made. The following experiment was carried out to solve the task. Samples of PZT films with 100 nm of thickness and with formed perovskite structure (the formation occurred in air during 30 min at 580 °C) were subjected to further heat treatment for 180 min. Auger spectra samples of PZT films have been identified: immediately after the HF magnetron sputtering, after formation of the perovskite phase and after further annealing (table 1).

Table 1. The intensities of the signals of the components of the PZT films, determined from Auger spectra.

PZT films	Pb	O	Ti	Zr
After deposition on the "cold" substrate	3	9.75	2.25	1
After formation of perovskite phase	2.5	10.5	3	1
After repeated annealing in air	1.5	8	3.25	1
After annealing in vacuum	1.5	7	3	1

According to results in table 1, the amount of lead oxide on the film surface was reduced more intensely than at the additional heat treatment in air or in vacuum during the formation of the perovskite phase. This indicates that during the formation of the perovskite structure, the diffusion of lead oxide to the surface of the film occurs more intensively than after its formation.

4. Conclusions

It is shown that the content of lead oxide has a significant effect on the structure and properties of PZT films obtained by the high-frequency magnetron sputtering. The fixation of polarization processes occurs by charges located in interlayers of lead oxide, which in turn leads to a decrease in the dielectric constant and to the growth of coercive fields. It was shown that the presence of a sublayer of lead oxide leads to a more dense formation of the crystallization centers of the perovskite phase and to the appearance of a charge at the lower interface. The built-in charge at the interface between the PZT film and the lower electrode causes the self-polarized state of the PZT film. During the formation of the perovskite structure, the diffusion of lead oxide to the surface of the film occurs more intensively than after its formation.

Acknowledgment

This work was financially supported by the grant of The Ministry of Education and Science (Minobrnauka) of the Russian Federation (project Goszadanie № 3.3990.2017/4.6).

References

[1] Dawber M and Bousquet E 2013 *MRS bulletin* **38**(12) 1048–55
[2] Afanasjev V P, Chigirev D A, Mukhin N V and Petrov A A 2016 *Ferroelectrics* **496**(1) 170–6
[3] Aman A, Majcherek S, Hirsch S and Schmidt B 2015 *Journal of Applied Physics* **118** 164105
[4] Levitskii V S, Redka D N and Terukov E I 2016 *Ferroelectrics* **496**(1) 163–9
[5] Schmidt M-P, Oseev A, Lucklum R, Zubtsov M and Hirsch S 2016 *Microsystem Technologies* **22**(7) 1593–9
[6] Oseev A, Schmidt M-P, Hirsch S, Brose A and Schmidt B 2017 *Sensors and Actuators B: Chemical* **239** 1213–20
[7] Sigov A S, Vorotilov K A and Zhigalina O M 2012 *Ferroelectrics* **433**(1) 146–57
[8] Sanjeev A and Ramesh R 1998 *Annu. Rev. Mater. Res.* **28** 463–99
[9] Pronin V P, Senkevich S V, Kaptelov E Yu and Pronin I P 2013 *Phys. Solid State* **55**(1) 105–8
[10] Mukhin N V 2014 *Glass Phys. Chem.* **40**(2) 238–42
[11] Mukhin N V 2016 *Glass Phys. Chem.* **42**(1) 64–9
[12] Maraeva E V, Moshnikov V A and Tairov Yu M 2013 *Semiconductors* **47**(10) 1422–5
[13] Afanasjev V P, Petrov A A, Pronin I P et al. 2001 *J. Phys.: Condens. Matter* **13**(39) 8755–63

The influence of the discharge current axial component on the magnetic field distribution in the cathode region of magnetron sputtering system

V D Goncharov[1], K S Sorokin[1] and E M Fiskin[2]

[1] Saint Petersburg Electrotechnical University "LETI", Laboratory of "Electric impulsive technologies", 197376, Saint Petersburg, Russia
[2] Irkutsk state technical university, Russia

E-mail: vdgoncharov@rambler.ru

Abstract. The paper includes the description of the mathematical model, which simulates magnetic field distribution in the cathode area of the sputtering system discharge considering the current created by electrons moving along the surface of cathode-target. Presented results of the numerical modeling were received using the developed model. It is also shown that the current component, directed along the cathode surface significantly impacts total magnetic field distribution in the magnetron discharge.

1. Introduction

The properties of the discharge ignited in the crossed electric and magnetic fields depend on the complicated character of the electron motion in the cathode area of this discharge. Special conditions in the area are created in the way to force the electrons overcome the distance between cathode and anode couple thousand times. As a result of low pressure (meaning longer free path for electrons and ions) a significant energy flow might be reached on the surface of the cathode. It allows using the discharge to sputter the target more effectively. This principle is the main idea of magnetron sputtering systems [1]. These systems are widely used in manufacturing, though a number of the problems of maintaining such type of discharge is still unsolved. First problem is the mechanisms of forming the current density on the surface of the cathode, especially at high-power pulsed sputtering [2]. Another problem is the possibility of decreasing the operating pressure in the system lower than 0.1 Pa [3]. Such a decrease improves the flexibility of the system, allows controlling the coatings thickness, modifying the sputtering process with ion bombardment of the growing coating [4]. It all opens the new perspectives for applying magnetron sputtering systems for creating protective coatings and coatings with special properties [5].

In general, to simulate the processes in magnetron sputtering system the expressions describing electric and magnetic field distribution with free electrons and their motion for every moment of the simulation must be solved. Also we have to consider the expressions describing the electrons incidents with gas molecules, expressions of the ion component drifting, etc. This problem is very complicated, but it might be solved in a number of separate steps.

And the first one in our opinion is the description of the magnetic field distribution near the sputtering system target. This paper is dedicated to calculating this field.

2. Model description

The most common construction of magnetron system is taken into consideration in the paper. This model includes the plane target (with electric field directed normal to the surface) and “arc” type of the magnetic field distribution.

To maintain such form of glowing discharge [6] the emitted from the cathode electrons get trapped by crossed electric and magnetic fields. Being trapped the electron moves along the cycloid trajectory until the collision with gas molecule happens (figure 1). The cycloid radius should be enough to make the electron reach a certain energy level to make the gas molecules ionization possible (condition defining the longitudinal component of the magnetic field) [7]. After inelastic (with energy loss) or elastic (only direction of movement changes) collision with molecule, the electron accelerated by electric field moves away from the cathode surface to a distance equal to Larmor radius. In the area where electron drifted the field decreases and the cycloid radius increases. It lasts until electron gets “trapped”, so before it reaches anode it moves along the surface of the target. Such motion causes axially directed current, which exceeds the discharge current many times (figure 1).

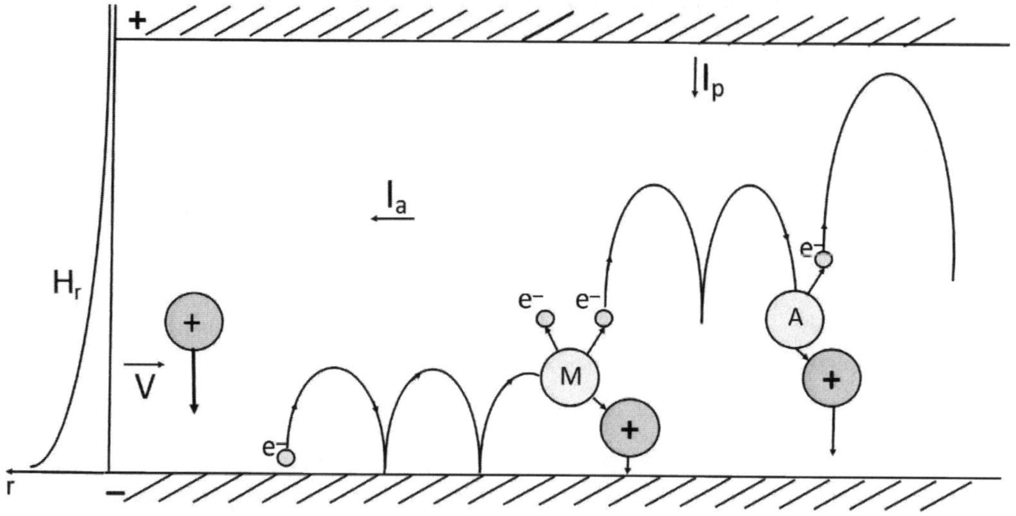

Figure 1. The scheme of the cathode processes of the magnetron sputtering system, forming the most of discharge.

The ratio between these currents is equal to the ratio of the distances, which electron passes between cathode and anode. This ratio might be equal from couple hundreds to couple thousands. For example, when the discharge current is 5 A, the axial one is 10 000 A. The direction of the current is so that it’s magnetic field decreases the field of the permanent magnets at the target and increases it when moving away from it. Studying the summary of distribution of the magnetic field in the cathode area experimentally is very complicated. That’s why we made theoretical researches. The paper is dedicated to this mathematical model.

The geometry of the studied area and qualitative distribution of the lines of the magnetic field are presented in the figure 2. The half of the section of the studied area is presented. Permanent magnetic field is created by SmCo magnet *1*. The lines of the magnetic field are commutated through the magnetic core made of electrotechnical steel *2*. The vector of the magnetic field might be presented as vector sum of longitudinal (H_r) and transversal (H_z) *3*. Over the target surface the current loop is placed *4*, where the axial current flows. It’s width was defined experimentally according to a sputtering zone on the target surface.

Simulation of magnetic field distribution [8, 9] requires very complicated calculations, and the problem doesn’t have any particular analytical solution (especially considering real magnetic materials characteristics). That’s why this simulation was provided numerically using mathematical model created in ANSYS. The geometry of the model copied a real device. The field distribution was defined

IOP Conf. Series: Journal of Physics: Conf. Series **872** (2017) 012046 doi:10.1088/1742-6596/872/1/012046

by fourth Maxwell equation solution $\text{div}\vec{B} = 0$. Current loop with magnet fields were defined by solving first Maxwell equation solution $\text{rot}\vec{H} = \vec{J}$, where $\vec{H}$ is a vector of a magnetic field, $\vec{J}$ is a current density in the loop. We presumed the axial current to fill the loop evenly.

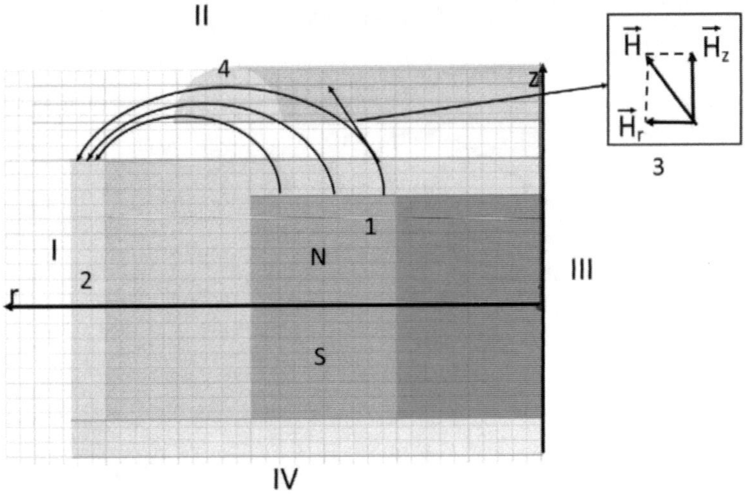

Figure 2. The geometry of studied area and qualitative character of magnetic field distribution.

Also in the figure 2 the boundary conditions are shown: I, II – zero magnetic field on a significant distance from the studied area; III, IV – no normal component of the magnetic field.

The results of the simulation for clarity are presented in 2-dimentional simplified plots. In the figure 3 two images of magnetic field simulation are presented (left is a case when there's no axial current, right one is a case when axial loop current is 1 kA).

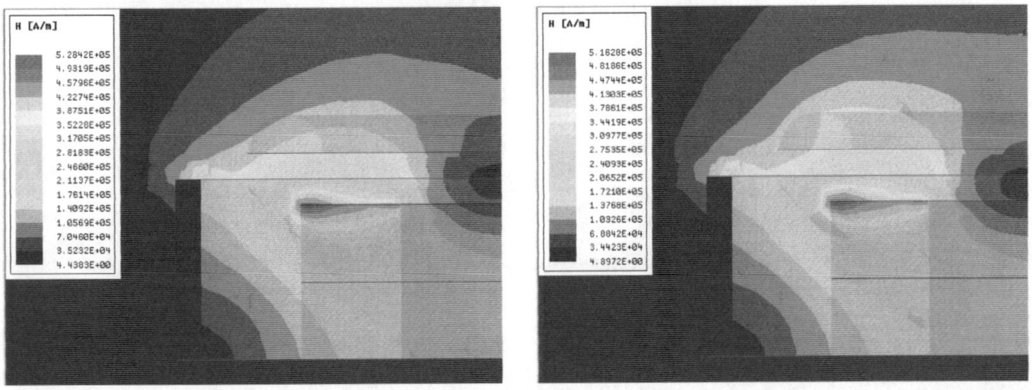

Figure 3. Magnetic field distribution (left) without axial current and (right) with axial current influence on field.

However, in simulating the magnetic field of the sputtering system, the highest interest is in tangent component of *H*. directed along the target. In the figure 4 the tangent component of the magnetic field is presented, calculated fordifferent distances from the megnetic system section. These results allow defining Larmor radius and choose the right target thickness for the megnetic system configuration.

The less gas pressure, the more free path distance of the electron hence the higher axial current. That's why the results received when increasing I_a might be treated as field distribution when the gas pressure decreases with the same other conditions.

Consideration the axial current component influences the field distribution dramatically. So, in the figure 5 depicted the results of calculating the tangent component of the magnetic field on the surface of the target for different cases: without the axial current, with I_a = 1 kA (200 times more than discharge current) and with I_a = 2.5 kA (500 times more than discharge current). Wherein the maximal H value decreases when axial current increases, the zone, where this value exists becomes wider.

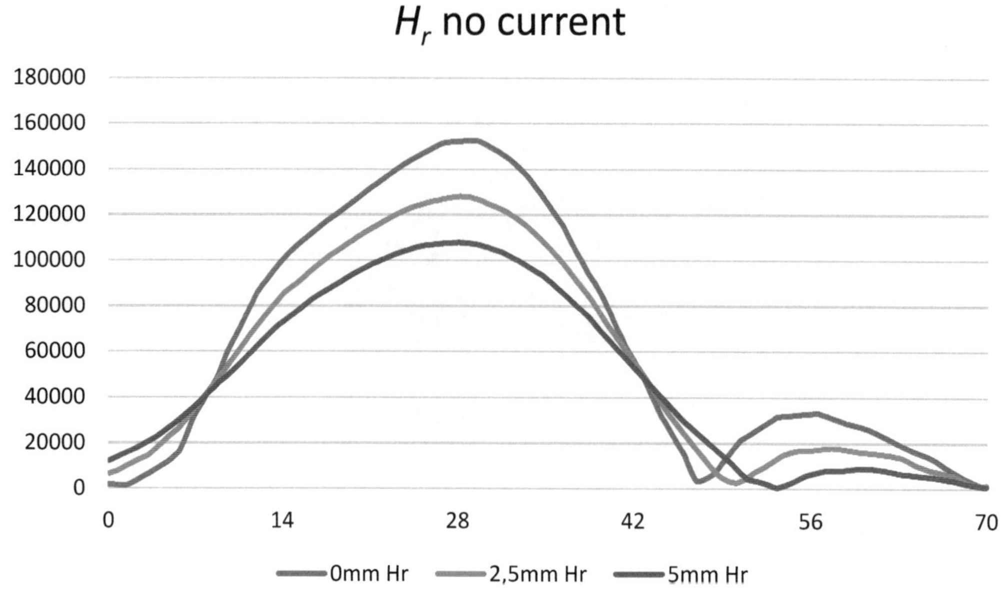

Figure 4. The results of calculating the tangent component of magnetic field at different distances from the magnetic system.

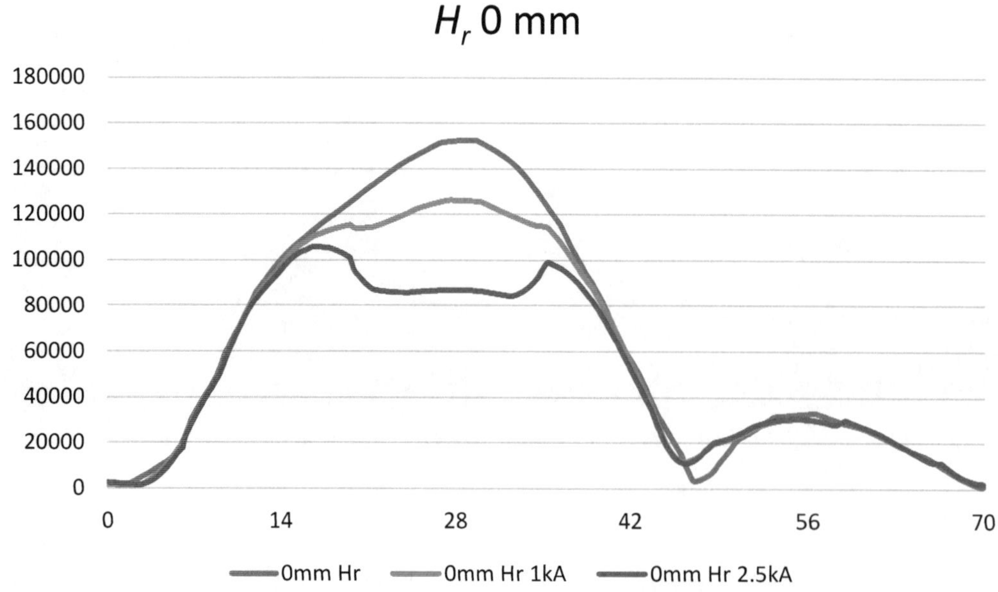

Figure 5. Calculation results of tengent component of H on the surface of the target (7 mm from the magnet system) for the ceses withous axial current, with I_a = 1 kA (200 times more than discharge current) and with I_a = 2.5 kA (500 times more than discharge current).

Another important plot (figure 6) is tangent component of the magnetic field distribution at different distances from the magnetic system at 1 kA axial current. It appeared to be that considering this current allows alligning the tangent component of magnetic field at different distances from the target. In other words, the cyclotrone radius doesn't change at the distances up to the 5 mm above the target. This is the area of localization of the cathode area of magnetron discharge. Beyond this area the tangent component decreases significantly. It casues the cyclothrone radius insrease and electron emmision from the cathode area.

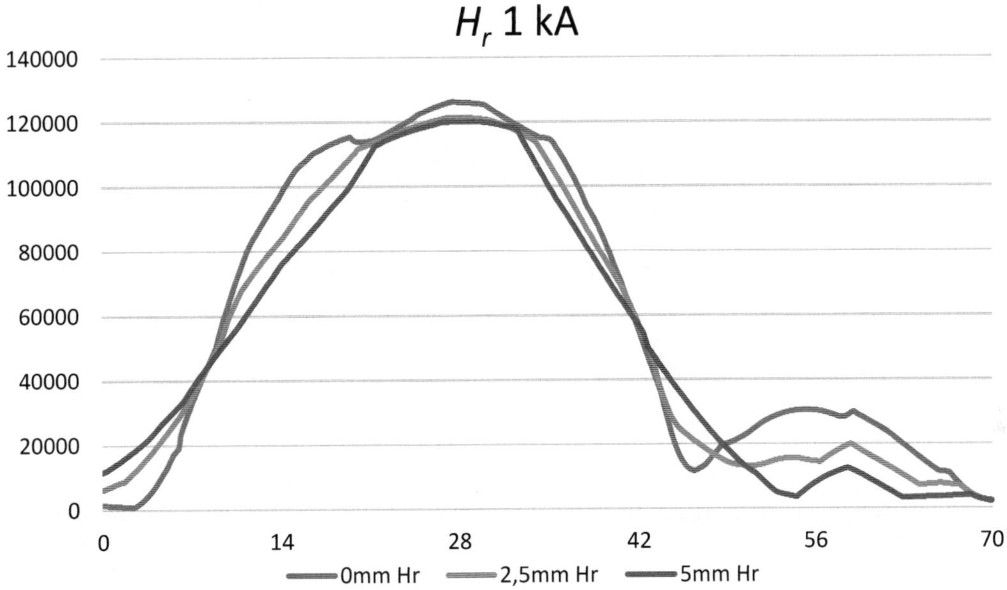

Figure 6. The results of tangent magnetic field calculation at different distances from the magnetic system. The curent I_a = 1 kA.

3. Conclusions

1. The axial component of current influences the magnetic field in the area of the discharge dramatically.

2. The value of the magnetic field on the surface of the target decreases in the discharge area.

3. Increasing the magnetic field at the boundaries of the discharge might cause the forming additional "magnetic traps" there, preventing electrons from leaving the cathode area.

References

[1] Kostrin D K, Lisenkov A A and Uhov A A 2016 *Journal of Physics: Conference Series* **735** 012055
[2] Horwat D and Anders A 2010 *J. Appl. Phys.* **108** 123306–11
[3] Danilin B S 1989 *Primenenie nizkotemperaturnoj plazmy dlja nanesenija tonkih plenok* (Moscow)
[4] Golosov D A, Eungsun B and Zavadskij S M 2014 *Zhurnal tehnicheskoj fiziki* **9**
[5] Svadkovskij I V 2007 *Doklady Bguir* **2** 112–14
[6] Ramazanov A N, Kostrin D K, Goncharov V D and Lisenkov A A 2016 *Journal of Physics: Conference Series* **729** 012004
[7] Kostrin D K, Trifonov S A and Lisenkov A A 2017 *Journal of Physics: Conference Series* **857** 012022
[8] Martsinukov S A, Kostrin D K, Chernigovskiy V V and Lisenkov A A 2016 *Journal of Physics: Conference Series* **729** 012023
[9] Chernigovskiy V V, Kostrin D K, Martsinukov S A and Lisenkov A A 2016 *Vakuum in Forschung und Praxis* **6** 34–7

IOP Conf. Series: Journal of Physics: Conf. Series **872** (2017) 012047 doi:10.1088/1742-6596/872/1/012047

The properties of ZnO films obtained by high-frequency magnetron deposition with subsequent vacuum annealing and plasma treatment

D N Redka[1], **K G Elanskaia**[1] **and L D Bakhchova**[2]

[1] Saint-Petersburg State Electrotechnical University "LETI", St. Petersburg, 197376, Russia
[2] National Technical University "Kharkov Polytechnic Institute", Kharkov, 61000, Ukraine

E-mail: dnredka@gmail.com

Abstract. The method of high-frequency magnetron sputtering allows to obtain thin films of zinc oxide with nanostructured surface morphology and, consequently, with unique optical characteristics. Such coating can provide an increase in the number of particles penetrating the photovoltaic structure and the length of their optical path. Vacuum-plasma processing of ZnO films allows to change the morphology of their surface, and annealing largely influences the optical transmission.

1. Introduction

Zinc oxide films are attractive for use in photovoltaics (and in solar converters in particular) as transparent electrodes for current collection [1], as a heterocontact in oxide heterostructures [2–4] and other oxide photovoltaic media [5] as well as in sensorics, for example, as an active layer in sensors on surface acoustic waves [6–9], etc. One of the problems of oxide film structures is their instability under various external influences [2, 10]. Zinc oxide can possess an electronic type of conductivity due to a high concentration of intrinsic defects such as interstitial zinc or oxygen vacancies. These defects are responsible for an increase in *n*-type conductivity and a decrease in crystallinity [11]. If a certain morphology of the structure can be created in ZnO films, then they can perform the functions of the diffusive light diffuser in the solar energy converter. When passing through a layer of such material, the ray of solar radiation changes its trajectory, i.e. the radiation scattering occurs that leads to an increase in the optical path of the particle in the photoactive structure. Due to this, a more effective conversion of solar energy into electrical energy is carried out, and the short-circuit current is increased. The annealing of the material leads to the formation of a more uniform structure, the reduction of hardness, the removal of stresses, the equalization of chemical inhomogeneity and the improvement of optical characteristics [12, 13]. On the other hand, the annealing temperature mainly affects the orientation in the plane. Processing in plasma allows changing the physical properties of the material, hardness and morphology of the surface [14].

2. Preparation of zinc oxide films by high-frequency magnetron sputtering

As targets for the deposition of zinc oxide films, ZnO powders that were uniformly distributed over the entire surface of the magnetrons (figure 1(a)) were used. A complex of experiments showed that the studied oxide is characterized by low sputtering rates. Dependences of the growth rate between zinc oxide and the power of the high-frequency discharge are shown in figure 1(b). It should be noted that further increase in the power of the HF discharge to increase the growth rates of the films is undesirable, because this will lead to overheating of the magnetrons and their possible failure. The gas mixture of the process contained 76 % Ar and 24 % O_2. The operating pressure of the gas mixture was 10^{-2} Torr. The substrate temperature (self-heating) was 260–270 °C.

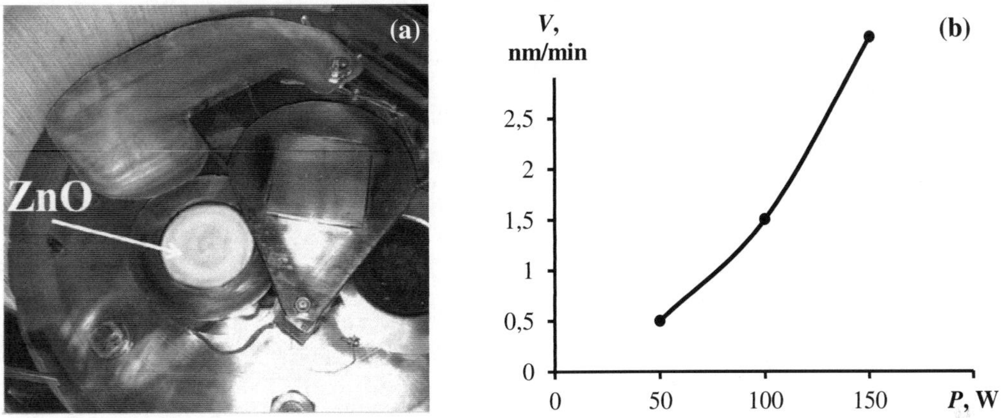

Figure 1. The internal equipment of the vacuum chamber **(a)**, the dependence of the growth rate of zinc oxide on the power of the HF discharge **(b)**.

3. Investigation of the optical properties of zinc oxide

The transmittance and reflection spectra of the films of zinc oxide (ZnO) were studied, which are intended for use as transparent electrodes in thin-film solar cells; parameters of the samples are given in table 1. The data of films thickness were obtained by an interference method, the surface resistance was measured by a four-probe method.

Table 1. Characteristics of ZnO film samples.

Sample №	Thickness, nm	Surface resistance, Ohm/kV
ZnO 4	192	–
ZnO 3	510	155
ZnO 2	982	43
ZnO 1	1609	18

The spectra were measured on a Cary-5000 double-beam spectrophotometer in the wavelength range of 400–1000 nm. The integrating sphere was additionally installed in the instrument, because a significant light scattering, which is necessary to increase the proportion of absorbed radiation in subsequent semiconductor layers, is observed in the studied zinc oxide. The sphere allows to measure both the radiation flux passing through the sample or reflected from it (flux into the hemisphere) and diffusely scattered radiation flux (flux into the hemisphere excluding flux along the normal to the sample), which makes it possible to determine the coefficients of total (TT) and diffuse (TD) transmission and reflection. Figures 2, 3 show the spectra of the total transmission and reflection of the studied samples. According to the results of the transmission measurement, the spectral dependences of the HF (haze factor) parameter were calculated for each film. It determines the fraction of diffusely scattered radiation in the transmitted flux (figure 4).

According to the graphs in figures 3, 4, a high haze factor (a significant proportion of the diffuse-scattered radiation) and a relatively small losses because of optical absorption (10–15 %) are achieved only for a 1.6 μm film (sample ZnO 1), which should probably be considered optimal. For films of smaller thicknesses, the loss of absorption decreases slightly (figure 2), but the haze factor falls sharply (figure 4), and the surface resistance increases (table 1).

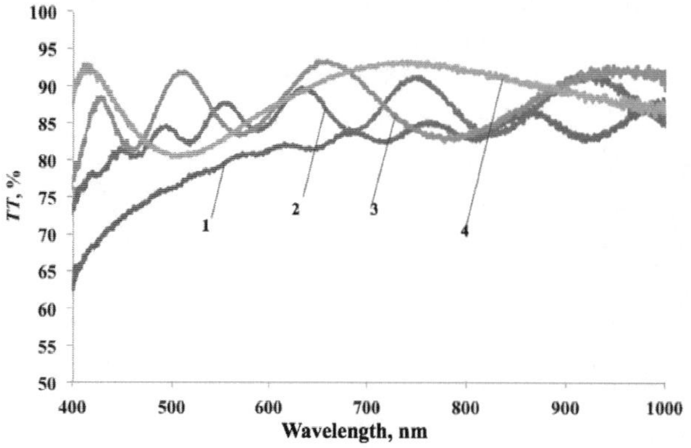

Figure 2. Spectra of total transmission for ZnO obtained by the method of magnetron sputtering (*1* – sample ZnO 1; *2* – ZnO 2; *3* – ZnO 3; *4* – ZnO 4).

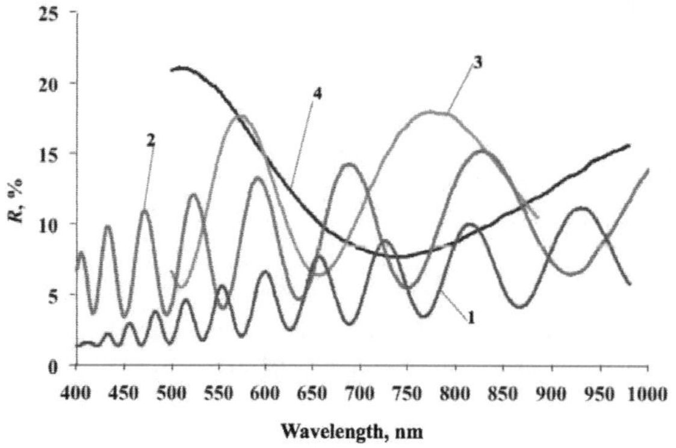

Figure 3. Spectra of total reflection for ZnO obtained by the method of magnetron sputtering (*1* – sample ZnO 1; *2* – ZnO 2; *3* – ZnO 3; *4* – ZnO 4).

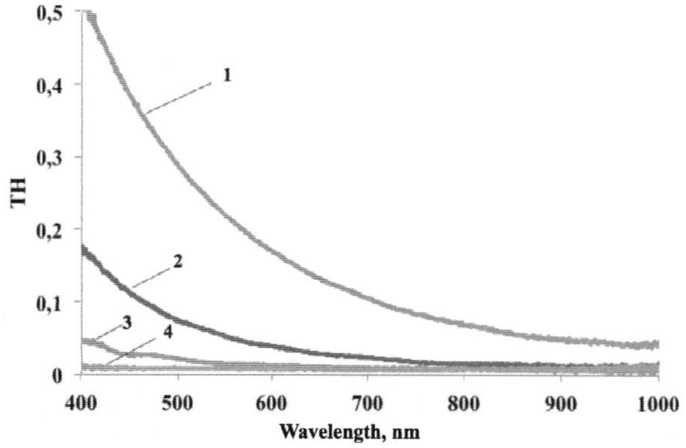

Figure 4. Effects of light scattering of ZnO films (*1* – ZnO 1; *2* – ZnO 2; *3* – ZnO 3; *4* – ZnO 4).

4. Effect of annealing in vacuum and processing in plasma on the properties of ZnO films

Further, the samples were annealed in vacuum at different temperatures and different annealing times. The results of the total transmission measurement are shown in figure 5. According to the graphs, the annealing time and temperature affect the transmission considerably at the annealing temperature of 300 °C, but there is no such strong effect at the annealing temperature of 250 °C.

Plasma treatment of ZnO that is aimed at the surface modification of films was performed in magnetron sputtering systems at ambient temperature and at a typical pressure of about 60 Torr. For processing, the samples were placed as a target directly above the magnet with the sputtered target. This position of the samples provides a strong magnetic field near the processed sample, which increases the efficiency of ionization of argon atoms (Ar) by electron pulses and, consequently, increases the plasma density and the Ar ion flux to the treated surface.

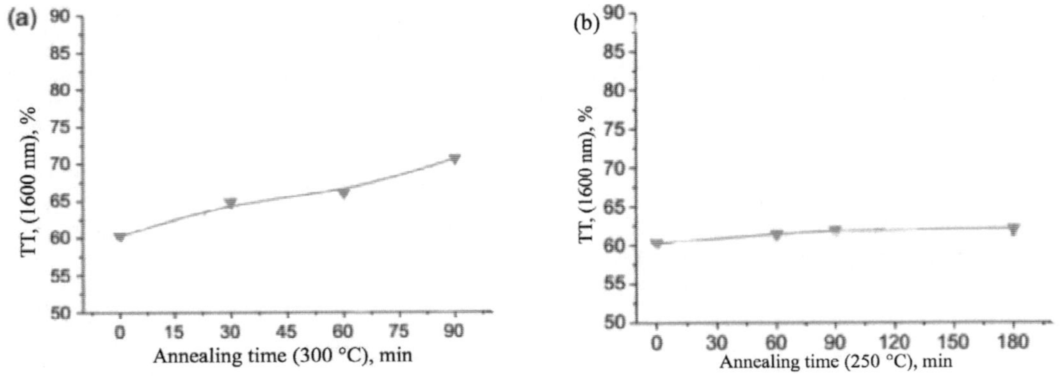

Figure 5. Change in the total transmission at a wavelength of 1600 nm with a variation of temperature (250 °C and 300 °C) and the annealing time.

As a result, the processing time can be expected to decrease without increasing the power of the HF signal. The processing time was 5 min at a power density of 1.3 W/cm^2. We have shown that plasma treatment of boron-doped ZnO layers allows to transform the V-shape surface into a U-type morphology. Plasma treatment leads to a decrease in the turbidity values in the entire range from 400 to 900 nm for all samples. The turbidity value for the sample treated for 5 min has a value of 22 % at 600 nm, and an increase in the treatment time leads to a decrease in the turbidity values. The light-scattering factor also decreases, as can be seen in figure 6.

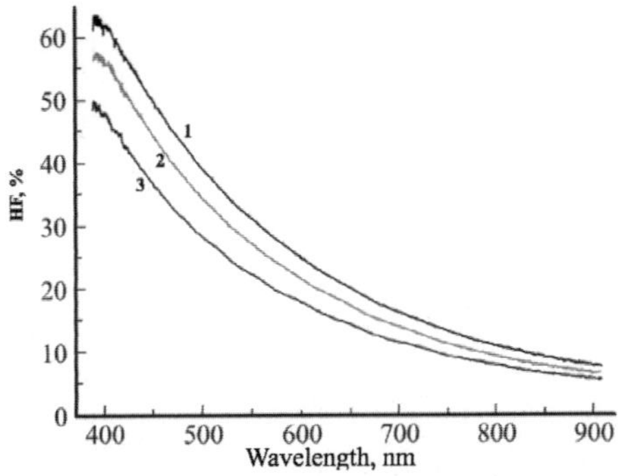

Figure 6. Light scattering effects of ZnO film: *1* – before plasma treatment; *2* – after 5 min of plasma treatment; *3* – after 10 min.

To verify the electronic properties of the films, we performed measurements of the electrical conductivity of the samples, carrier concentration and mobility. The results show that there is no significant change in the values of mobility and carrier concentration for samples with a short treatment period (5 min), while a longer exposure to radio-frequency magnetron plasma leads to a deterioration in the electronic properties. This result can be explained either by the formation of defects in the bombardment with massive Ar ions or by overheating of films deposited on glass substrates with a low thermal conductivity during prolonged treatment with intense plasma and low pressure. It can be expected that under these conditions the structural properties of the films can also deteriorate. As we have shown, processing for five minutes is sufficient to transform the surface morphology from V-type to U-type without decreasing the electronic properties of the films.

5. Conclusion

In the framework of this paper, we obtained samples of zinc oxide films with a nanostructured surface on substrates made from special photovoltaic glass with increased light transmission (reduced iron oxide content). The optimal film thickness was 1.6 μm, since in this case relatively small losses for optical absorption (10–15 %) and optimal values of the light scattering factor are obtained. It was also found that the effect of annealing in vacuum at 300 °C significantly influences the value of optical transmission, namely, it increases it, which indicates an improvement in the structural parameters of the film. By vacuum-plasma treatment it is possible to change the structure of the surface from the V-shape to the U-type morphology, which is more advantageous for use in photovoltaics. Also, the optimal plasma processing time is 5 min, since this is sufficient to reduce the turbidity values without disturbing the electronic properties of the material.

Acknowledgment
This work was supported by the grant of The Ministry of Education and Science (Minobrnauka) of the Russian Federation (project Goszadanie № 3.3990.2017/4.6).

References

[1] Redka D N, Mukhin N V and Zakharov I G 2016 *Technical Physics* **61**(11) 1744–6
[2] Adilov Sh R, Afanasiev V P, Kashkul I N, Kumekov S E, Mukhin N V and Terukov E I 2017 *Glass Phys. Chem.* **43**(3) 272–5
[3] Afanasjev V, Bazhan M, Klimenkov B, Mukhin N and Chigirev D 2016 *Journal of Physics: Conference Series* **729** 012013
[4] Afanasjev V P, Konoplev G A, Chigirev D A, Mukhin N V, Terukov E I and Terukova E E 2016 *Proc. of SPIE* **9896** 98980V
[5] Afanasjev V P, Chigirev D A, Mukhin N V and Petrov A A 2016 *Ferroelectrics* **496**(1) 170–6
[6] Fernández M J, Fontecha J L, Sayago I, Aleixandre M, Lozano J, Gutiérrez J, Gràcia I, Cané C and Carmen Horrillo M 2007 *Sensors and Actuators B: Chemical* **127**(1) 277–83
[7] Schmidt M-P, Oseev A, Lucklum R, Zubtsov M and Hirsch S 2016 *Microsystem Technologies* **22**(7) 1593–9
[8] Lucklum R, Zubtsov M, Oseev A, Schmidt M-P, Hirsch S and Hagemann F 2013 *European Frequency and Time Forum & International Frequency Control Symposium (EFTF/IFC)* pp 69–72
[9] Oseev A, Schmidt M-P, Hirsch S, Brose A and Schmidt B 2017 *Sensors and Actuators B: Chemical* **239** 1213–20
[10] Afanas'ev V P, Vorotilov K A and Mukhin N V 2016 *Glass Phys. Chem.* **42**(3) 295–301
[11] Damiani L R and Mansano R D 2012 *Journal of Physics: Conference Series* **370** 1–7
[12] Iakovlev A, Redka D, Afanasjev V, Mukhin N, Grishkanich A, Terukov E and Hirsch S 2017 *Proc. of SPIE* **10091** 100911B
[13] Mukhin N V 2016 *Glass Phys. Chem.* **42**(1) 64–9
[14] Afanasjev V P, Mukhin N V, Redka D N, Rudenko M V, Terukov E I, Oseev A and Hirsch S 2017 *Ferroelectrics* **508**(1) 124–9

Water repellent properties of dispersed metals containing low-dimensional forms of ammonium compounds on the surface

A G Syrkov, V R Kabirov and M O Silivanov

Department of general and technical physics, Saint Petersburg Mining University, 199106, Saint Petersburg, Russia

E-mail: syrkovandrey@mail.ru

Abstract. For the first time the change of the water repellent properties of dispersed copper, modified using quaternary ammonium compounds on 24 h time scale in saturated water vapours was studied. Exponential time dependences of the water repellent properties of dispersed copper with adsopted QAC were derived and characterized. It was established that the samples modified in mixed and consistent modes by both modifiers reach the saturation state faster than others, due to the small number of hydrophilic centers on the surface of metals. The last conclusion was confirmed by the distribution spectra of centers of adsorption, which were obtained by the adsorption of acid-base indicators for more dispersed samples based on aluminum powder.

1. Introduction

Dispersed metals are widely used in various fields of science and technology, including electronics, tribotechnology and heterogeneous catalysis. Due to the high chemical activity of the particles, dispersed metals can be oxidized in air, primarily by water vapours. Therefore, to stabilize the functional properties of solid materials containing dispersed metals, it is important to create the protective films on the surface of particles of such metals to improve their water repellent properties. One of the promising methods is layering of different-sized molecules of quaternary ammonium compounds (QAC). The advantages of the method are the possibility of applying submonomolecular coatings, low vapor pressure of QAC, gas-phase modification at room temperature, and environmental friendliness of the method [1–3].

Previously, it was shown that the chemisorption of QAC-based cationic surfactants has a positive effect on water repellent and antifriction properties of metals, such as aluminum [2, 3]. However, there are more interesting objects for modification such as dispersed copper powders, not only from the point of view of surface chemistry, but also in case of variety of practical use.

In this research, the water repellent properties of dispersed copper with adsorbed QAC on the surface were studied in saturated water vapours (5–24 h). This topic is significant not only for regulation of different properties of surface of dispersed metals, but also for establishment of quantitative regularities between synthesis, composition, structure and properties of various heterogeneous materials, containing nanostructured metals, which is a basic goal of modern physics and chemistry of low-dimensional systems.

2. Materials and methods

Copper powder PM-1 (GOST 4960-75) with specific surface area of 0.34 ± 0.02 m^2/g and with specific volume of monolayer of nitrogen of 0.08 ml/g was chosen as the initial sample of dispersed metal. More dispersed non-porous aluminum powder PAP-2 was also used to analyze the distribution spectra of adsorption centers on the surface of metal samples. According to electron microscopy, the modification of the surface of metals does not lead to a significant change of the shape and size of the particles.

Metal powders were treated in vapours of alkamon (A), triamon (T) and hydrophobic silicone-organic liquid (HSL-94) at room temperature using various adsorption programs. The composition of the obtained samples with adsorbed ammonium or silicon-organic compounds was determined by EDX spectroscopy (analytical attachment EDAX/TSL, shooting mode – 6 kV). The binding energies of the electrons of chemical elements on the surface were determined by the X-ray photoelectron spectroscopy. Measurements of the X-ray photoelectron spectra were carried out on Escalab 220iXL (University of Leipzig). The binding energy of the characteristic level was determined with accuracy of 0.1 eV. The water repellent properties of the samples were evaluated by parameter $1/a$, which was calculated from the value of adsorption of water vapours (a). Adsorption of water vapours was measured gravimetrically (g/g) at relative vapour pressure of p/p_0 = 0.98 ± 0.02 and temperature of 20 ± 2 °C. The fact of adsorption of water vapours on metal was proved by the presence of the O 1*s* peak with binding energy of 532.5 eV in the X-ray photoelectron spectra of samples [4]. The mathematical description of the process was obtained using programs such as MathCad and MS Excel.

Alkamon and triamon are widespread surfactants based on quaternary ammonium compounds with the general formula of $[R_1R_2R_3R_4N^+]X^-$, where R_1, R_2, R_3, R_4 – organic radicals on a nitrogen atom; X^- – is a polar group, usually an inorganic anion. It should be noted that alkamon has larger organic radicals (C_{16}–C_{18}) than triamon (C_1, C_2). The composition of technical triamon correspond to the formula: tris- (β-oxyethyl) methyl ammonium methyl sulfate $[(HOC_2H_4)_3N^+CH_3][CH_3SO_4^-]$. The industrial hydrophobizator HSL-94, which was used in the research, in fact, was a reference model.

3. Results and discussion

Modification of samples of dispersed copper was carried out according to the methods [2, 3, 5] and confirmed by EDX-spectroscopy. Due to the fact that the initial Cu powder did not contain any noticeable amount of nitrogen and sulfur, it can be assumed that their presence in modified powder is a result of chemisorption of QAC. Therefore the fact of chemisorption can be proved by the chemical composition of the samples. The content in initial powder is: copper at least 91.6 at. %, oxygen is 3.1 at. %, and carbon is 4.7 at. %. After treatment in alkamon vapours, the copper sample with the specific surface area of 0.35 m^2/g contained 0.2 at. % of nitrogen, 0.3 at. % of sulfur, 3.7 at. % of oxygen and 5.8 at. % of carbon. The similarly prepared Cu/A (treated in triamon vapours) sample with specific surface area of 0.32 m^2/g contained 0.4 at. % of nitrogen, 0.4 at. % of sulfur, 3.7 at. % of oxygen and 5.0 at. % of carbon. The Cu/T/A sample (prepared via sequential treatment in the triamon and alkamon vapours) with the specific surface area of 0.36 m^2/g contained 0.7 at. % of nitrogen, 0.8 at. % of sulfur, 3.1 at. % of oxygen and 6.3 at. % carbon [5].

Based on the experiments of adsorption of water vapors by various samples, the parameter a (value of adsorption of water vapours) was converted into parameter $1/a$, which characterizes the water repellent properties of the samples (table 1).

The kinetic equation of mass transfer (1) for adsorption processes was used as the basis to describe the process of adsorption of water vapor on metal in time [6, 7], where n (t) – is the function of the amount of adsorbed water; k – is the kinetic coefficient of adsorption; t – is the time

$$\frac{dn(t)}{dt} = -k \times n(t) \Rightarrow n(t) \sim e^{-kt}, \text{where } n(t) \sim a. \quad (1)$$

After elementary mathematical transformations

$$1/a \sim 1/n(t) \sim e^{kt}. \quad (2)$$

Table 1. Examples of conversion of value of adsorption *a* into parameter 1/*a*.

Sample	Exposure time, h							
	3		6		12		24	
	a, rel. un.	1/*a*	*a*, rel. un.	1/*a*	*a*, rel. un.	1/*a*	*a*, rel. un.	1/*a*
Cu-initial	0.00437	229	0.00438	229	0.00655	153	0.00657	153
Cu/A+T	0.00197	509	0.00295	339	0.00393	255	0.00394	255
Cu/A	0.00598	167	0.00498	201	0.00497	201	0.00499	201
Cu/T	0.00401	250	0.00404	250	0.00501	200	0.00509	200
Cu/T/A	0.00399	251	0.00499	201	0.00598	167	0.00599	167
Cu/HSL	0.00293	341	0.00294	341	0.00391	256	0.00488	205

The correlation dependencies, based on the experimental data were plotted in the MathCad (figure 1). The part of the graph, which is parallel to the *OX* axis (time) corresponds to the stage of saturation of the sorbent. A mathematical description of the function $1/a = f(t)$ with sufficiently high coefficient of determination R^2 (table 2) was obtained.

Table 2. Approximated equations $1/a = f(t)$ and coefficient of determination R^2.

Sample	Equation $1/a = f(t)$	Coefficient of determination R^2
Cu-initial	$f(t) = 104 + 201e^{-0.126t}$	0.838
Cu/(A+T)	$f(t) = 241 + 543e^{-0.229t}$	0.982
Cu/T/A	$f(t) = 166 + 217e^{-0.31t}$	0.997
Cu/T	$f(t) = 191 + 95e^{-0.126t}$	0.838

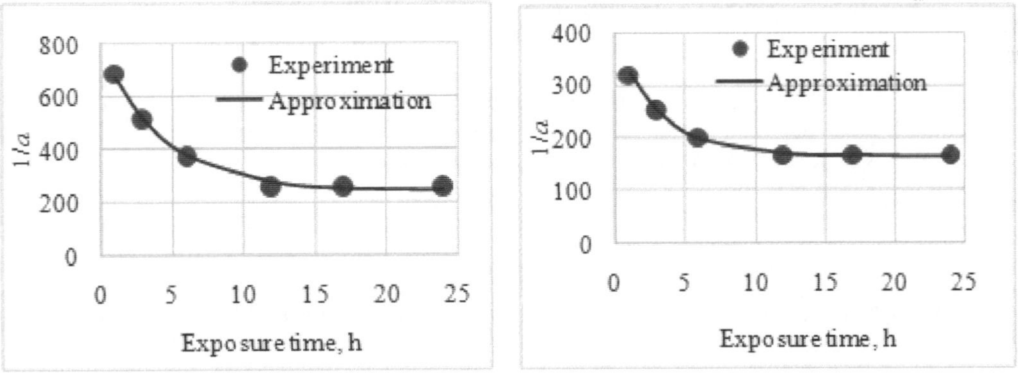

(a) **(b)**

Figure 1. Dependency graphs $1/a = f(t)$ for Cu/A+T **(a)** and Cu/T/A **(b)** samples.

The shape of the curves of the dependences and equations are similar to relaxation curves for other types of hydrophobic adsorbents, which were studied in Institute of general physics in Moscow [7]. In our previous researches of dispersed aluminum, where the exposure time was significantly greater (more than 320 h), the dependences of water repellent properties were approximated using a set of elementary functions (ln *t*, t^2 and Gaussian function) [8].

The sample Cu/(A+T) after 24 h of interaction with saturated water vapor remains the most hydrophobic (the maximum values are 1/*a* (table 1)). According to figure 1 and table 2 samples treated with T and A in mixed and sequential modes (the maximum value of the coefficient k), reach the saturation state of the surface faster than the original copper or the sample treated with only one modifier (T).

It is possible to trace the change in the nature of the active centers on the surface of metal powders during the layering of A and T by plotting the distribution curves of the adsorption centers. The curves can be plotted by adsorption of acid-base organic indicators with constant of ionization (acid force pKa) from –0.3 to 14.2 [9, 10]. The most informative and relief spectra of the distribution of adsorption centers were obtained for aluminum powder (PAP-2), its specific surface (2.6 ± 0.1 m^2/g) almost 8 times greater than the specific surface area of copper powder.

According to figure 2 mixed chemisorption of A and T on the metal leads to the decrease of the number of Lewis acid centers (pKa = 14.2) by an order, in comparison with the original aluminum powder. These centers correspond to the vacant orbitals of the Al atom, which can accept an electron pair, including the donor-acceptor interaction of QAC with a metal [2, 9, 10]. In addition, the intensity of the peaks, which are responsible for the hydrate-hydroxyl coating decreases for pKa = 5 and pKa = 9.5. This fact, confirms the thesis of reducing of the number of hydrophilic centers in case of mixed treatment with A and T. In case of layering of T or A separately, as seen from figure 2, the effects of the hydrophobization of the surface and the screening of electron-acceptor centers on the metal (the pKa range around 14), are weaker.

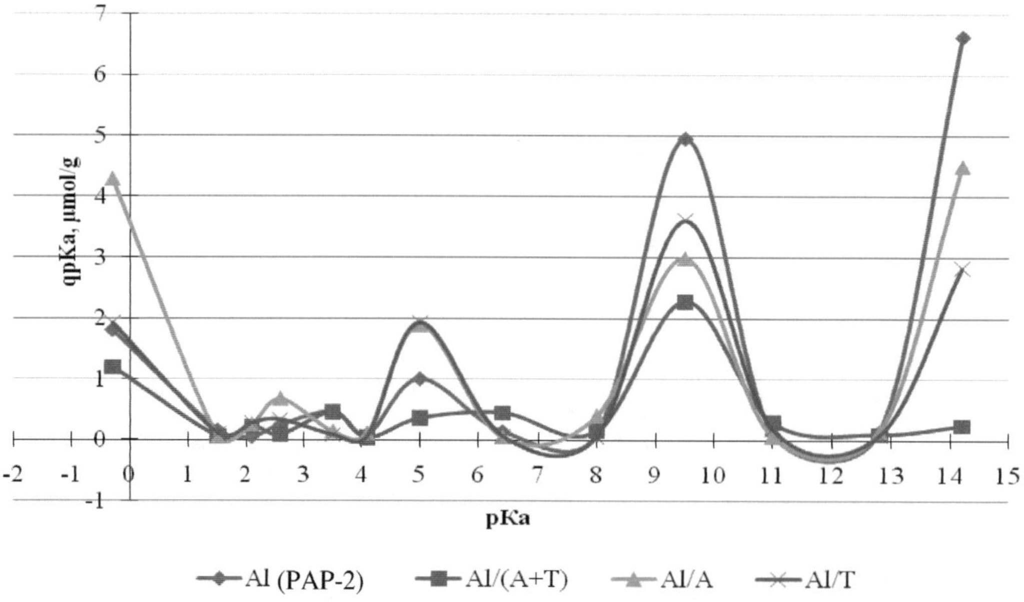

Figure 2. Distribution of acid-base centers on the surface of modified Al-powders.

4. Conclusion

For the first time the change of the water repellent properties of dispersed copper, modified using quaternary ammonium compounds (QAC) on 24 h time scale in saturated water vapours was studied.

Sufficiently accurate time exponential dependences of the water repellent properties of dispersed copper treated with QAC were derived and characterized. It was established that the samples, prepared by chemisorption of both modifiers in mixed and consistent regimes, reach the saturation state on the curves ($1/a$ – time) faster than others, due to the small number of hydrophilic centers on the surface of metals. The last conclusion was confirmed for similarly modified Al-powders by the effect of halving of the intensity of the peaks (pKa = 5 and pKa = 9.5) in the distribution spectras of the adsorption centers of acid-base organic indicators.

References

[1] Syrkov A G, Silivanov M O and Kushchenko A N 2016 *Journal of Physics: Conference Series* **729** 012026

[2] Syrkov A G, Kamalova T G, Kabirov V R, Kavun V S and Levine K L 2015 *Smart Nanocomposites* **6** 213–5

[3] Syrkov A G 2013 *Russian Journal of General Chemistry* **83** 1392–3

[4] Roberts M and Mackie C 1989 *Chemistry of the metal-gas interface* (Moscow: Mir)

[5] Syrkov A G 2015 *Russian Journal of General Chemistry* **85** 1538–9

[6] Romankov P G 2009 *Methods of calculations of processes and apparatus of chemical technology* (Saint Petersburg: Khimizdat)

[7] Artemov V G, Kurmasheva D M, Kapralov P O, Travkin V D, Tikhonov V I and Volkov A A 2014 *Bulletin of the Russian Academy of Sciences. Physical series* **78** 245

[8] Syrkov A G and Kabirov V R 2017 *Materials of international forum-contest of young researches "Rational use of natural resources"* **2** pp 236–7

[9] Sychov M M, Minakova T S, Slizhov Yu G and Shilova O A 2016 *Acid-base characteristics of the solid surface and properties control of materials and composites* (Saint Petersburg: Khimizdat)

[10] Syrkov A G, Silivanov M O and Pleskunov I V 2015 *Smart Nanocomposites* **6** 171–80

X-ray diagnostics in non-specialized conditions

A Y Vasil'yev[1], V A Lyubimenko[2], Y N Potrakhov[3], N N Potrakhov[3], V B Bessonov[3], V V Klonov[3], I A Larionov[3], A Y Gryaznov[3] and K K Zhamova[3]

[1] Moscow State University of Medicine and Dentistry, 127473, Moscow, Russia
[2] SPb City Children's Hospital № 1, 198095, Saint Petersburg, Russia
[3] Department of electronic instruments and devices, Saint Petersburg Electrotechnical University "LETI", 197376, Saint Petersburg, Russia

E-mail: yus87@yandex.ru

Abstract. The results of research in the development of X-ray diagnostic equipment in non-specialized conditions are described. The design of the first domestic digital X-ray diagnostic complex for neonatology in portable design is presented. Examples of X-ray images obtained "at home" from a patient are given.

1. Introduction

Currently, all the more urgent becomes the problem of X-ray studies in an unspecialized conditions, such as patient "at home" [1, 2]. Reasons for conducting X-ray diagnostic studies at home may be different, including:

- somatic "serious" condition in elderly patients who are unable to get to the health facility for suspected pneumonia, fractured limbs, other trauma, etc.;
- availability of financial well-off citizens, which ready to pay for X-ray diagnostic "at home".

However, in many cases, X-ray radiography in unspecialized conditions cannot be regarded as a method of definitive diagnosis, since there is a number of specific restrictions in carrying out such researches. At first, often medical condition of patient does not allow to perform X-ray examination in the standard (at least 2) projections; secondly, the characteristics of the technical means of X-ray radiology, do not allow conduct the X-ray studies.

Obviously, the use of traditional stationary tripod devices providing positioning X-ray apparatus relative to the patient, at home is not possible. Therefore, staff use portable tripods or takes the picture without using a tripod – "in hand". Therefore, special importance becomes the issue of radiation safety for the staff carrying out the study, and to the others who may also be involved in the study, such as the laying of the patient. However, the use of stationary means of protection against the unused X-ray radiation, as well as remove others at a safe distance in the home is almost impossible. Therefore it is necessary to use such a technique of X-ray photography, which can significantly reduce the exposure dose of radiation than when using stationary devices. In this case, is extremely important to increase the image quality of the test authority, the necessary and sufficient for a decision of the presence or absence of pathology.

Of particular importance to the issue of X-ray diagnostics in unspecialized conditions gives the important challenges of modern clinic – radiological surveys in neonatology. Additional difficulties in its decision are caused by the small size and weight of the newborn (up 400 g); conditions in which there is a child (incubator, open resuscitation system, operating), as well as the severity of his condition (reanimation, intensive care, rehabilitation). Often, because of the condition of the patient: artificial ventilation, cardiac or infusion therapy, active aspiration – X-ray study can be conduct only in the reanimation.

According "City Children's Hospital № 1" (Saint Petersburg) only in its resuscitation units annually around 4500 radiographic procedures in 2 or 3 projections. In total, it is up to 10 thousand X-rays in a year. Currently, these studies are carried out on the domestic or foreign X-ray machines, practically not adapted to the problems and conditions of the resuscitation of newborns with pathologies. Accordingly, the additional radiation load subjected to medical staff and other patients.

Research carried out in collaboration with experts from Federal Almazov North-West Medical Research Centre and "City Children's Hospital № 1" showed that virtually all requirements to the organization of the X-ray studies in unspecialized environment and, above all, neonatology, satisfied by method of microfocus X-ray radiography. Its main features are:

- using X-ray apparatus with a size of focal spot of X-ray tube is less than 0.1 mm;
- 1.5–2 fold increase voltage on X-ray tube and 3–5 times decrease focal distance as compared to standard X-ray technique.

To implement a methodology was developed specialized digital X-ray diagnostic complex "PARDUS-Neo" composed of the X-ray apparatus "PARDUS" in a portable version and X-ray visualisation system on the basis of the screen with the photoluminophore. The construction of the first domestic portable apparatus "PARDUS-R" allows to obtain X-ray images without the use of a stationary tripod. X-ray lab assistant takes a picture while holding the device in hand, while being in the radiation safety conditions (figure 1).

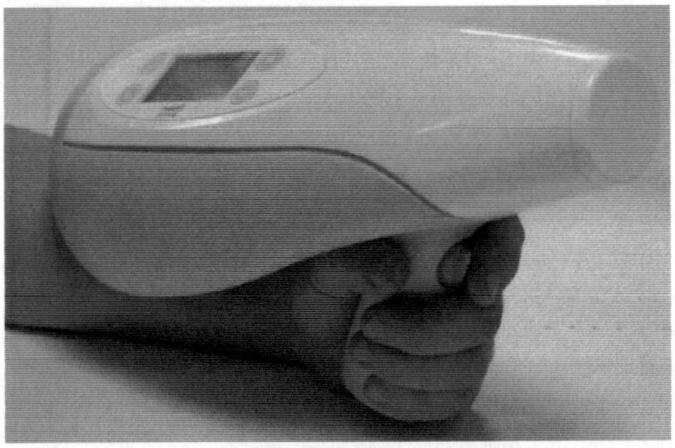

Figure 1. Portable X-ray apparatus "PARDUS-R".

Figure 2 shows an X-ray image of the preparation of the newborn body obtained by the method of microfocus X-ray radiography in unspecialized conditions on the X-ray diagnostic complex "PARDUS-Neo".

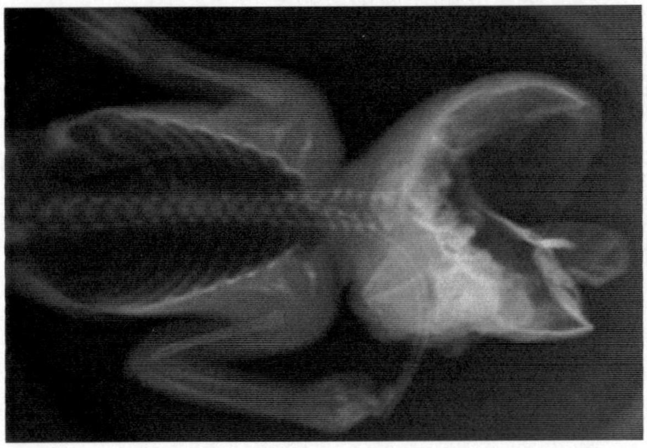

Figure 2. X-ray image of the preparation of the newborn body.

On the X-ray image is good visualized structure of bone elements: bone of the skull base, thorax skeleton, spine, pelvis and long bones, and soft tissues, including non-contrast nasotracheal tube. Figure 3 shows the X-ray images of the thoracic department (patient P., age 7 months, weight 3.5 kg), obtained in a stationary X-ray room in “City Children's Hospital № 1” (figure 3(a)), and obtained in unspecialized conditions “at home” of the same patient (figure 3(b)) using the complex “PARDUS-Neo”. An radiograph of pelvic colon of the same patient is shown in figure 4. According to experts the informativeness of the X-ray image obtained at home, almost as good as X-ray image obtained in the hospital.

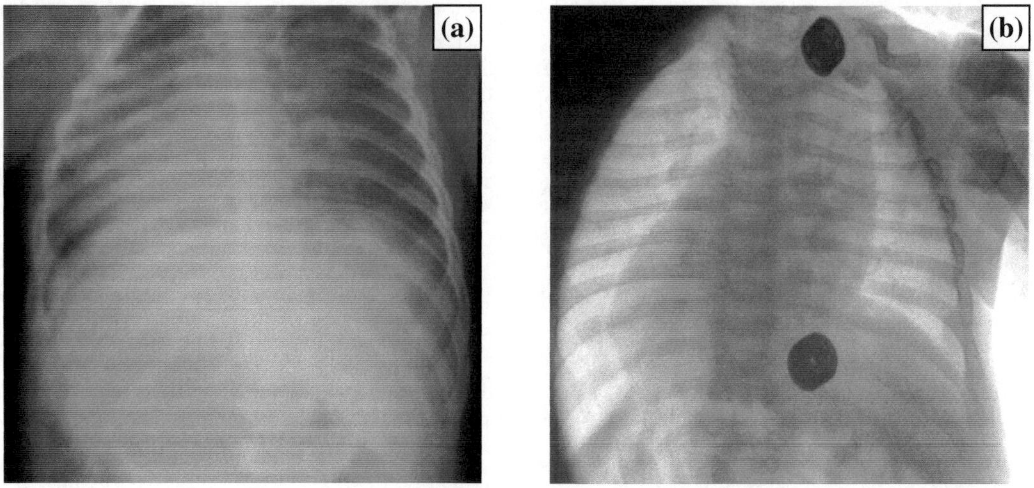

Figure 3. X-ray images of the thoracic department: **(a)** – stationary X-ray room in “City Children's Hospital № 1”; **(b)** – unspecialized conditions “at home” using the complex “PARDUS-Neo”.

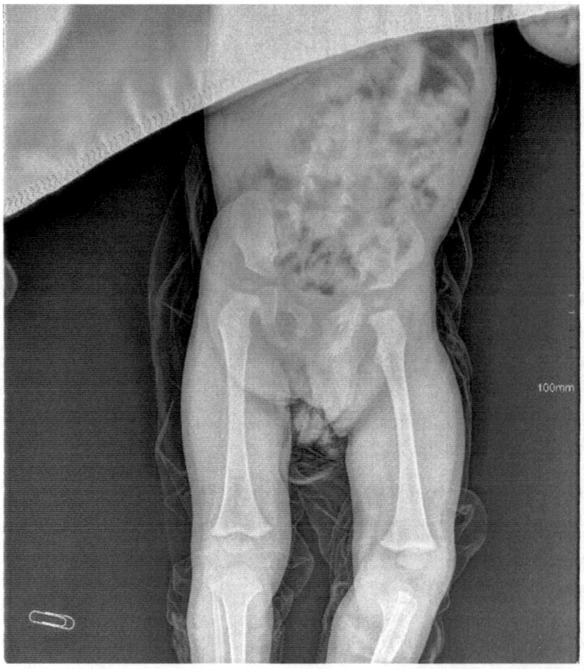

Figure 4. X-ray image of the lumbar spine and pelvic colon, obtained at the “home” of the patient.

Thus, the described results of research open the prospect of the use of techniques of microfocus X-ray portable technical equipment for newborns research.

Acknowledgments
The work was supported by the Russian Science Foundation in the framework of the project "Creating portable systems for microfocus radiography with the purpose of operational control of the microstructure, physicochemical properties and determining the residual life of aircraft parts and assemblies made of composite materials". Number of project 15-19-00259.

References

[1] Potrakhov N N 2015 *Biomedical Engineering* **48** 237–40
[2] Potrakhov N N and Potrakhov Yu N 2017 *Biomedical Engineering* **50** 406–9

Determination of the plasma impedance of a glow discharge in carbon dioxide

A S Kiselev and E A Smirnov

Department of electronic instruments and devices, Saint Petersburg Electrotechnical University "LETI", 197376, Saint Petersburg, Russia

E-mail: alex.kiselev.epu@gmail.com

Abstract. In this work an expression for the dynamic resistance of a glow discharge flowing in long tubes is obtained and analyzed. The expression describes the physical processes occurring in the positive column of a glow discharge. The frequency dependences of the active and reactive components as well as the dynamic resistance module for the discharge conditions corresponding to CO_2–lasers have been calculated. Based on the simulation results developed a computer program in the C# programming language for modeling the dynamic resistance discharge of glow discharge lasers.

Gas-discharge devices are used widely in various fields of science and technology [1]. The high prevalence of such devices is due to the wide variety of design solutions associated with the type of discharge flowing in the device. Among of such devices a special place is occupied by instruments with an extended positive column of a glow discharge [2, 3]. Glow discharge lasers belong to such devices. Glow discharge lasers are used actively for technological and metrological purposes. Due to physical and constructive features, the laser radiation has a temporary instability due to the influence of various factors. Therefore glow discharge lasers are used in along with radiation power stabilization systems [4].

Methods for stabilizing the power of laser radiation are classified on passive and active. In practice are used often active methods. Passive methods stabilizing do not take into account all the factors affecting the instability of the laser radiation power. Active stabilization methods assume negative feedback. One of the options for stabilizing the power of laser radiation is adjustment the pump level. In glow discharge lasers this process reduces to modulating the discharge current. Let us assume that at the time t the laser had a power level P_0 at a current I_0, which corresponds to the point A (figure 1). Due to external influences, the energy characteristic of the laser can shift and this current will correspond to the power $P_0 + \delta P$ (point B). The principle of current stabilization of power consists in forced changing the discharge current by such a value δI so that the radiation power level again is equal to P_0 (point C).

The discharge conditions realized in glow discharge lasers correspond to the region of sharp drop in the CVC of the discharge gap. In such a case there is a transition stage from the electrons free diffusion to the ambipolar. To obtain expressions for the total dynamic resistance, one can use the method of small perturbations. In accordance with this method, the current and voltage of the discharge will, respectively, have the form: $U = U_0 + u$, $I = I_0 + i$. In these expressions I_0, U_0 are stationary values of the current and voltage at the discharge i, u are the variable components of the discharge current and voltage, where $u \ll U_0$, $i \ll I_0$.

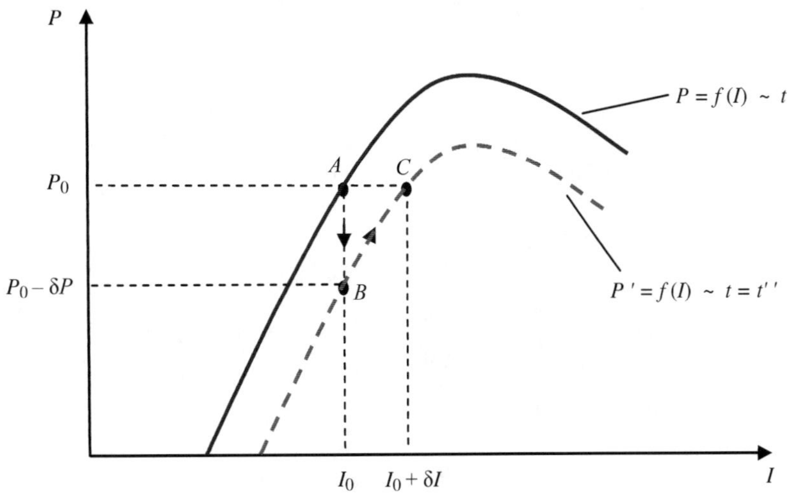

Figure 1. Explanation of the operation principle of the radiation power stabilization system.

Then the dynamic resistance will be determined by solving:

$$\frac{1}{I_0}\frac{\partial^2 i}{\partial t^2}+\frac{1}{I_1}\frac{di}{dt}+\frac{i}{I_2}=\frac{1}{U_0}\frac{\partial^2 u}{\partial t^2}+\frac{1}{U_1}\frac{du}{dt}+\frac{u}{U_2}. \quad (1)$$

Solution (1) was sought in the form:

$$Z_d=\frac{u_0}{i_0}e^{j\varphi}=\frac{I_0^{-1}(j\omega)^2+I_1^{-1}j\omega+I_2^{-1}}{U_0^{-1}(j\omega)^2+U_1^{-1}j\omega+U_2^{-1}}. \quad (2)$$

After transformation (2), an expression for the dynamic resistance of the glow discharge lasers is obtained, as the total positive column resistance

$$Z_d=R+jX=\frac{(I_2U_2)^{-1}+\omega^2\left[(I_1U_1)^{-1}-(I_2U_0)^{-1}-(I_0U_2)^{-1}+\omega^2(I_0U_0)^{-1}\right]}{U_2^{-2}-2\omega^2(U_0U_2)^{-1}+\omega^2U_1^{-2}+\omega^4U_0^{-2}}+$$
$$+j\frac{\omega\left[(I_1U_2)^{-1}+\omega^2(I_0U_1)^{-1}-(I_2U_1)^{-1}+\omega^2(I_1U_0)^{-1}\right]}{U_2^{-2}-2\omega^2(U_0U_2)^{-1}+\omega^2U_1^{-2}+\omega^4U_0^{-2}}, \quad (3)$$

where R and X – active and reactive components of the impedance of the positive column; ω – angular frequency. The remaining parameters in (3) depend on the discharge conditions [5].

An important role in the analysis of the dynamic discharge resistance is played by a static current-voltage characteristic. For the glow discharge lasers conditions the CVC has a falling character [6]. An experimental study of the CVC was conducted out on the basis of the serial laser LGN-705. The laser was connected to a vacuum system, allowing pumping out and filling the working volume. The length of the positive column was L = 400 mm, the diameter of the discharge gap was d = 9 mm. Figure 2 shows the current-voltage characteristics for the filling of CO_2:N_2:He = 1:1:6, depending on the pressure of the gas mixture. In this case the CVC can be approximated by an expression that includes the length of the discharge gap L and the longitudinal potential gradient E_z:

$$U_0 = U_k + aE_zLI^b,$$

where U_k – the cathode voltage drop, which is determined by a pair of gas-cathode material; a, b – constants depending on the nature of the gas and the discharge conditions. In this case can introduce a static slope of the current-voltage characteristic [7]:

$$\rho = dU/dI = abLE_zI^{b-1}.$$

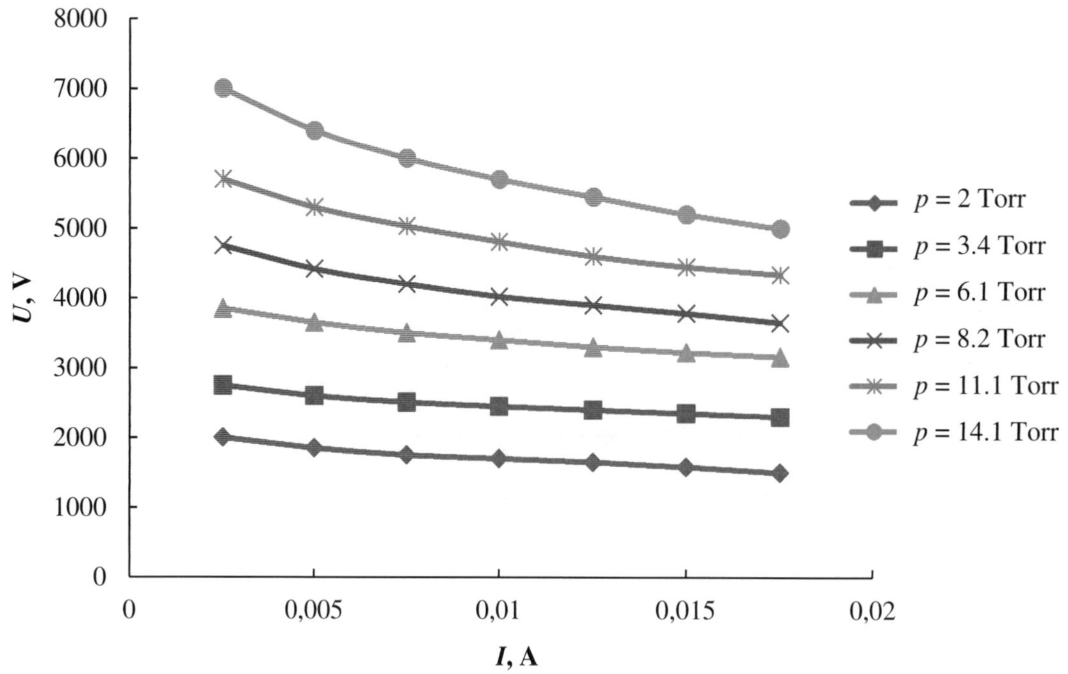

Figure 2. The set of current-voltage characteristics of the discharge in a gas mixture CO_2:N_2:He = 1:1:6.

The anode voltage drop was then assumed to be zero. The longitudinal gradient of the potential E_z in the positive column of envy from the genus of gas, working pressure, and also the diameter of the discharge gap. It can be determined taking into account the electron accommodation coefficient χ [8]:

$$E_z = \frac{3kT_e\sqrt{\chi}}{2e\lambda_e}.$$

Rigorous calculation of χ for the conditions of discharges in long narrow tubes, realized in the gas discharge lasers, is difficult because of the diversity and complexity of the processes occurring in the plasma. In the literature there are experimental data about χ relating to a number of pure gases. They reflect the dependence $\chi = f(T_e)$ for a narrow range of the discharge conditions, so the use of the reference values in the calculation is difficult [9–11].

To automate the analysis of the discharge impedance behavior, a program was developed in the programming environment of MS Visual C# (figure 3). The program allows calculating the active, reactive components of impedance, the impedance module, and also building an impedance hodograph. Before calculating the user, it is possible to select the type of gas mixture, set the discharge conditions, the range of the discharge current and the frequency of the signal. The result of the calculation is the family of dependences of the impedance components on the frequency for different discharge currents.

The dynamic properties of the discharge in the range of measurement parameters corresponding to the working conditions of glow discharge lasers are described quite adequately by a model that takes into account, as the main factors, the inertia of the direct ionization process, the finite time for establishing the space charge, and the falling character of the CVC. The resulting expressions for impedance can be used in analyzing the stability of power supplies of gas-discharge lasers based on the principles of current, passive or active power stabilization, when knowledge of discharge behavior in the nonstationary regime is required.

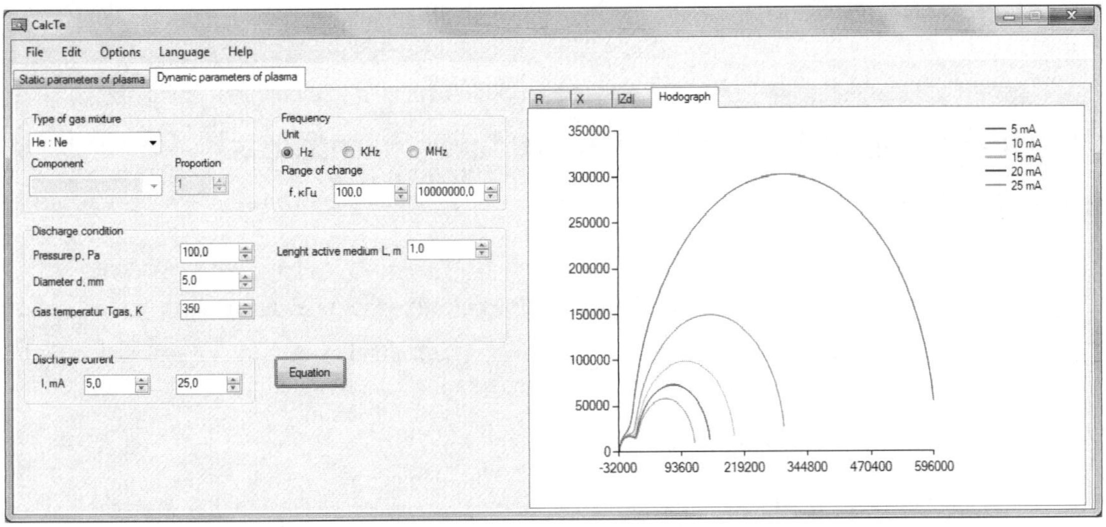

Figure 3. Example of calculating the plasma impedance for a mixture CO_2:N_2:He.

References

[1] Martsinukov S A, Kostrin D K, Chernigovskiy V V and Lisenkov A A 2016 *Journal of Physics: Conference Series* **735** 012049

[2] Chernigovskiy V V, Kostrin D K, Martsinukov S A and Lisenkov A A 2016 *Vakuum in Forschung und Praxis* **6** 34–7

[3] Martsinukov S A, Kostrin D K, Chernigovskiy V V and Lisenkov A A 2016 *Journal of Physics: Conference Series* **729** 012023

[4] Smirnov E A, Morozov S S and Timoshichev A A 2008 *Petersburg Journal of Electronics* **2-3** 39–48

[5] Smirnov E A 2013 *Proceeding of Saint Petersburg Electrotechnical University "LETI"* **10** 6–10

[6] Smirnov E A 2005 *Proceeding of Saint Petersburg Electrotechnical University "LETI"* **2** 43–8

[7] Kiselev A S and Smirnov E A 2015 *Vacuum technique and technology* **1** 172–4

[8] Granovsky V L 1971 *Electrical current in gas* (Moscow: Science)

[9] Kiselev A S and Smirnov E A 2016 *Proceeding of Saint Petersburg Electrotechnical University "LETI"* **8** 13–5

[10] Kiselev A S and Smirnov E A 2014 *Vacuum technique and technology* **1** 56–9

[11] Kiselev A S, Kostrin D K, Lisenkov A A and Smirnov E A 2017 *Journal of Physics: Conference Series* **789** 012027

IOP Conf. Series: Journal of Physics: Conf. Series **872** (2017) 012051 doi:10.1088/1742-6596/872/1/012051

Physical basis of the device for surface cleanliness measurement

S S Gavriushin[1], V B Poliakov[2], E P Prokhorov[1] and E A Deulin[1]

[1] Bauman Moscow State Technical University, 105005, Moscow, Russian Federation
[2] OOO "SKTB ELPA", 152613, Uglich, Russian Federation

E-mail: deulin@bmstu.ru

Abstract. The physical model of the device for surface cleanliness measurement and its technical realization are being considered in this paper. The number of gas monolayers being sorbed on the experimental surfaces is considered as a cleanliness parameter of these surfaces. The calibration graphs for the device under variable conditions of gases residual pressure, temperature and relative humidity are shown.

1. Actuality of cleanliness measurement

The problem of surface cleanliness measurement in nuclear and space technologies that is calculated as a number of sorbed gases monolayers is in the absence of sensors or even methods known by nowadays that could provide it direct quantitative estimation. Nevertheless the authors have realized it through some indirect and labour-intensive methods.

Scientific and social importance of the surface cleanliness sensor that is under development best of all may be illustrated with IVC-20 Secretariat decision [1] and e-mail letter from Prof. Jay Hendricks to Prof. Deulin (Chair VSTD IUVSTA): "We are working both with accelerators/R&D and industries. As result of our work we started to define purity classes. First purity classes are fixed for UCV (ultra clean vacuum similar to UCV purity classes we are working on purity classes for UHV/XHV".

2. Theoretical foundation of the sensor working

The first theoretical estimation of coverage Θ parameter in region of high vacuum we call [4] Henry equation:

$$\Theta = K_G P, \tag{1}$$

where K_G – Henry gas constant; P – residual pressure.

The more complex equation for theoretical coverage Θ parameter estimation in high and middle vacuum we call Langmuir [8] equation:

$$u = \frac{bP}{1+bP}, \tag{2}$$

where b – gas constant; P – residual pressure.

The most complex equation for theoretical coverage parameter Θ estimation in wide vacuum region, close to atmosphere, we call BET (Brunnauer Emmet Teller) equation [11], where P – residual pressure over the surface; P_L – sorbed gas saturation pressure; E_a – sorption heat; E_L – evaporation (condensation) heat of gas being sorbed; R – gas constant; T – surface (gas) temperature:

$$\Theta = \frac{\frac{P}{P_L}\exp\left[\frac{E_a - E_L}{RT}\right]}{\left(1-\frac{P}{P_L}\right)\left(1+\exp\left[\frac{E_a - E_L}{RT}\right]\right)\frac{P}{P_L}}, \quad (3)$$

The next equation was done [3] for coverage parameter Θ estimation in process of surfaces contacting in vacuum mechanisms:

$$\Theta \le \frac{Pd_0\alpha}{4\sqrt{\frac{2mkT_a}{\pi}}}\left[\frac{L}{V} - \tau_0\exp(E_d / RT)\right], \quad (4)$$

where T_a – mean temperature of contacting surfaces after contact, K; L – mean distance between micro roughness on contacting surfaces, m; V – friction speed, m/s; α – sticking coefficient; k – Boltzmann constant; τ_0 – adhesion constant, s; d_0 – gas molecule diameter $d_0 \approx 3\cdot10^{-10}$, m; m – sorbed molecule mass, kg.

Well known physics position says that energy (force) of links between sorbed molecule and a surface depends on "molecule–surface" distance and may be described with Lennard-Jones equation [8]. That equation shows that the largest energy (force) exists in a zone of first monolayer where "molecule–surface" distance is equal to sorbed molecule diameter $d_0 \approx 3\cdot10^{-10}$ m (zone of "chemistry sorption"):

$$E(X) = 4\varepsilon\left[\left(\frac{r_0}{r}\right)^{12} - \left(\frac{r_0}{r}\right)^{6}\right], \quad (5)$$

where ε – potential linking energy; r – distance between centers of molecules; r_0 – distance where the energy takes zero value.

Figure 1 shows an equation (5) graphical interpretation where distance r (marked with symbol X) increasing leads to sorption energy variation till minimal values that we call "chemical" and "physical" sorption energy.

Taking into account that mentioned energy influences on friction force, we can use this effect for coverage Θ value detection. In figure 1 "sorbed molecule–surface" linking energy $E(X)$ takes negative values, but in engineering practice the friction force F_T as a function of $E(X)$ cannot be negative.

To eliminate this contradiction let's represent diagram from figure 1 in inverted form (figure 2(a)) with a friction force F_T as a function of distance X. The "inverted" diagram from figure 2(a) comparison with engineering experimental results in figure 2(b) shows their identity.

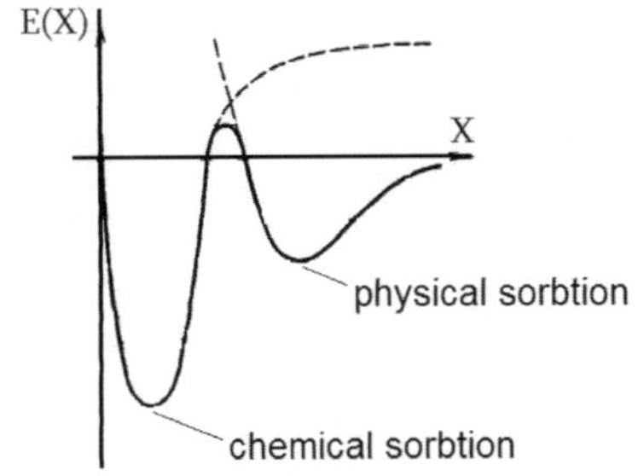

Figure 1. Lennard-Jones diagram of the sorption energy as a function of "sorbed molecule–surface" distance.

We can see mentioned "negative" energy from figure 1 can be presented as a positive friction force F_T and may be used for the sensor calibration. The results of static friction force coefficient measurement may be presented (figure 2(b)) as a function of distance X ("sorbed molecule–surface") or a function of coverage Θ value after recalculation [2]. Taking into account the physical sense of Lennard-Jones diagram we understand that F_T parameter value that is equal to "zero" in figure 1 and figure 2(a) was obtained theoretically and couldn't be measured that is reflected in graph (figure 2(b)).

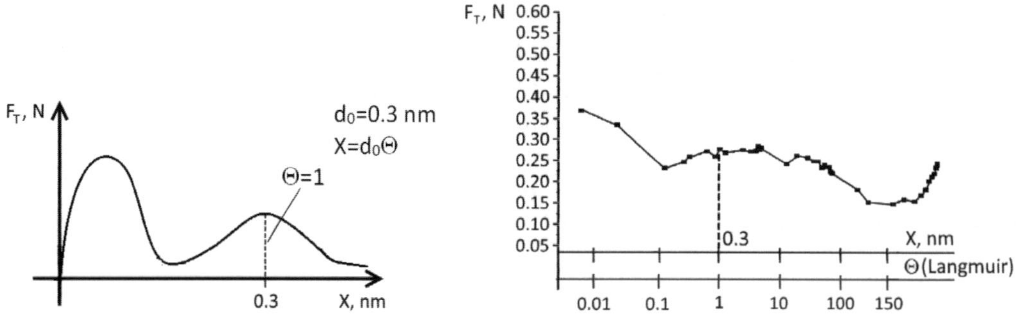

(a) **(b)**

Figure 2. Lennard-Jones diagram presentation: into "engineering" inverted form **(a)**; into friction force F_T as a function of distance X (coverage Θ value) **(b)**.

3. Sensor design description

The cleanliness sensor patented principle [5] may be realized in different design variants. The presented in figure 3 design is based on three polished plates usage, one of which is movable (*5*) and is clamped between two another ones (*4*, *6*). Piezo drive *1* presses plate *4* to a plate *5* with the normal force F_N. Piezo drive *10* provides the tangential force F_T by moving plate *5* with element *7*. Sensors *3* and *9* being fixed on elastic elements are used for normal force F_N and tangential force F_T measurement respectively. Base *11* with a frame *2* are fixed on UHV (Ultra High Vacuum) flange CF40.

The design idea shown in figure 3 was realized in prototype that may be seen in figure 4. The sample of cleanliness sensor model was fixed on UHV flange Conflat CF40. Electric feedthrough fixed on CF40 UHV flange provides the signal transmission between control units that are in atmosphere and piezoelectric drives with strain gauges that are in vacuum.

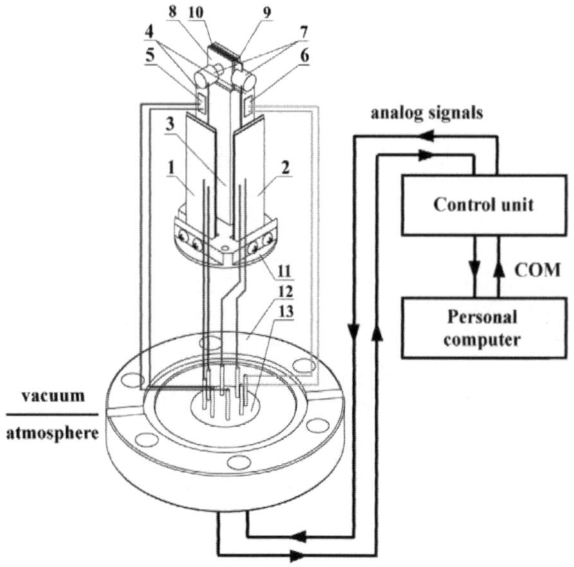

Figure 3. Design scheme of the surface cleanliness sensor: *1* – piezoelectric drive for the force F_N creation; *2* – piezoelectric drive for the tangential force F_T creation; *3* – frame; *4* – elastic elements; *5* – sensor for the normal force F_N measurement; *6* – sensor for the tangential force F_T measurement; *7* – plate holders; *8* – clamping plate; *9* – movable plate; *10* – static plate; *11* – base, *12* – UHV flange CF40; *13* – UHV electric feedthrough.

Figure 4. The surface cleanliness sensor specimen on UHV flange CF40.

The sensor novelty is that it makes possible to measure the coverage coefficient Θ directly on plates *8–10* (figure 3). The "Control unit" (figure 3) receives analog signals from F_N and F_T measuring strain gauges, converts them to digital format and transmits to the personal computer COM port. The PC program collects the data from control unit in form of a static friction force coefficient:

$$\mu_s = F_T / F_N . \tag{6}$$

Finally the program processes the accumulated data of static friction force μ_s coefficient values using calibrating diagrams and presents the result of measurement in form of a key parameter – coverage Θ coefficient.

4. Experimental base

The method of the cleanliness parameter measurement was patented in 2005–2008 [4, 5]. The experimental data based on used methods [3] show the ways for the sensor realization, as we can see in figures 2, 5.

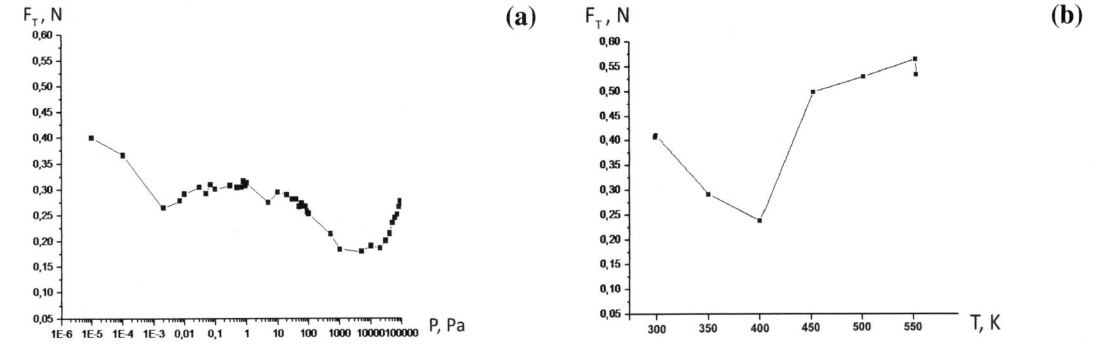

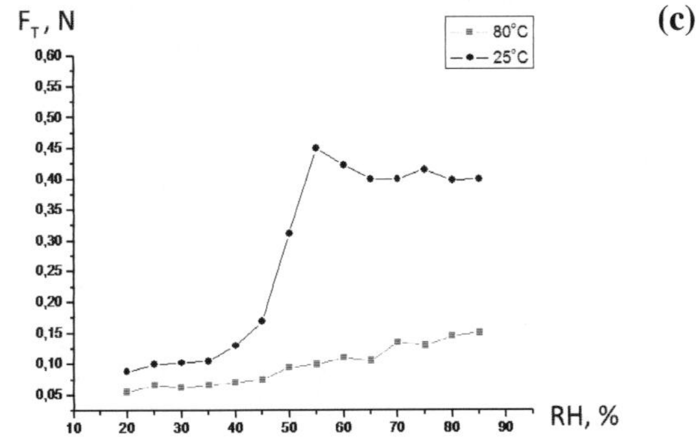

Figure 5. Experimental data of static friction coefficient as a function of: pressure [1] **(a)**, temperature [6] **(b)**, relative humidity [2] **(c)**.

The experimental data of static friction coefficient measurement at variable parameters *P*, *T*, *RH* (figure 5) shows the influence of these parameters on the coverage Θ parameter and may be used for the cleanliness sensor calibration. Thus measuring the static friction force coefficient with our sensor we can transform its values into coverage Θ parameter [2, 10] that would describe the number of sorbed gases monolayers on the surface in specific conditions of pressure, temperature and relative humidity of the sensor ambient. Also we must remember that we consider this Θ parameter as a cleanliness factor.

5. Summary

The physical base of device for the surface cleanliness measurement was done and used in practice. The sensor design main idea to consider "cleanliness" of the object surface as a number of sorbed gas monolayers is being realized and show its productivity. Theory shows that all the experimental results at residual pressure, temperature, humidity and materials variation may be presented as a function of coverage coefficient that may be used for cleanliness estimation by the sensor being designed in BMSTU, Russia. The sensor calibration graphs show the sensor marketability for using it in conditions

IOP Conf. Series: Journal of Physics: Conf. Series **872** (2017) 012051 doi:10.1088/1742-6596/872/1/012051

of residual pressure, temperature, relative humidity variation, i.e. in the conditions being demanded for nuclear, vacuum and space technology.

Acknowledgments
The authors would like to acknowledge Ministry of Science and Education of Russian Federation for Grant P24-21 "Creation of a New Methods based on Dry Friction Physics for Catastrophic Results in Electronics, Nuclear and Gas Transporting Systems Prediction and Prevention" that has helped in the process of work with this paper and with the experimental results transformation understanding and presentation.

References
[1] Deulin E A, Gatsenko A A and Loginov B A 1999 *Surface Science* **433-435** 288–92
[2] Deulin E A, Mikhailov V P, Panfilov Yu V and Nevshupa R A 2010 *Mechanics and Physics of Precise Vacuum Mechanisms. Fluid Mechanics and Its Applications* **91** (Springer)
[3] Deulin E A 2010 *Trenie i smazka v mashinakh i mekhanizmakh* **9-10** 35–43, 43–7
[4] Deulin E A 2005 *Friction vacuum gauge. Patent RU 2263886*
[5] Deulin E A 2008 *Method of measuring vacuum. Patent RU 2316744*
[6] Gladyshev I V 2002 *Proceedings Nordtrib, Keynotes and Abstracts* p 176
[7] IVC-20 Secretariat – People-X Inc.: contact@ivc20.com
[8] Langmuir I 1913 *Phys. Rev.* **1** 337
[9] Lennard-Jones J E 1924 *Proc. Roy. Soc.* **A106** 463
[10] Rozanov L N 1990 *Vakuumnaya tekhnika* (Moscow: Vysshaya shkola)
[11] Dushman S 1962 *Scientific Foundations of Vacuum Technique* (New York-London: John Wiley & Sons)

Modeling of thin films growth processes in the early stage for atoms with covalent bonds

V A Tupik, V I Margolin and Chu Trong Su

Department of Radio Microelectronics and Radio Equipment Technology, Electrotechnical University "LETI", Saint Petersburg 197376, Russia

E-mail: chusu171@mail.ru

Abstract. Computer simulation for obtaining thin film's growth process at an early stage with the proposed model of atoms with isotropic and anisotropic interactions been considered. Carrying out the procedure for analyzing the problem on the basis of the program being implemented, computer simulation of thin film growth processes has been carried out on several examples. The results of computer simulation of the growth process of thin film on a given substrate and an aggregate in a vacuum condition are shown. Some characteristic distributions of the obtained structure have been carried out to evaluate the proposed adequate model and to reflect the high complexity of thin films growth process.

1. Introduction

Nowadays, with the advancement in nanotechnology, one of the high promising directions is the application of computer modeling with an approach that allows taking into account features at the atomic level, which for human cognition, is always indirectly accessible. And using the computer approach for the optimization and synthesis of the processes of formation of these thin films at the early stage also plays a big role. The work was designed so that from the beginning the optimization of configuration setting and the process macroparameters are based on the macroparameters of the final function of the resulting thin film, like the film thickness, deposition rate, distribution of thin film thickness etc., then we proceed to optimize microparameters thin film such as surface roughness, atomic structure, defect structures, etc. Based on previous studies [1–4, 6, 8], many particular feature processes were shown for simple atoms which had an isotropic bond which is described by a pair potential interaction.

In this study, we would take a look on most of the described computer approach for more complex atoms wherein the bond is anisotropic, which means we can expand the range of applications, for example, the semiconductor material atoms.

2. Description of new models

As shown in [1, 7] we can observe that at present it is more important to use an empirical approach in the description of interatomic interaction. Also, due to restrictions on the available power of computing techniques and the number of simulated atoms (in the order of 100 atoms in the approach based on the first principles of quantum mechanics), the new idea, using in this work, was such that all the particles (atoms) were represented as described in figure 1(a), (b), (c) [1].The atom has two parts, namely the external electron orbitals (often referred to as electron, and denoted as *e* in figure 1) and the nucleus (more precisely, the inner part of the atom, excluding the outer electronic orbital). The electrons repel each other and form an angle φ_0^0 (figure 1), they also attract the electrons to the optimal distance r_1 in figure 1.

Atoms are related to each other depending on the distance between them (r_2), and if this distance is greater than the threshold value, we then use an isotropic potential bond to simplify the computational load and accelerate the modeling process. Otherwise, when atoms related to each other at a close distance is less than the threshold value; we use an anisotropic potential bond, separating the electrons into bound (connected) and unbound (unconnected) ones. A more detailed description is given in [1].

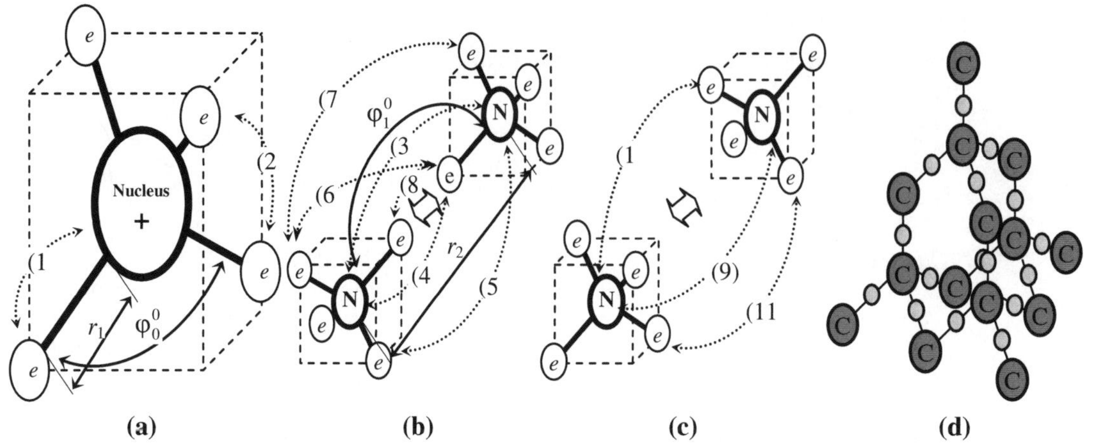

Figure 1. A new model of one atom **(a)** and two cases of interatomic interaction: $r \geq d_{\text{threshold}}$ **(b)** and $r < d_{\text{threshold}}$ **(c)**. There are shown 11 various types of bonds in this model [1]. The atomic diamond structure **(d)** used in this work as a substrate.

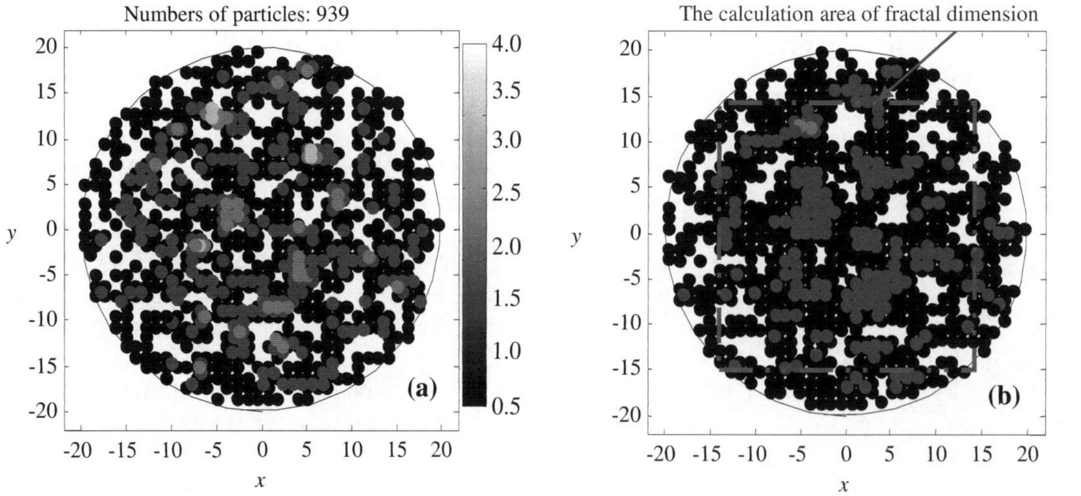

Figure 2. Computer simulation of thin film growth process without thermal noise **(a)** and with thermal noise **(b)** [2].

As shown in this paper [2, 3], we also considered the same problem, but for complex atoms with sp^3 type of hybridization. For example, figure 2 shows the result of computer simulation of thin film growth on a substrate with a body-centered cubic lattice with an equal lattice parameter. The left diagram, figure 2(a) shows the process without thermal noise on the substrate, and in the other diagram figure 2(b) we took into account the parameter of thermal noise in the average energy order of 10^{-1}, and also the optimal interaction energy of two particles in Lenard-Jones pair potential [2]. As shown in figure 2, the thermal noise will affect the size of the nucleus and the roughness of the resulting thin film. We want to note here that all the linear dimensions used in this paper were normalized by $r = r_{real} / \sigma_0$. Where $d_{threshold}$, σ_0 are adjustable parameters for the working material and each type of bonding (in Lenard-Jones pair potential), as usually offered as the distance is longer than $3\sigma_0$.

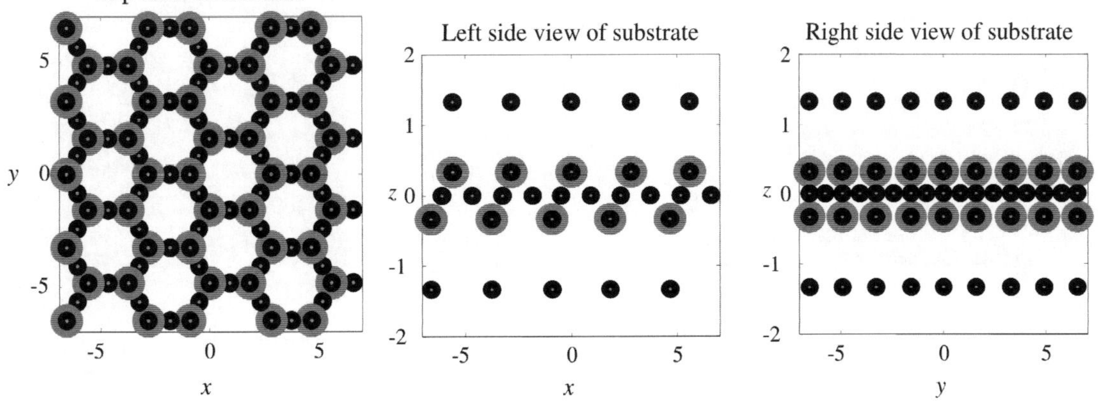

Figure 3. Using model of substrate.

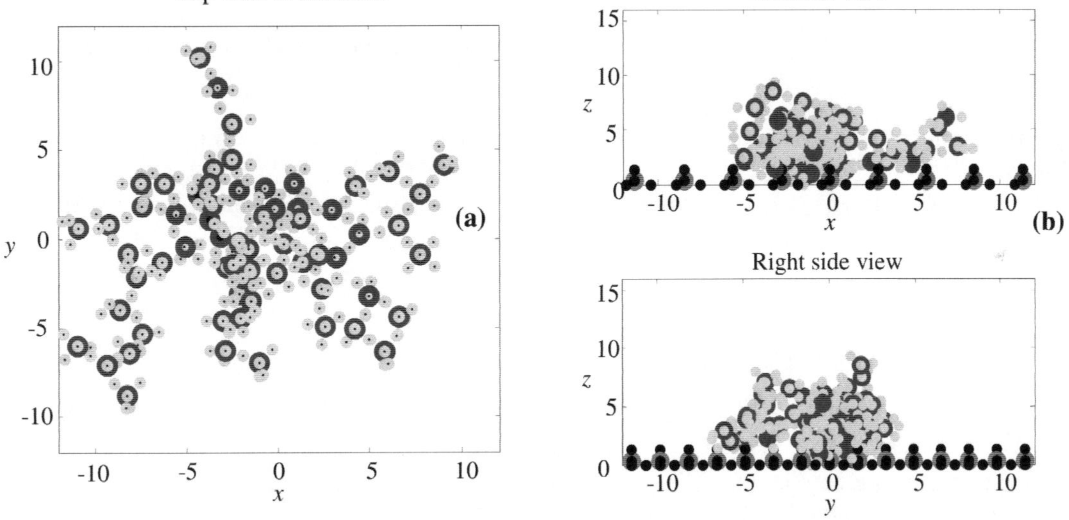

Figure 4. 3D result of computer simulation of the growth process of 64 atoms (nucleus – blue) with sp^3 hybridization type (green colours), on the left – in the vacuum condition, on the right – on the used model of the cover (with diamond structure – red and black).

3. Two computer simulations with new algorithm

Using the new approach (anisotropic model), we conducted two experiments, in the first computer experiment, the deposited structure is obtained without a substrate in the vacuum condition (figure 4(a)), and in the second experiment, on the selected structure of the substrate as shown in (figure 4(b)) as the substrate parameter, we use a structure where $r_1 = 1$ and it consists of two layers (figure 3) of diamond structures with the orientation shown as in figure 1(d). This structure only creates a potential impact and is a fixed structure in the thin films growth process. The growth process proceeds according to this algorithm, the particle is randomly generated at a distance r from the substrate with random coordinates (x, y), and then the growth process, the system of deposited particles behaves this way so that the energy of the system (deposited particles with the substrate) becomes minimum. Numerical methods were used by the Monte Carlo method for random particle selection and a quasi-Newtonian method was used for selecting the optimal direction of particle motion [3, 4]. After that, we calculated its analytical statistical characteristics and we compared it with the structure of the film deposited on the substrate.

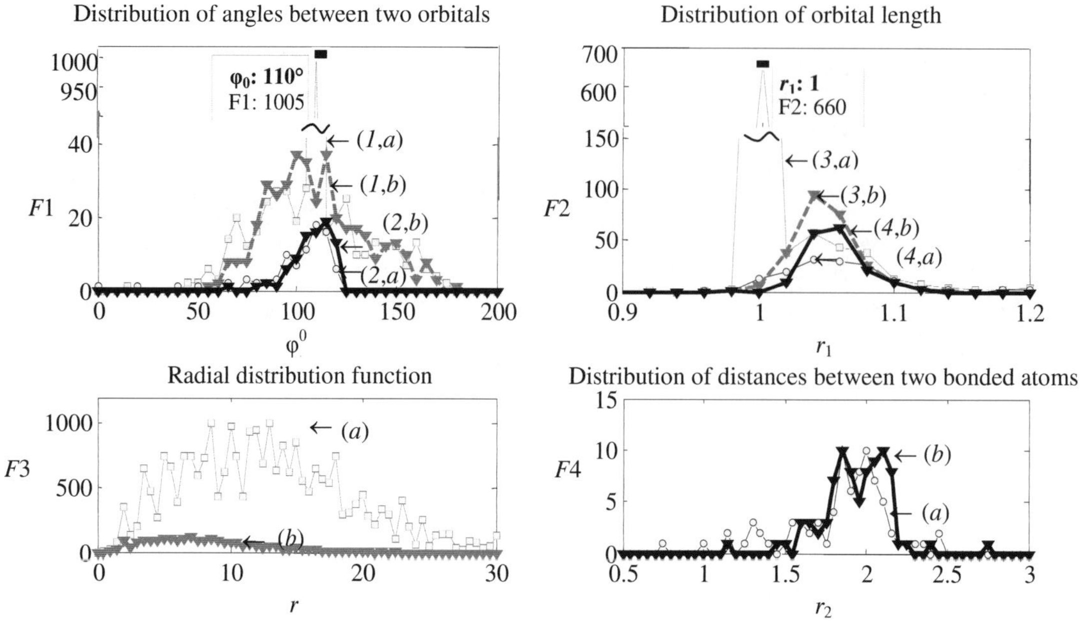

Figure 5. Some characteristic distributions of the obtained structure (figure 4). F(1–4) – the number of atoms or orbitals having a given distance or angle between two orbitals: *1* – analysis for internal atoms (φ_0^0); *2* – analysis only for bonded atoms (φ_1^0); *a* – analysis for obtained film on a substrate (figure 4 (b)); *b* – analysis for the obtained aggregate, which was growth in a vacuum condition (figure 4 (a)).

4. Results and discussion

Based on the results obtained with the aid of the calculation model in [1], in figure 5, $F1$ – it shows the angle between two orbitals φ_0^0 inside one particle it also changes (not constantly), and most particles have an optimal angle of 109° (acos(–1/3)), the angle between two connected particles φ_1^0 also changes, as shown on the image. In figure 5, $F2$ it is shown that the orbital length inside one particle (r_1) and the distance between two particles (r_2) vary. And in figure 5, $F3$ – the character of the crystallization of the obtained structure is shown (if the number of simulated particles is sufficient), as a result, it is seen that the structure obtained is likely to have an amorphous structure, as shown in figure 4.

As for the process of obtaining single crystals of SiC, it shows there is a lot of different polytypes and polymorphs [9, 10], which implies that the system which consists of two parts could go one in a lot of directions of system development, and the distribution of sp^2, sp^3 [5] reflects the complexity (nonlinear dynamics) occurring at this stage. Our model which will assist in showing our understanding at this stage is really the first approximate calculation closest to reality. But the use of this approach from first principles is still a bit far from implementation due to the complexity and loaded work [11].

5. Conclusions

In this paper, we presented the result of studying the process of condensation of thin films on the potential field of an ideal substrate structure for complex atoms with sp^3 type of hybridization. The procedure for the movement of computer simulation into synthesis and optimization of simple to complex models were suitably shown for different types of interatomic potential. It has been estimated that in spite of the first approximate results, the model can account for an anisotropic bond, change bond lengths and angles between the two particles in the condensation process. These are very significant steps to the modeling of atoms with an anisotropic bond as discussed and shown.

References

[1] Tupik V A et al. 2017 *IEEE Conference of Russian Young Researchers in Electrical and Electronic Engineering (EIConRus)* 1209–12
[2] Tupik V A et al. 2016 *J. Phys.: Conf. Ser.* **729** 012025
[3] Chu Trong Su 2016 *Proceeding of the Russian Universities: Radioelectronics* **6** 22–31
[4] Tupik V A et al. 2016 *IEEE NW Russia Young Researchers in Electrical and Electronic Engineering Conference (EIConRusNW)* 93–6
[5] Osswald S et al. 2006 *Journal of the American Chemical Society* **128** (35) 11635–42
[6] Tupik V A et al. 2016 *IEEE V Forum Strategic Partnership of Universities and Enterprises of Hi-Tech Branches (Science. Education. Innovations)* 42–5
[7] Tersoff J 1988 *Phys. Rev. B* **37**(12) 6991–7000
[8] Tupik V A et al. 2015 *Proceeding of the Russian Universities: Radioelectronics* **5** 15–9
[9] Ching W Y et al. 2006 *Materials Science and Engineering: A* **422**(1–2) 147–56
[10] Wasyluk J et al. 2010 *Materials Science Forum* **645-648** 359–62
[11] Lee J G 2016 *Computational material science* (CRC Press)

Dynamic characteristics of the plasma of a helium-cadmium laser

A S Kiselev and E A Smirnov

Department of electronic instruments and devices, Saint Petersburg Electrotechnical University "LETI", 197376, Saint Petersburg, Russia

E-mail: alex.kiselev.epu@gmail.com

Abstract. In this work the dynamic characteristics of a helium-cadmium laser are investigated in the process of entering the working regime. The study of the spectra of oscillations of the coherent radiation power in combination with the study of the spectrum of discharge current oscillations is carried out. The spectra of lateral spontaneous emission were also studied using a small-size diffraction spectrometer in the process of entering the laser into operating mode.

The most typical representative of metal vapor lasers is a helium-cadmium (He-Cd) laser. It is able to generate continuous coherent radiation in the blue (441 nm) and ultraviolet (325 nm) regions of the spectrum. He-Cd lasers are used in printing, holographic, spectroscopy and so on. But a serious drawback of He-Cd lasers is the high level of instability of the coherent radiation power. The level of radiation oscillation depends on the discharge current, the helium pressure, the geometry of the discharge gap, the concentration of cadmium vapor [1].

The construction of helium cadmium He-Cd lasers is far more complex than that of other helium based lasers. The laser tube contains a reservoir for cadmium and a heater to vaporize the metal. As a result, a heated filament cathode is often used in place of the cylindrical tube that comprises a He-Ne laser. Additionally, the laser itself needs to sustain a higher level of internal pressurization allowing the vaporized cadmium to remain in the tube. The lifetime of a specific helium cadmium laser is dictated by the amount of cadmium in the reservoir [2].

The active medium of He-Cd lasers contains an inert gas He under pressure 500...700 Pa and Cd vapor (0.1 Pa). Excited cadmium ions are radiating particles. Because of the low concentration of Cd atoms their excitation due to direct collisions with electrons plays an insignificant role. The main process of pumping the active medium of a He-Cd laser is ionization of Penning – process of formation of an excited ion of an easily ionized particle (Cd atom) upon collision with an excited metastable atom of a He buffer gas having a large excitation cross section:

$$He^* + Cd = He + (Cd^*)^+ + e.$$

The aim of the study was to search for and analyze the dynamic parameters of the plasma of a helium-cadmium laser, leading to a short-term and long-term instability of its radiation power. The method of investigation is reduced to the study of the spectra and dependences of the coherent fluctuating power for various physical parameters of the laser and to reveal the causes of the appearance of noise of He-Cd lasers, which makes it possible to make assumptions about the occurrence of noise in the laser radiation.

In this work studied the spectra of the oscillations of the coherent radiation power in combination with the study of the spectrum of discharge current oscillations, as well as the spectra of the side spontaneous emission in the process of entering the laser into operating mode. The spectra of spontaneous emission were studied with the help of a small-sized spectrometer [3]; the radiation was fed through the optical fiber. At the initial stage working of laser the discharge proceeds in pure helium (figure 1). This can be seen from the comparison of the spectrum of the lateral spontaneous study combined with the helium emission lines (figure 2). It was found that the intensity of the emission lines of cadmium and helium over time varies, a transient process is observed. In the first minutes of working the laser the intensity of the helium lines is approximately constant. Approximately at the fifth minute of working the laser the emission lines of cadmium begin to appear in the spectrum. Then there is a decline corresponding to the time of the onset of evaporation of cadmium. It can be noted that the oscillations in the intensities of the emission of helium and cadmium are made in antiphase (figure 3).

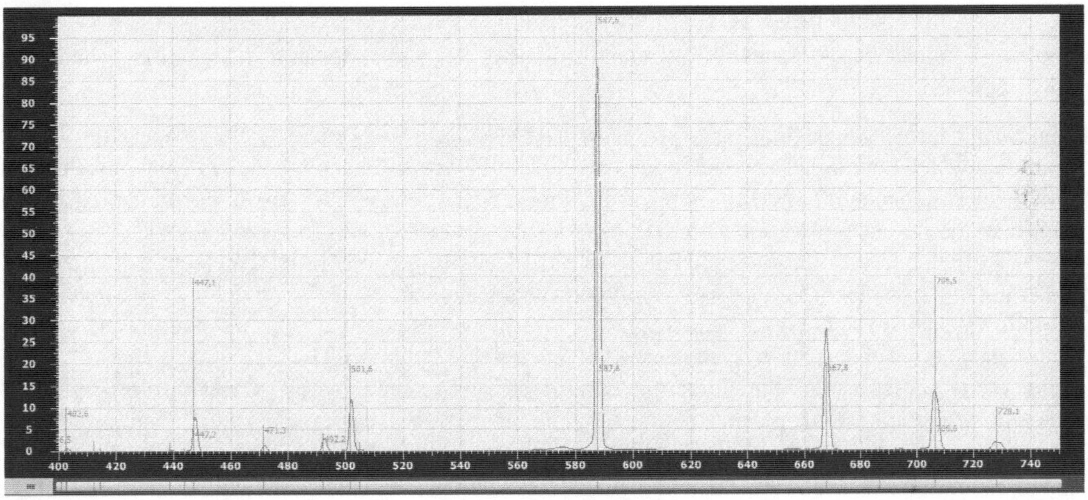

Figure 1. The spectrum of the lateral spontaneous emission of a He-Cd laser at the moment of switching.

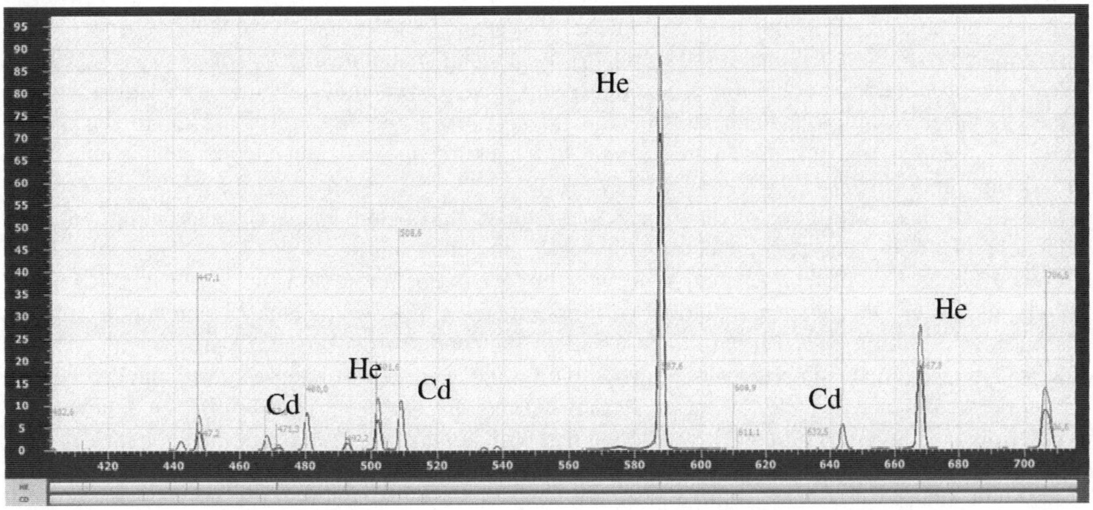

Figure 2. Comparison of the spectrum of the lateral spontaneous study of the He-Cd laser at the moment of switching on and after 25 min.

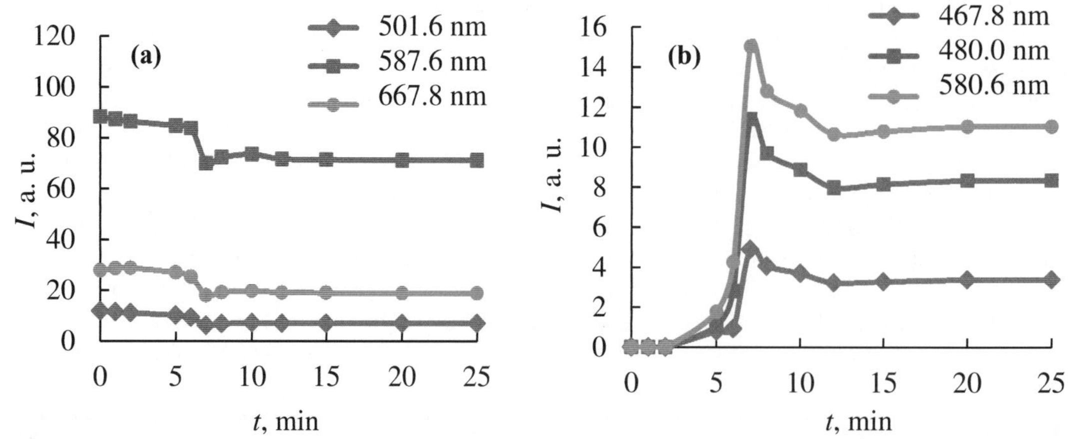

Figure 3. Time dependences of emission intensity of some spectral lines: **(a)** – helium; **(b)** – cadmium.

A significant influence on the output power of a He-Cd laser is exerted by the temperature of the evaporator, which determines the vapor pressure Cd in the positive column of the discharge. Initially the increase in concentration with increasing temperature of the evaporator is accompanied by an increase in the output power due to an increase in the number of emitting particles Cd^{+*}. A further increase in the concentration of readily ionisable Cd leads to a decrease in the electron temperature and the potential gradient in the positive column. The mechanism for reducing the energy of electrons is as follows. As the electron energy increases, so does the number of Cd^{+} ions involved in the process of compensating for the negative space charge of electrons in the plasma [4].

The own oscillations of the discharge current are determined by the variety of processes of interaction of particles in the plasma, and also by the interaction of the external electric circuit and the discharge as a whole. Such oscillations include reactive (relaxation) oscillations, stratification of discharge, double layer oscillations and cathode oscillations. Reactive oscillations (tens–hundreds kHz) are manifested at small discharge currents corresponding to the cessation of its existence. The steepness of the dropping current-voltage characteristic of the discharge has large values. Reactive oscillations increase with increasing parasitic capacitance "cathode–anode" [5]. The stratification of discharge, or layered oscillations, is due to the discrepancy in the length of the discharge of the ionisation and excitation maxima of the atoms [6]. As a result spatial layers (ionization waves) with an increased concentration of either ions or excited atoms arise. Layers move at high speed along the axis of the discharge, forming running stratification layers, increasing from the cathode to the anode. Stratification layers not only modulate current and radiation in time, but also make the discharge spatially non-uniform [7].

In a He-Cd laser are more often present non-regular stratification layers – oscillations not one frequency, but noise ones, the spectrum of which includes a large number of harmonics. The difference frequencies of the harmonics fall into the low-frequency region (up to 80...100 kHz) and strongly modulate the coherent radiation with amplitudes from one to tens of percent, forming a Poisson noise $1/F$. In the discharge of a He-Cd laser, the conditions ensuring maximum amplification of the active medium are simultaneously favourable for the existence of intense stratification layers. In this work studies were made of the spectrum of the discharge current oscillations during the evaporation of cadmium (figure 4). It was found that with an increase in the concentration of cadmium in the discharge, the spectrum shifted to low frequencies, and at the same time the increases amplitude of current oscillations (by an order of magnitude).

The mechanisms described above significantly influence the output power of the He-Cd laser radiation. During the evaporation of cadmium occurs inertia in active medium. Also in the laser there are natural oscillations of the discharge. This leads to modulation of the discharge current. An obligatory condition for the efficient use of lasers based on cadmium vapor is the presence of systems for stabilizing the power of their radiation.

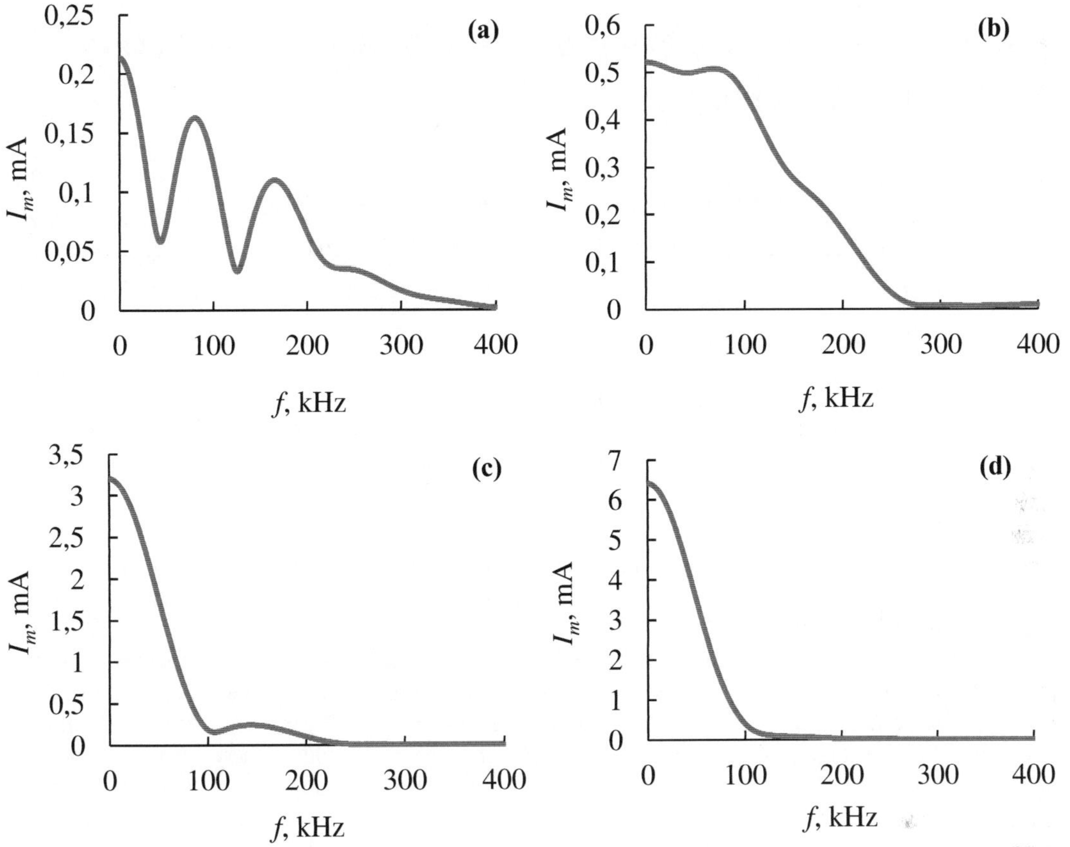

Figure 4. The spectra of discharge current oscillations: **(a)** – in pure helium; **(b)** – 2 min after the evaporator has been switched on; **(c)** – 7 min; **(d)** – 25 min.

References

[1] Pikhtin A N 2015 *Quantum and optical electronics* (Moscow: Abris)
[2] Smirnov E A, Morozov S S and Timoshichev A A 2008 *Petersburg Journal of Electronics* **2-3** 39–48
[3] Uhov A A, Gerasimov V A, Kostrin D K and Selivanov L M 2014 *Journal of Physics: Conference Series* **567** 012039
[4] Granovsky V L 1971 *Electrical current in gas* (Moscow: Science)
[5] Kiselev A S and Smirnov E A 2015 *Proceeding of Saint Petersburg Electrotechnical University "LETI"* **6** 3–6
[6] Kiselev A S, Kostrin D K, Lisenkov A A and Smirnov E A 2017 *Journal of Physics: Conference Series* **789** 012027
[7] Kiselev A S and Smirnov E A 2014 *Vacuum technique and technology* **1** 56–9

An investigation of radiation instability on reconstruction quality in tomography

V B Bessonov, V V Klonov, I A Larionov and A V Obodovskiy

Saint Petersburg Electrotechnical University "LETI", 197022, Saint-Petersburg, Russia

E-mail: vbbessonov@yandex.ru

Abstract. The paper describes the experience of detecting defects in layered structures using microfocus tomography. The basic features of the equipment for microfocus tomography are given. The sizes of defects (pores and inclusions) and their location in the object are analyzed.

Modern tomograph is a software and hardware complex in which the obtaining of necessary information for the reconstruction of the object's section is associated with its consistent radioscopy from a large number of angles [1].

Obtaining projections is associated with such parasitic phenomena as digital noise and scatter in the intensity of exposure of the X-ray receiver in a series of tomographic projections. In this connection, it seems expedient to investigate the effect of these phenomena on the quality of the rebuilt section [2].

To solve this problem, a specialized computer program was developed in the Qt Creator environment. The program allows to recreate the digital noise effect of an ideal tomographic projection for further investigation of the influence of this parameter's value on the reconstructed tomographic section quality.

The addition of digital noise (figure 1) was carried out using the following relationship:

$$Delta = (RND \cdot k) - k/2,$$

where *Delta* – brightness added to the original pixel of the image; *RND* – random variable in a given range; *k* – coefficient that allows to vary the overall level of digital noise in the image.

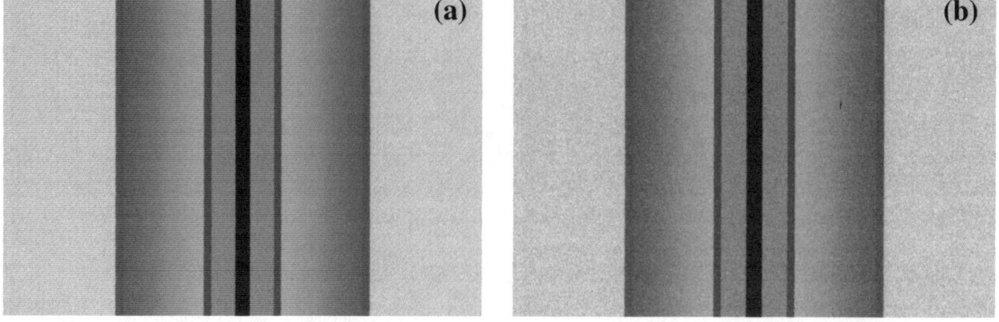

Figure 1. Adding to the original image of digital noise: **(a)** – initial projection; **(b)** – projection with artificially added noise.

The program also implements the possibility of creating a set of tomographic projections with a spread in the intensity of exposure of the X-ray receiver with a given value of the instability of the X-ray source. The variation of the spread (figure 2) was carried out according to the following relationship:

$$Delta = (RND - 0.5) \cdot k + 1.$$

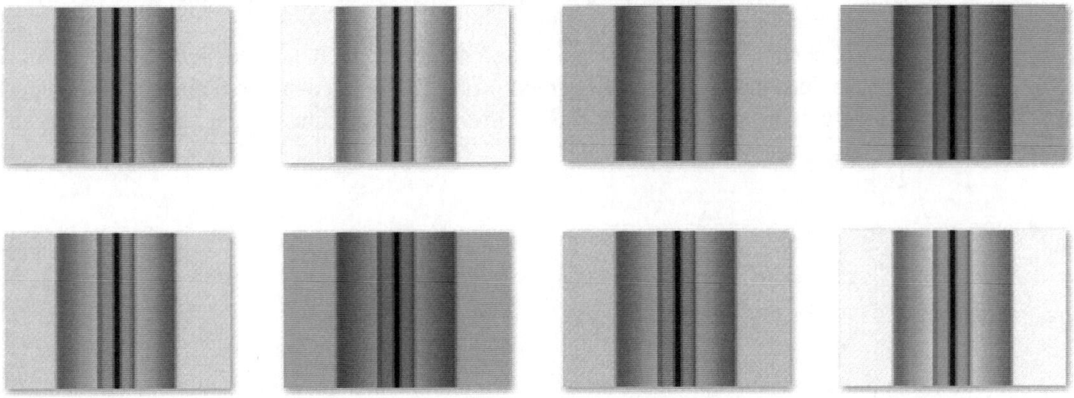

Figure 2. A projection series with an imitation of the scatter of radiation intensity.

With the help of the written program, a set of tomographic projections with different levels of digital noise was created. Superposition of noise was carried out using relationship represented above, wherein the noise level is determined by the coefficient *k*. Projections with different values of *k* are shown in further sections. Also under each section there is an example of the projection of the corresponding level of noise pollution (figure 3).

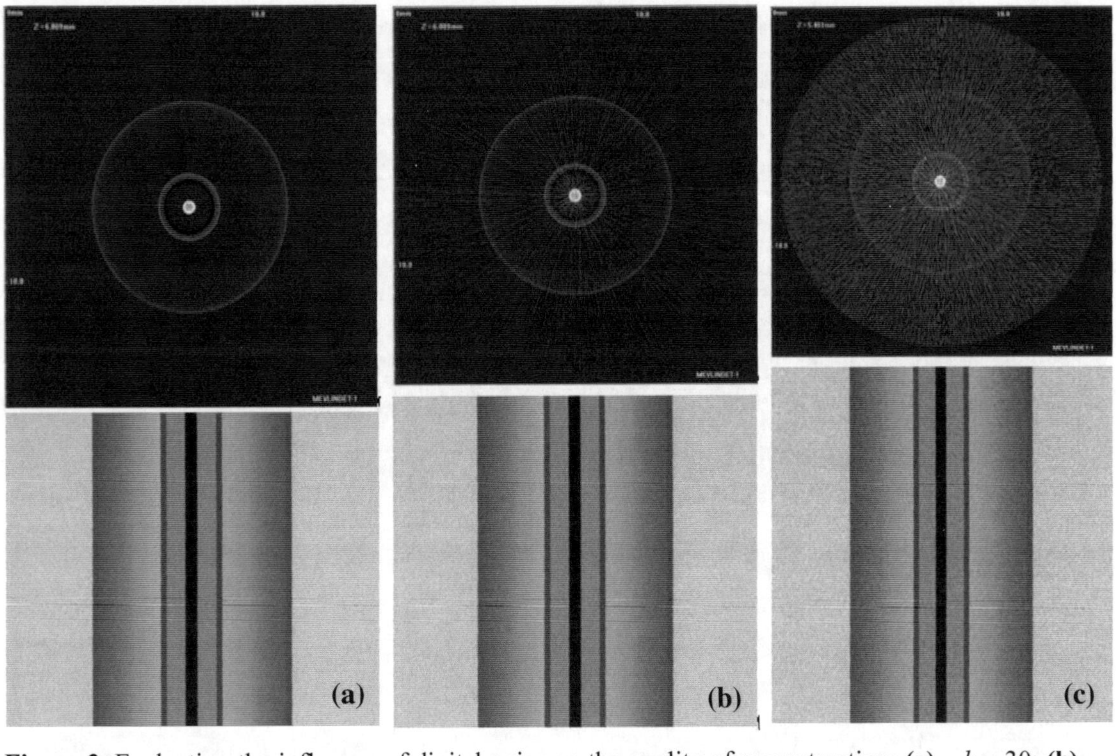

Figure 3. Evaluation the influence of digital noise on the quality of reconstruction: **(a)** – $k = 30$; **(b)** – $k = 50$; **(c)** – $k = 100$.

IOP Conf. Series: Journal of Physics: Conf. Series **872** (2017) 012054 doi:10.1088/1742-6596/872/1/012054

From the images shown above, it can be seen that even a small amount of digital noise in the image influences on the quality of the reconstruction, so it is worthwhile to pay the greatest attention to its filtering in the set of projection data.

When evaluating the obtained data, it is worthwhile to note that the noise component appearing during the reconstruction takes different values for the slices reconstructed from the same number of projections and increases with the increase in the noise component due to the imperfection of the X-ray receivers. Hence, we can conclude that the noise of the reconstruction depends on the noise of the projections, which in turn indicates that the effect of projective noise on the quality of the reduced cut increases nonlinearly as its magnitude increases. The graph in figure 4 most clearly reflects this relationship.

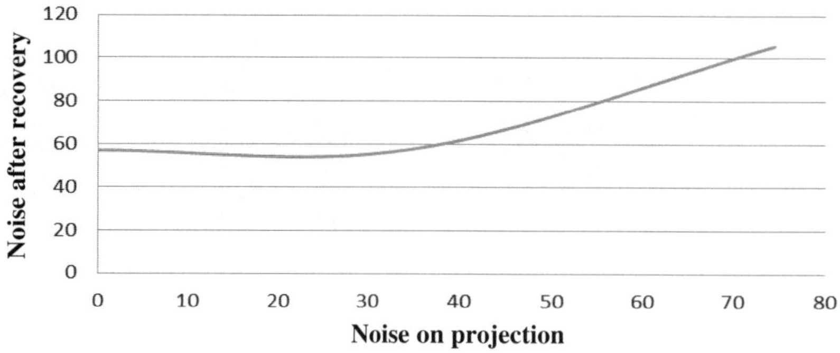

Figure 4. Dependence of the projection noise on the noise of the shear recovery.

The results of a simulation describing the non-uniformity of radiation during a set of projection data are presented in the figure 5.

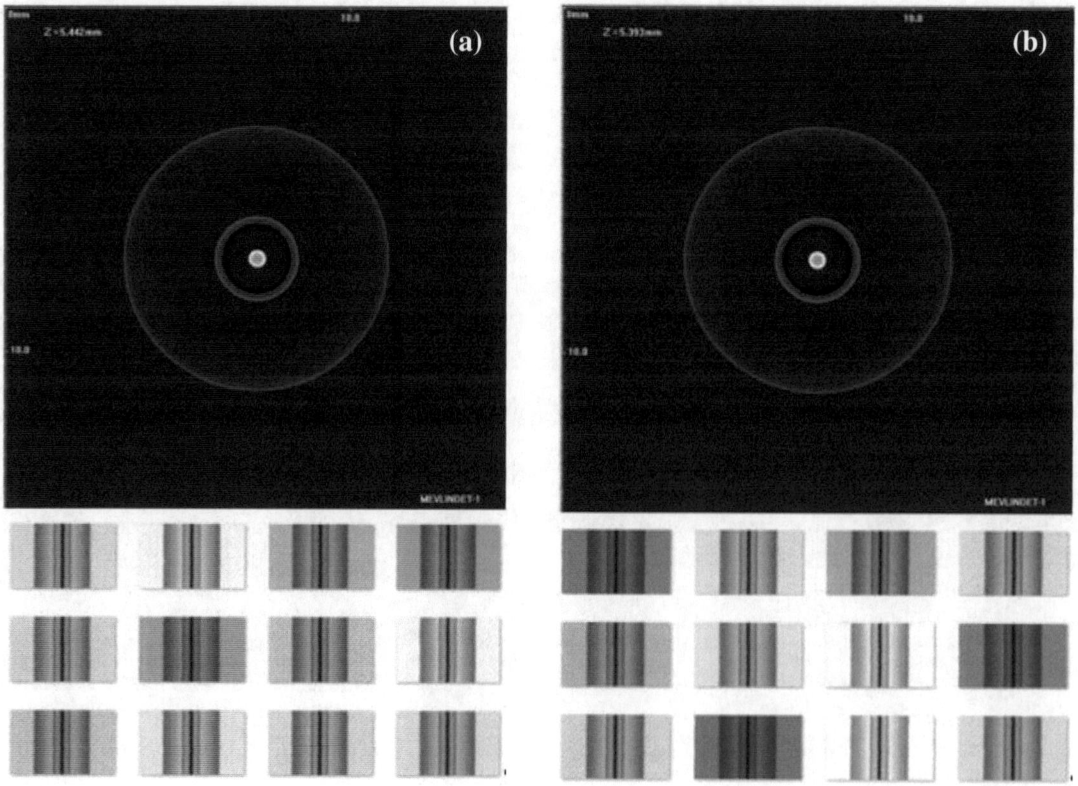

Figure 5. Evaluation of the radiation effect unevenness on reconstruction.

The data presented in figure 5 clearly illustrate that the non-uniformity of radiation does not have a significant effect on the reconstruction quality. However, in this case it should be noted that imitation of non-uniformity was made within the linear area of the dynamic range curve of the radiation receiver.

Speaking about microtomography like a method of research (usually, inanimate objects), that makes it possible to obtain tomograms with a resolution at the level of one micrometer, due to the peculiarities of the used physical and technical conditions of shooting(small anode currents of the X-ray tube and, consequently, large exposition time, which is necessary for obtaining one projection), it is advisable to analyze, how much projection data or how small the angle of rotation of an object is required to achieve the required quality of reconstruction. The tomographic section was reconstructed using 200, 800 and 1600 projections. The restored tomographic sections are presented in figure 6.

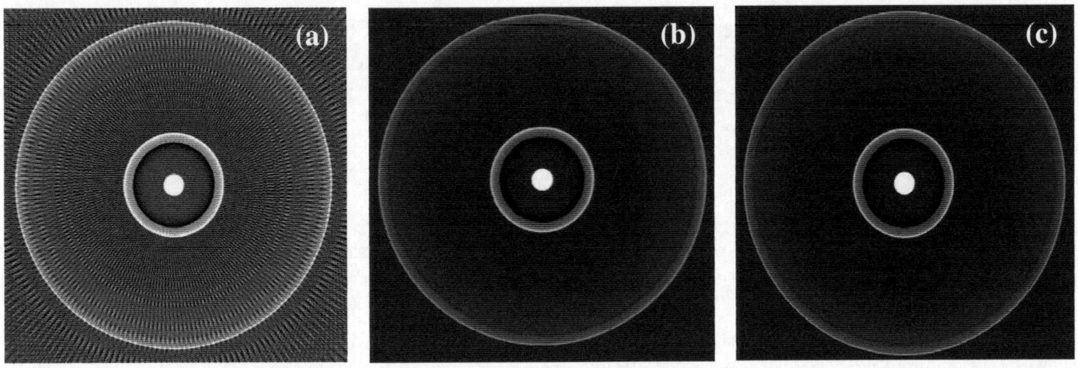

Figure 6. To explain the effect of the number of projections on the quality of reconstructions in tomographic images: **(a)** – section reconstructed from 200 projections; **(b)** – section reconstructed from 800 projections; **(c)** – section reconstructed from 1600 projections.

As can be seen in figure 6(a), on a section reconstructed from 200 projections, the artifacts from the boundaries of the object have a significant effect on the quality of the reconstruction. With 800 projections, artifacts are almost eliminated. On a cut all details of the investigated object are visible, and there are no defects. The use of 1600 projections does not introduce any noticeable changes, which indicates the inexpediency of further increases in the number of projections in this study. Thus, it can be concluded that for this simulated test object, using the condition of "necessity and sufficiency" of information for diagnosis, it is possible to significantly reduce the study time from 2 to 8 times.

Based on the results of the conducted research, it seems expedient to develop a specialized mathematical model that includes small-sized and low-contrast details, allowing to estimate the number of projections necessary to achieve the required quality of reconstruction.

Acknowledgments

The Russian Science Foundation have supported this work as the project entitled "The robotic system is multi-view X-ray microfocus aircraft parts and assemblies made of polymer composite materials with complex action". Project number 16-19-00155.

References

[1] Bessonov V B, Obodovskiy A V, Klonov V V and Kostrin D K 2014 *The Eurasian Union of Scientists* **5-3** 12–5

[2] Blinov N N, Vasilyev A Y, Bessonov V B, Gryaznov A Y, Zhamova K K, Potrakhov E N and Potrakhov N N 2014 *Biomedical Engineering* **2** 58–61

IOP Conf. Series: Journal of Physics: Conf. Series **872** (2017) 012055 doi:10.1088/1742-6596/872/1/012055

Modelling emission spectra of the gas-discharge plasma for the aims of its analysis and parameters estimation

D K Kostrin[1], A A Lisenkov[2], A A Uhov[1] and A N Ramazanov[1]

[1] Department of electronic instruments and devices, Saint Petersburg Electrotechnical University "LETI", 197376, Saint Petersburg, Russia
[2] Institute of Problems of Mechanical Engineering, Russian Academy of Sciences, 199178, Saint Petersburg, Russia

E-mail: dkkostrin@mail.ru

Abstract. In this paper modelling of the emission spectra of gas-discharge plasma using the data about the electronic transitions in substances under the assumption of local thermodynamic equilibrium is discussed. The possibility of using these spectra to simplify the analysis of the plasma composition during the coating deposition process and determine the parameters of the gas-discharge plasma are shown.

1. Introduction

Methods of optical spectroscopy are widely used to control ion-plasma processes, in particular during the application of films and coatings from the material atomized by gas discharge [1–4]. Atoms and molecules of the deposited substances due to collisions with electrons are partially transferred to the excited state, while the reverse transition to the ground state is accompanied by the emission of optical radiation. The number of all electronic transitions and hence the number of lines in the emission spectrum of a substance is determined by the number and placement of the outer electrons. Spectra of atoms with few valence electrons (e.g. alkali metals) have relatively few lines. Atoms with complex outer shells, especially the elements of side subgroups of the periodic system, have radiation spectra with a very large number of lines [2].

Lines, emerging due to electron transfer to the main level, are called resonance. Special analytical value have all the resonance lines, corresponding to transitions from the first excited level. For excitation of the resonance lines in the spectra of alkali metals a minor energy is necessary, while in the case of non-metals, this energy is very high.

Ions have their own emission spectra, which differ significantly from the spectra of neutral atoms of a given element. As a result, in the spectrum appear new ion lines, and the intensity of atomic lines is reduced. Thus, with increasing degree of ionization intensity of neutral atoms lines will be lowered by reduction of their quantity. In the case of the plasma diagnostics methods of optical spectroscopy allow obtaining information about such properties as chemical composition, density, potential, electron temperature, and some others [5].

2. Properties and parameters of gas-discharge plasma

Possibilities of application of methods of optical spectroscopy depend on the parameters of the system under study. The state of homogeneous plasma in a closed volume with isothermal walls can be described using a number of macroscopic parameters (temperature, pressure and concentration of components) without detailed consideration of the microscopic processes [6]. In this case, plasma is in thermodynamic equilibrium. Upon reaching equilibrium to describe the states of matter and properties of radiation in the volume can be used the laws of statistical mechanics: the laws of Maxwell, Boltzmann and Saha describe the distribution function of particle energies, processes of excitation and ionization, respectively. Planck's law describes the spectral distribution of radiation.

It should be noted that full thermodynamic equilibrium is observed almost exclusively in stellar bodies, where large amounts of substances are at higher temperatures. In turn, in the laboratory conditions it is customary to speak of local thermodynamic equilibrium. In the case of a small plasma volume, a significant part of the radiation leaves the volume without absorption (optically thin plasma), spectral irradiance does not obey Planck's law, and the processes of excitation can be fully described by the Boltzmann distribution. However, in the case of high electron density, when the processes of excitation are mainly determined by the collisions with electrons, relaxation and recombination assumptions of local thermodynamic equilibrium can be used. In this case, to describe the spectral distribution of the intensity can be applied the Boltzmann distribution [6]:

$$I(\lambda_{mn}) = \frac{1}{4\pi}\frac{hc}{\lambda_{mn}} A_{mn} g_m \frac{N_i}{Z(T_e)} \exp\left(-\frac{E_m}{kT_e}\right), \tag{1}$$

where $h = 6.626 \cdot 10^{-34}$ J·s – Planck's constant; $c = 2.998 \cdot 10^{8}$ m/s – speed of light; λ_{mn} – wavelength of the radiation corresponding to the transition from the upper level m to the lower level n; A_{mn} – transition probability; g_m – statistical weight of the upper level; N_i – total concentration of particles of type i; $Z(T_e)$ – statistical sum; $k = 1.38 \cdot 10^{-23}$ J/K – Boltzmann's constant; T_e – electron temperature; E_m – energy of the upper level. The statistical sum that contains information about the statistical properties of the system in a state of thermodynamic equilibrium is determined as follows:

$$Z(T_e) = \sum_m g_m \exp\left(-\frac{E_m}{kT_e}\right).$$

It should be taken into account that real spectral lines are broadened due to the influence of various factors. The broadening can be described by Lorentz, Gaussian or Voigt profile. Type of broadening depends on the factors leading to it. Real broadening of spectral lines is determined by the total influence of various factors.

One of the methods of determining the electron temperature of the plasma is the calculation of the ratio of experimental intensities of two spectral lines [7]:

$$T_e = \frac{E_{m2} - E_{m1}}{k \ln\left(\dfrac{I(\lambda_{mn1}) A_{mn2} g_{m2} \lambda_{mn1}}{I(\lambda_{mn2}) A_{mn1} g_{m1} \lambda_{mn2}}\right)}. \tag{2}$$

where E_{m1}, λ_{mn1}, A_{mn1}, g_{m1} – parameters corresponding to the first of the considered spectral lines; E_{m2}, λ_{mn2}, A_{mn2}, g_{m2} – to the second. Instead of one of the lines may, if it exists, be considered a continuous component of the spectrum. It should be noted that obtained according to the expression (2) results are often not accurate, so computations over a large number of pairs of lines, followed by statistical analysis, are used in this case.

3. Results and discussion

Let us consider modelling spectra of plasma radiation for the visual determination of its parameters. To build the spectra in this work, we will use expression (1) under the assumption of local thermodynamic equilibrium. Later the data were multiplied by a function simulating real broadening of spectral lines. Before proceeding to complex emission spectra of the plasma during the coating deposition process, let us determine the most suitable methodology for a further use. The best source of radiation for pre-testing of the methods of modeling spectra of radiation is a mercury low-pressure lamp, the spectrum of which contains intense, spaced from each other at a great distance lines of atomic mercury.

The real emission spectra of the plasma, with which were compared all the modelled spectra in this work, were obtained using a universal optical spectrometer ISM3600 [8–9]. For the instrument control and spectra analysis was used Aspect2010 software. In addition to obtain more reliable experimental results, all spectra were processed using digital filters and averaged over multiple measurements. Before the experiments the spectrometer was calibrated using the spectrum of a reference source to correct its spectral characteristic.

For modelling the spectra of radiation can be used a variety of data about possible electronic transitions in atoms. Let us consider three variants: data from tables of spectral lines, as well as from the Kurucz [10] and NIST [11] databases. Experimental spectra were built from points corresponding to the pixels of the spectrometer's photodetector; all the modelled radiation spectra were obtained using the Mathcad software package. Figure 1 show fragments of plasma radiation spectra of mercury for a real low-pressure mercury lamp and modelled using three mentioned above variants.

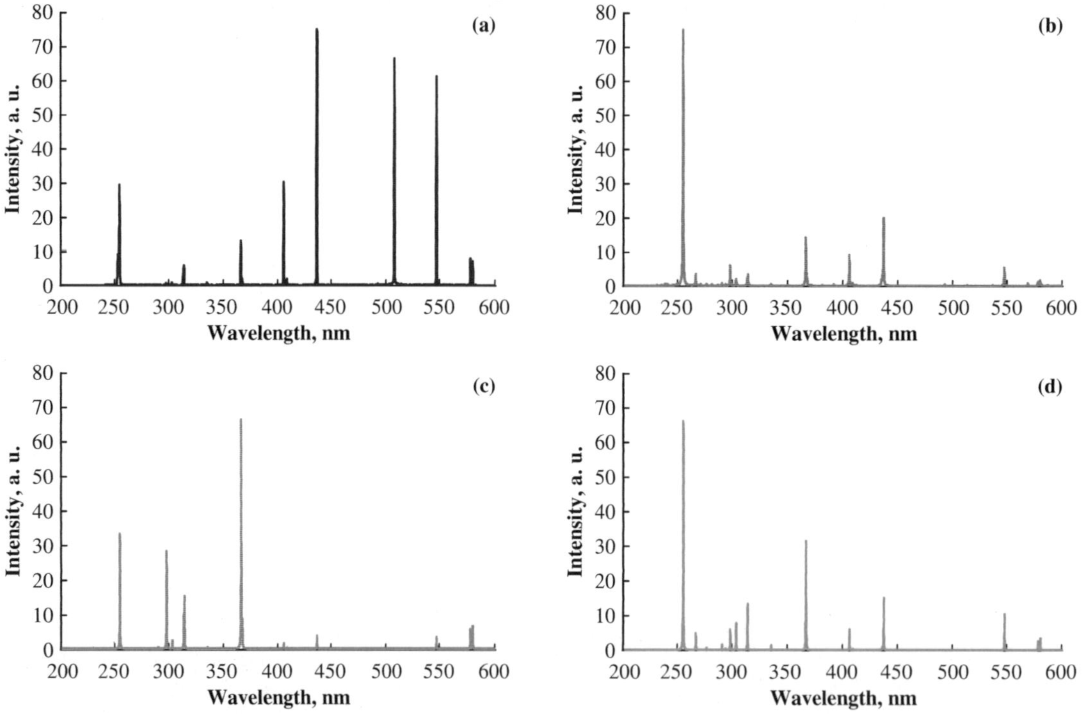

Figure 1. Plasma radiation spectra of mercury: **(a)** – experimental for a mercury low-pressure lamp; **(b)** – modelled using tables of spectral lines; **(c)** – modelled according to Kurucz database; **(d)** – modelled according to NIST database.

The experimental emission spectrum is built according to 3648 points corresponding to the pixels of a sensor with charge transfer used in the optical scheme of the spectrometer. Tables of spectral lines for mercury in this wavelength range contain 139 lines. In the Kurucz and NIST databases there are 24 and 28 allowed electronic transitions, causing the appearance of spectral lines, respectively.

It is obvious that the modelled spectra do not fully match with the experimental spectrum. Of the three modelled spectra for further work we have selected built using data from the NIST database (figure 1(d)), as most accurately reflecting the pattern of the spectral distribution of radiation.

Shown in figure 1 modelled emission spectra built using the databases were obtained for the electron temperature of 11 600 K or 1 eV. Let us consider the modelled emission spectra of atomic mercury at different electronic temperatures according to the NIST database (figure 2).

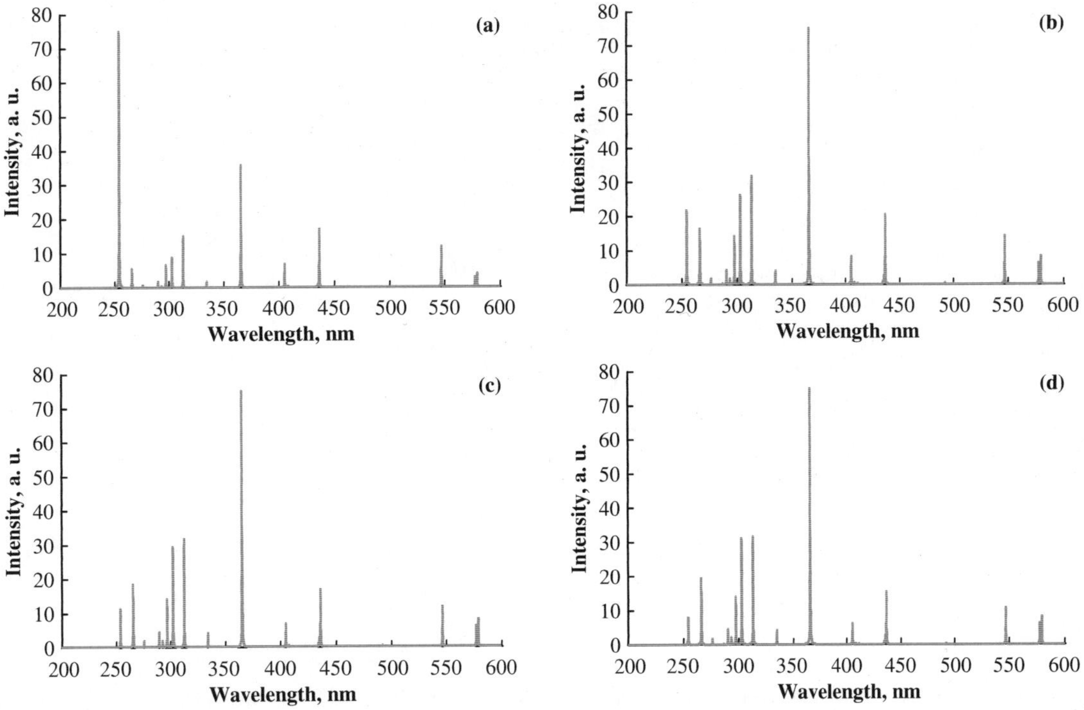

Figure 2. Modelled emission spectra of atomic mercury at different electronic temperatures built according to the NIST database: **(a)** – $T_e = 1$ eV; **(b)** – $T_e = 2$ eV; **(c)** – $T_e = 3$ eV; **(d)** – $T_e = 4$ eV.

It is obvious that the modelled spectra with the electronic temperatures of 3 and 4 eV are closer to the experimental spectrum than the earlier obtained spectrum for a temperature of 1 eV. It should be noted that according to the literature the electron temperature for a mercury low-pressure lamps is 1–2 eV. In this case, either used custom-made source of radiation has an increased electronic temperature or modelled spectra do not fully correspond to the declared electronic temperature.

Nevertheless the following fact is obvious: choosing the parameters in equation (1) for modelling the spectrum its form most matching with the real spectrum can be handpicked and thus the parameters of the real gas-discharge plasma can be estimated.

Estimation of the approximate electron temperature using the experimental spectrum by formula (2) gave unreliable results – for some pairs of lines the results are quite close to the expected, while for a number of pairs results differ from the ordinary by 2–3 orders of magnitude.

Let us consider the plasma radiation spectrum during the deposition of titanium coating using vacuum-arc plasma source [12–14]. When applying the titanium coating in the spectrum of the plasma radiation are detected spectral lines corresponding to atomic titanium and its ions. The degree of ionization of the plasma in vacuum-arc discharge is high enough; the plasma flow of titanium plasma contains singly and multiply charged ions [15].

Figure 3 shows the real emission spectrum of the plasma and also modelled using data from the NIST base radiation spectra of atoms and ions of titanium for the electron temperature 1 eV, which is considered close to the real value for the vacuum-arc plasma source.

According to the data presented in figure 3 it can be noted that the emission spectrum of atomic titanium is close to the total experimental spectrum of radiation of titanium plasma. The spectral lines of titanium ions with different charge, given the low spectral resolution of the used spectrometer, are superimposed on the lines corresponding to the atomic titanium.

In the NIST database in the considered wavelength range to the atomic titanium correspond 360 lines, to Ti^{+} – 205 lines, Ti^{2+} – 121 line and Ti^{3+} – 8 lines. Spectral lines of the tetravalent titanium ion lie in the ultraviolet region and do not belong to the considered range.

IOP Conf. Series: Journal of Physics: Conf. Series **872** (2017) 012055 doi:10.1088/1742-6596/872/1/012055

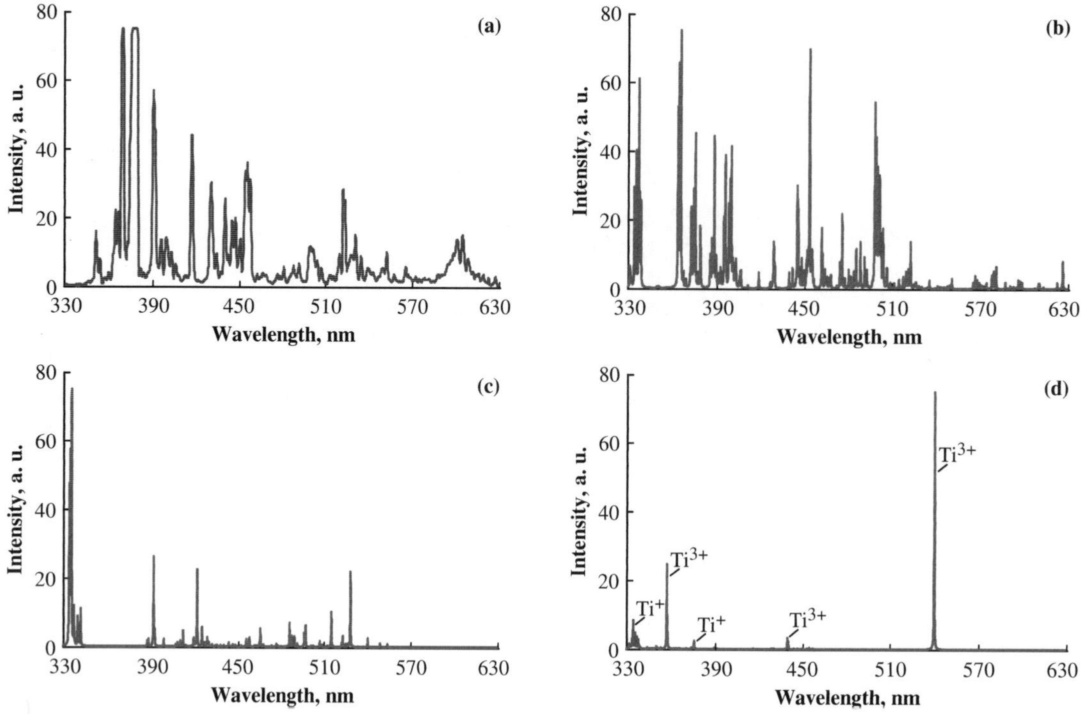

Figure 3. Emission spectrum of the vacuum-arc discharge plasma when sputtering a titanium cathode **(a)**, modelled emission spectrum of atomic titanium **(b)**, modelled spectrum of double-charged ions of titanium **(c)**, modelled spectrum of the total radiation for single- and triple-charged titanium ions **(d)**.

4. Conclusion

Studies show that the use of modelled spectra to determine the composition of plasma and its parameters is of a limited use – determination of parameters is possible in principle, but difficult and unpractical. The plasma in a coating deposition process is a very complicated system – while for mercury lamp such technique can be effectively used, for vacuum-arc discharge estimation is too difficult, even taking into account one substance and excluding other chemical elements, which can also participate in the process. However it can be noted that with sufficient ease of modelling, these spectra can greatly simplify the identification of the elements composing the plasma.

Also can be mentioned the fact that data for spectral lines from Kurucz and NIST databases correspond to astrophysical measurements and is primarily intended to assess the spectra of stars. Tables of spectral lines in turn contain information for some ideal case. In a real system there is a complex interaction of the particles forming the plasma, thus, the modelled spectra do not fully reflect the real distribution of radiation.

References

[1] Devia D M, Rodriguez-Restrepo L V and Restrepo-Parra E 2015 *Ingenieria y Ciencia* **21** 239–67
[2] Kostrin D K, Uhov A A and Lisenkov A A 2016 *Vakuum in Forschung und Praxis* **3** 34–7
[3] Kostrin D K, Lisenkov A A, Uhov A A and Ramazanov A N 2016 *Journal of Physics: Conference Series* **729** 012030
[4] Kostrin D K, Lisenkov A A and Uhov A A 2016 *Journal of Physics: Conference Series* **735** 012055
[5] Kiselev A S, Kostrin D K, Lisenkov A A and Smirnov E A 2017 *Journal of Physics: Conference Series* **789** 012027

[6] Andre P, Ondet J, Pellet R and Lefort A 1997 *Journal of Physics D: Applied Physics* **30** 2043–55
[7] Mirapeix J, Cobo A, Gonzalez D A and Lopez-Higuera J M 2006 *Optics Express* **4** 1884–97
[8] Uhov A A, Gerasimov V A, Kostrin D K and Selivanov L M 2014 *Journal of Physics: Conference Series* **567** 012039
[9] Simon V A, Kostrin D K, Uhov A A, Ramazanov A N and Lisenkov A A 2017 *Proceedings of the 2017 IEEE Russia Section Young Researchers in Electrical and Electronic Engineering Conference* 342–5
[10] Smith P L, Heise C, Esmond J R and Kurucz R L *Atomic spectral line database built from atomic data files from R L Kurucz CD-ROM* (http://www.pmp.uni-hannover.de/cgi-bin/ssi/test/kurucz/sekur.html)
[11] Kramida A, Ralchenko Yu and Reader J *NIST Atomic Spectra Database* (http://physics.nist.gov/PhysRefData/ASD/lines_form.html)
[12] Bystrov Yu A, Vetrov N Z, Lisenkov A A and Kostrin D K 2014 *Vakuum in Forschung und Praxis* **5** 19–23
[13] Bystrov Yu A, Kostrin D K, Lisenkov A A and Vetrov N Z 2015 *Vakuum in Forschung und Praxis* **2** 22–5
[14] Kostrin D K and Lisenkov A A 2016 *Materials Science Forum* **843** 278–83
[15] Kostrin D K, Trifonov S A and Lisenkov A A 2017 *Journal of Physics: Conference Series* **857** 012022